Forecasting Demand for U.S. Ground Forces

Assessing Future Trends in Armed Conflict and U.S. Military Interventions

MATTHEW LANE, BRYAN FREDERICK, JENNIFER KAVANAGH,
STEPHEN WATTS, NATHAN CHANDLER, MEAGAN L. SMITH

Prepared for the United States Army
Approved for public release; distribution unlimited

For more information on this publication, visit **www.rand.org/t/RR2995**.

About RAND

Research Integrity

Published by the RAND Corporation, Santa Monica, Calif.

Library of Congress Cataloging-in-Publication Data is available for this publication.

ISBN: 978-1-9774-0450-3

Cover: U.S. Air Force photo/Tech. Sgt. Erik Gudmundson; ABC Vector/Adobe Stock..

Limited Print and Electronic Distribution Rights

Preface

This report documents research and analysis conducted as part of a project titled *Forecasting Demand for U.S. Ground Forces*, sponsored by the Office of the Deputy Chief of Staff, G-3/5/7, U.S. Army. The purpose of this project was to identify potential future demands for U.S. ground forces, including size, location, and capabilities, to inform decisions regarding planning, posture, and investments.

This research was conducted within RAND Arroyo Center's Strategy, Doctrine, and Resources Program. RAND Arroyo Center, part of the RAND Corporation, is a federally funded research and development center (FFRDC) sponsored by the United States Army.

RAND operates under a "Federal-Wide Assurance" (FWA00003425) and complies with the *Code of Federal Regulations for the Protection of Human Subjects Under United States Law* (45 CFR 46), also known as "the Common Rule," as well as with the implementation guidance set forth in DoD Instruction 3216.02. As applicable, this compliance includes reviews and approvals by RAND's Institutional Review Board (the Human Subjects Protection Committee) and by the U.S. Army. The views of sources utilized in this study are solely their own and do not represent the official policy or position of DoD or the U.S. Government.

Contents

Figures

Tables

Summary

The research reported here was completed in February 2019, followed by security review by the sponsor and the U.S. Army Office of the Chief of Public Affairs, with final sign-off in April 2022. The authors also wish to note that one of the scenarios modeled in this report, a hypothetical future Global Pandemic, differs in important ways from the COVID-19 pandemic that emerged in 2019–2020. As a result, the analysis of that scenario in this report does not reflect the forecasts of this research for the likely effects of the current pandemic.

Assessments of current and emerging threats inform a range of different decisions that are central to the U.S. military's ability to prepare for future operations: decisions about what types of equipment to invest in, how to train military personnel, where to station U.S. forces abroad, and how to manage units' deployment cycles. These assessments are informed by historical data, analyses of current trends, simulations, scenarios, and expert opinion.

To defend against potential threats, the U.S. Army devotes significant resources to strategic and operational planning. But U.S. Army planning is an exercise in risk management across the wide array of potential threats facing the United States. U.S. Army plans support the larger U.S. Joint Military Force, and Army resources are allocated with knowledge that there are likely gaps in how U.S. ground forces prepare for and respond to individual threats as policymakers prepare U.S. ground forces for the contingencies that pose the greatest strategic challenge.

The U.S. Army experience in Iraq and Afghanistan highlights the limitations of force-planning tools that are misaligned with a changing strategic landscape. To avoid these types of gaps in the future, military planners need better tools that leverage emergent trends in the global geostrategic environment to forecast future contingencies to preemptively build, shape, and prepare U.S. forces for the kinds of missions they are most likely to encounter in the future and for the contingencies that pose the greatest strategic risk to the United States. Furthermore, the retirement of the Support for Strategic Analysis planning process means that new tools are needed to assess likely future demand signals for ground forces in all theaters. There is an urgent need for better and more accurate forecasting methods that can provide more sophisticated and detailed inputs into the Army planning process. Fortunately, advances in forecasting tools and methodologies provide a starting point for the development of improved approaches to forecasting demand for ground forces.

Objective and Approach

In this report, we aim to provide empirically grounded assessments of future demands for U.S. ground forces by presenting a dynamic forecasting model that projects future U.S. ground interventions in a range of scenarios through the year 2040. The model we develop incorporates annual projections of opportunities for U.S. intervention—including armed conflicts and their aftermath—and U.S. ground interventions themselves for each year in the 2017–2040 time frame. We develop three main types of projections: trends in the future operating environment, including the incidence of interstate wars and intrastate conflicts; future U.S. ground interventions, including those involving deterrence, combat, and stabilization activities; and the anticipated average force requirements for those interventions.

The forecasts presented in this report will be useful to the U.S. military generally and the U.S. Army in particular in several ways. First, our analysis identifies key factors that can serve as early warning indicators of future conflicts or U.S. ground interventions to enable better planning and anticipation. Second, our forecasts provide an improved empirical basis for estimating the frequency, magnitude, duration, and overlap of future contingencies to help military personnel plan for a scalable and flexible force. Third, our forecasts may help increase the responsiveness of U.S. Army forces in the event of an overseas crisis by informing force posture decisions that most usefully position U.S. Army forces around the world. Finally, the methodology we present also provides an empirical process that can inform a replacement to the Support for Strategic Analysis to support steady state Army planning, filling an emerging gap with a more flexible and dynamic process.

Limitations

Our approach to forecasting future demands for U.S. ground forces comes with a number of limitations important to note at the outset. First, any set of forecasts has predictive error associated with it. Empirical models are only approximations of the world; they omit variables and rely on particular assumptions about the future and, in this case, about the way projected demands for U.S. ground forces stem from key factors. U.S. intervention decisions have historically been somewhat idiosyncratic, making them difficult phenomena to model because it is difficult to accurately capture all the factors that affect demands for U.S. ground forces. In addition, for some key factors we have relied on proxy measurements that only imperfectly capture the underlying factors, and data are inconsistently available across our historical series. Further, because the models presented in this report use one set of forecasts (conflict) and build another set of forecasts (interventions) on top, and then apply qualitative assessments of ground force characteristics using historical data, the potential error at each stage of our forecasting process may compound.

Estimating the sizes of projected future interventions is a particular challenge. Although we use dynamic models to forecast conflicts and the incidence of interventions, we lack fully developed theoretical models of the factors likely to drive future force-sizing decisions that would enable a similar dynamic forecasting process to project intervention sizes. As a result, we use historical patterns and descriptive statistics to identify how overall intervention size and the relative use of heavy and light forces in different interventions vary based on key criteria. We then categorize and assign a typical or average "force package" of overall intervention size and levels of heavy and light forces to each projected intervention. While this will give the Army a general estimate of expected size for each projected intervention, there has historically been substantial variation in size within these categories, and we expect similar variation in the future. Our projections of anticipated force demands should therefore be viewed with particular caution.

We have taken steps to minimize and account for the underlying uncertainty of our estimates, to the extent possible, in three main ways. First, we implemented 500 repeated simulations of each model in an effort to gradually reduce the uncertainty in any individual projection through a larger number of forecasts. Second, in our discussion of model results, we are clear and explicit regarding the degree of confidence attached to each individual result, so that readers can interpret the trends projected in the appropriate context. Third, alongside our baseline, most likely future, we consider a range of possible alternative futures to identify which trends are particularly sensitive—or not—to the nature of the anticipated future strategic environment.

Forecasting Model Architecture

Broadly, our forecasting model works by sequentially developing annual predictions of armed conflict and U.S. ground interventions for each year in the 2017–2040 time frame. The model then uses those predictions to inform the subsequent year's predictions of conflict and interventions. For example, for 2017, our model first predicts levels and locations of intrastate and interstate armed conflicts, followed by predictions about the types and locations of U.S. ground interventions, followed by estimates of the forces required by those interventions. Considering the results of those 2017 forecasts, it then follows the same process for the year 2018, and so on. More specifically, the overall architecture of our forecasting model is built around four main components, summarized in Figure S.1. These four components work interdependently to predict future trends in armed conflict, U.S. ground interventions, and U.S. ground force requirements across a broad range of future strategic environments. We use empirical models to assess the opportunities for intervention and the realized U.S. ground interventions and rely on analysis of historical trends to assess the size of these interventions. A single iteration of our model involves the full simulation of each model component

Figure S.1
Forecasting Architecture Conceptual Diagram

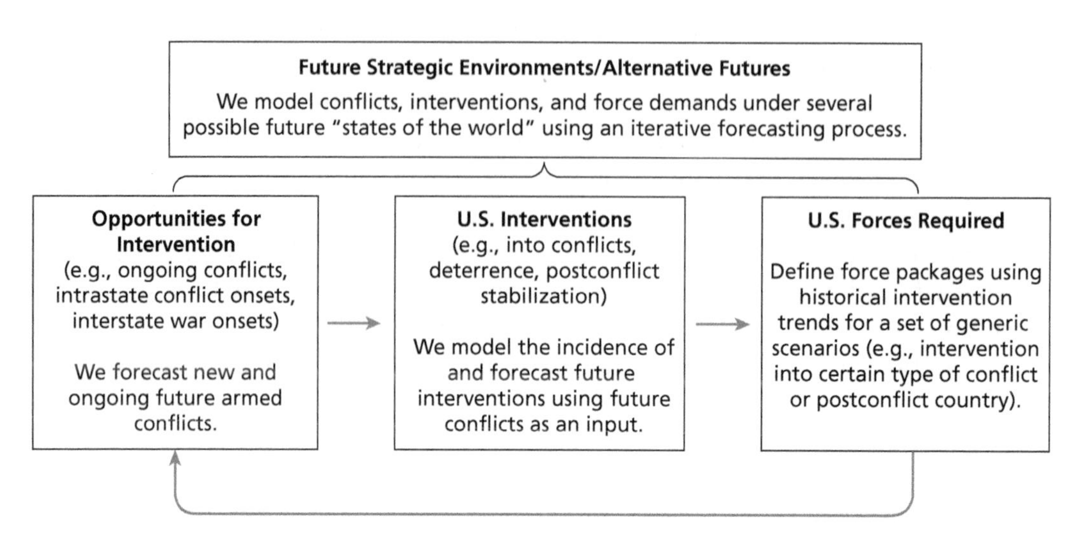

annually for each year from 2017 through 2040. Each subsequent iteration then re-simulates the entire 2017–2040 period using a different random seed. We consider five strategic environments—the baseline future and four alternative scenarios designed to consider possible shocks to the current international system and the effects such shocks would likely produce on potential demands for U.S. ground forces.

Future Demands for U.S. Ground Forces

We used our forecasting models to derive five sets of conflict and intervention projections, each with associated forecasts for future demand for ground forces at the global and regional level. These five scenarios are (1) a baseline scenario (the most likely future environment, which assumes gradual changes in key global factors); (2) a Global Depression scenario, which models the effects of a dramatic global economic collapse; (3) a Revisionist China scenario, which considers the effects of a dramatic expansion in Chinese efforts to revise the international system; (4) a Global Pandemic scenario, which considers the effects of a two-year pandemic and resulting global population and economic losses; and (5) a U.S. Isolationism scenario, which assumes that the United States substantially reduces its international engagement, including pulling out of all major alliances, multilateral trade agreements, and international institutions.

Summary of Projections

Our conflict and intervention projections vary across the five different scenarios, but we can make a number of cross-cutting insights. First, many of the results from our models have relatively high levels of uncertainty associated with their specific projections. This uncertainty illustrates the limitations of the models and the historical data surrounding conflict and U.S. ground interventions, but the uncertainty also accurately reflects the challenges involved in such an effort. Indeed, claims to provide highly precise estimates of future wars, conflicts, and U.S. interventions 10, 15, or 20 years from now would simply not be credible. Notwithstanding this uncertainty, the trends in our projections can still provide insights into possible future trends, which are of use to U.S. policymakers and the U.S. Army in particular.

Turning first to trends in armed conflict, our analyses suggest that, although future levels of interstate war are likely to remain low by historical standards, the risk of interstate war appears likely to increase, in both our baseline scenario and in several of our alternative scenarios. In the baseline scenario, this increase in risk is concentrated in Eurasia and the Middle East and corresponds roughly to an increase of one additional ongoing interstate war each year. This would effectively return the frequency of interstate war back to what it was on average during the Cold War period, while still remaining well below levels in the pre-1945 era.

In our alternative scenarios, the risk of interstate war is largest in the Global Depression scenario, in which new conflicts emerge at the friction points between separate geopolitical and trading blocs, and in the revisionist China scenario, in which conflict emerges between the group of states that become closely aligned with China and others that remain outside its orbit.

However, while our models consistently suggest some increase in the future likelihood of interstate war, our forecasts also consistently project a decrease in intrastate conflict from current levels, in the baseline scenario and across the alternatives. This decrease is generally spread across regions but is most pronounced in East/Southern Africa and the Middle East, the regions with the highest current levels of such conflict. Intrastate conflict remains prevalent, however, even with the anticipated declines, returning only to the level last seen in the late 1990s and early 2000s. Notably, this trend appears robust across scenarios, even the Great Depression and Global Pandemic scenarios, which one would expect to strain less-developed states.

Our projections of the number of future U.S. ground interventions and the troops anticipated to be employed in those interventions also vary across scenarios, but again some notable patterns emerge. In the baseline scenario, the total number of U.S. ground interventions is expected to decline slightly or remain the same, but this trend is accompanied by a projected increase in the forces required to meet the demands of these interventions, as detailed in Figure S.2. This increase is partially a result of assumptions made in the model regarding the force sizes of ongoing interventions, and the increase is reflective of a replacement of current interventions that have drawn down notably from their peak troop levels, such as in Afghanistan, with future interventions that are estimated to be notably larger. In the baseline scenario, for instance, our model suggests a demand that ranges from 100,000 (10th percentile of projections) to 425,000 (90th percentile of projections) troops, but with an average

Figure S.2
Baseline Forecasts of Demands for U.S. Ground Intervention Forces, 2017–2040

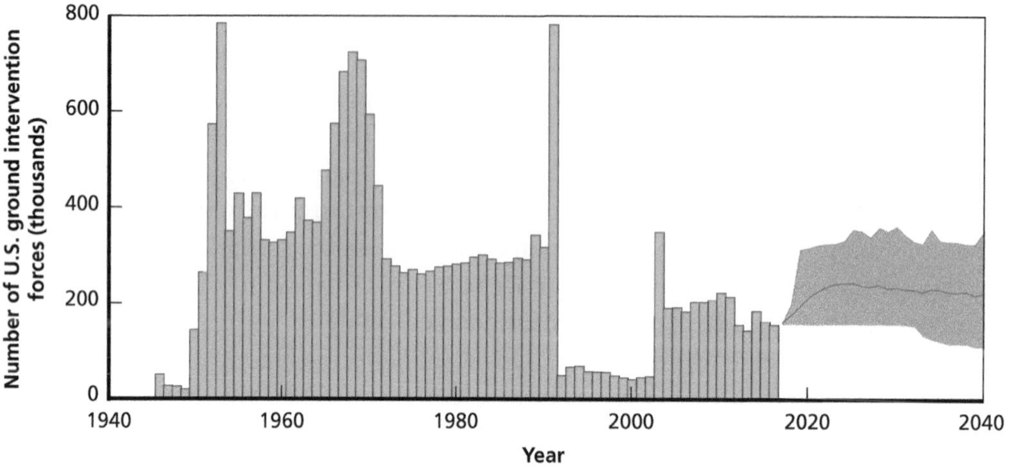

NOTES: The blue bars denote historical demands for U.S. ground intervention forces. The red line denotes the projected average number of U.S. ground forces required for interventions each year, based on 500 iterations of our forecasting model. The gray shaded area represents the range of forecasts bounded by the 10th and 90th percentiles of U.S. ground forces required for interventions each year, based on 500 iterations of our forecasting model.

number of projected troops employed of more than 200,000. This average projection represents a roughly 30 percent increase in U.S. ground forces employed in interventions between 2017 and 2040, largely driven by an increase in stability operations and combat missions. While this is a broad range, it notably does exclude both commitment levels at the Vietnam-era peak and the immediate post–Cold War valley. Further, the regional patterns of these anticipated interventions suggest that the deployed troops are most likely to be involved in interventions in the Middle East or Eurasia. We again underline, however, the difficulty of projecting specific troop demand numbers, for the reasons discussed previously.

We do not forecast a large increase in the number of interventions in any scenario. Even in the Global Depression scenario, the average number of U.S. ground interventions is projected to remain relatively constant with the number of interventions today. We do, however, observe some sizable increases in U.S. ground forces committed to interventions, particularly compared with the present, as detailed in Figure S.3. As in the case of conflict projections, however, our estimates of demand for future U.S. ground intervention forces have a high degree of uncertainty, and the 10th and 90th percentile projections underscore the wide range of plausible values. However, these projections are still valuable in a comparative sense, to understand how demand for U.S. ground intervention forces and the trends in this demand over time are likely to be affected by various assumptions in our five scenarios.

Figure S.3
Forecasts of Demands for U.S. Ground Intervention Forces Across Alternative Future Scenarios, 2017–2040

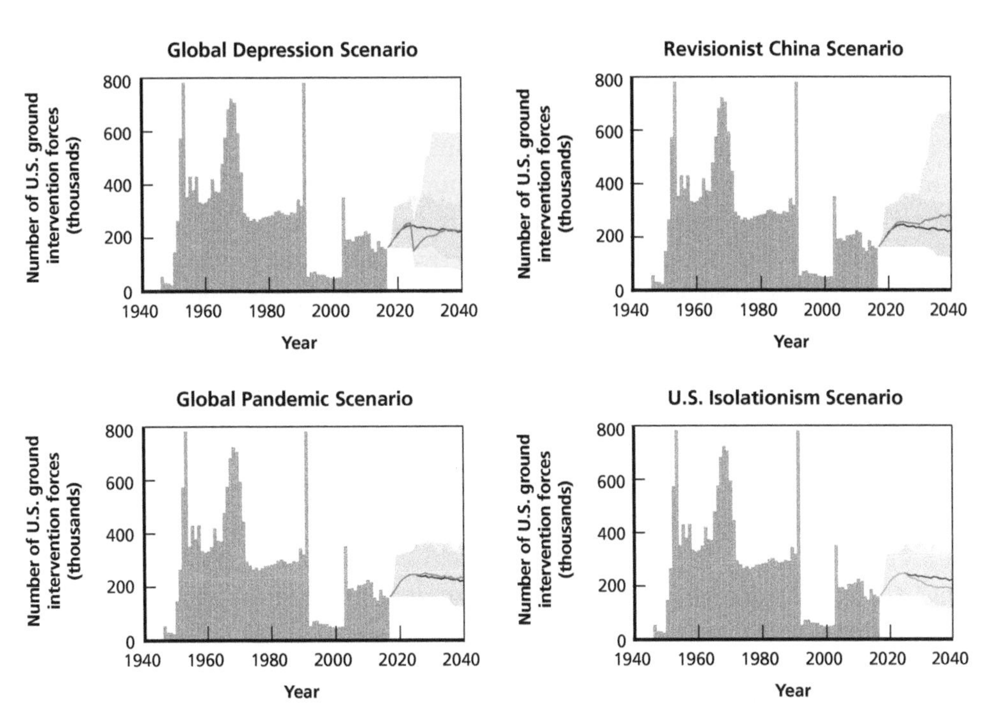

NOTES: The black bars in each chart denote historical demands for U.S. ground intervention forces. The black line in each chart denotes the projected average number of U.S. ground forces required for interventions each year in our baseline scenario. The gray shaded area in each chart represents the range of forecasts bounded by the 10th and 90th percentiles of U.S. ground forces required for interventions each year in our Baseline scenario. The colored line in each chart denotes the average number of U.S. ground forces required for interventions each year in each of our alternative future scenarios. The colored shaded area in each chart represents the range of forecasts bounded by the 10th and 90th percentiles of U.S. ground forces required for interventions each year in each of our alternative future scenarios.

We see the largest increases in demands for forces, for instance, with interventions into armed conflict for the Global Depression scenario and interventions into armed conflict in the Revisionist China scenario. For example, peak U.S. ground force demands in the Revisionist China scenario require roughly 250,000 troops, representing a roughly 60 percent increase over 2016 demands for U.S. ground intervention forces.

In terms of the regional patterns, despite the pronounced changes in other factors across scenarios, certain key regions persist in having the highest risk of larger U.S. ground interventions: the Middle East, Eurasia, and East/Southeast Asia. In the cases of the Middle East and East/Southeast Asia, such interventions would continue long-standing historical patterns. The model's suggestion that substantial U.S. forces could be deployed to Eurasia, however, would be a notable strategic departure for the United States and could involve heightened tensions with Russia, China, or both.

We can also compare across the different scenarios in a general sense to explore which types of external shocks are most likely to have significant implications for U.S. ground force demands. The scenario with the fewest implications for U.S. ground forces (or the fewest deviations from the baseline) is the Global Pandemic scenario. While we see a small increase in the projected number and size of deterrent interventions, force demand and conflict incidence do not vary as widely from the baseline as in other scenarios. Although there is much concern over the implications of a pandemic from economic and human welfare perspectives, our projections suggest limited implications for future levels of armed conflict and deterrent, conflict, and stabilization interventions.

Second, the effects of the economic collapse modeled in the Global Depression scenario are more severe. We often think about economic collapse in terms of its domestic effects on wages and employment, but our models clearly suggest that if severe economic downturn is accompanied by geopolitical dislocations and the establishment of rival trading blocs, it may also be associated with an increase in interstate war and a greater demand for U.S. combat forces. In a period of economic decline, there may be fewer resources to provide the military with the equipment, training, and personnel needed to be successful. This could complicate U.S. ground interventions in this context.

The Revisionist China scenario also suggests a more conflictual trajectory for the world and a higher demand for U.S. forces over the baseline. It is also especially relevant within the context of the National Defense Strategy and its focus on thinking about the potential for more direct great power competition. The practical implications for the U.S. military of the Revisionist China scenario are broadly similar to the Global Depression scenario—although the increases in interstate war and U.S. ground interventions are less significant.

Finally, the results of our U.S. Isolationism scenario have interesting implications. There is a long-running debate about the extent to which U.S. deterrent missions, participation in multilateral institutions, and building strong alliances are effective ways to reduce conflict. Although contributing to that debate was not an objective of this report, our U.S. Isolationism scenario suggests that, at least in some regions, these factors do provide a pacifying effect and that the risk of conflict may increase in their absence. In East/Southeast Asia, our results suggest that a drawdown in U.S. deterrence commitments may be more than compensated for by an increase in commitments to new combat missions in the region, such that U.S. forces in the region could actually increase despite the isolationism stipulated in the scenario. As such, while the overall numbers of U.S. ground intervention forces may slightly decline over the 2017–2040 period, the nature of missions for those forces committed abroad could significantly change, with greater numbers of U.S. ground forces undertaking combat missions brought about by an end to the United States' ongoing deterrence missions.

Implications for the Army

Planning for the future, whether in the U.S. Army or elsewhere, is necessarily an exercise in risk management. The important implications for the Army of the analyses presented in this report are therefore less about specific force estimates (although those are worth noting) and more about identifying and understanding how different aspects of the future operating environment may function as drivers of demand for U.S. Army forces. The Army can use this understanding to take steps to manage future risk and ensure that it is sufficiently prepared to respond to the most likely or high-consequence contingencies.

Future Stability Operations

Although our different scenario projections diverge in many respects, our results are relatively consistent in finding a high likelihood that the United States will conduct a sizable stability operation at some point between now and 2040. In the baseline and alternative scenarios, our projections consistently show the likelihood of an increase of at least one new stability operation in the projection period. Although the specifics of our projections, including the exact timing of such an intervention and where it will take place, remain highly uncertain, the recurrence at some point in the next two decades of conditions that have previously prompted the United States to undertake a stability operation seems quite likely.

Our expectation that a relatively sizable U.S. stability operation is likely to occur in the next 20 years is somewhat at odds with the current prevailing orientation of Army strategic thinking. In 2017, the Army published an updated version of its Field Manual 3-0, *Operations*, which shifted focus away from the counterinsurgency and stability operations and increased the emphasis on large-scale ground combat with near-peer adversaries such as Russia and China, based on an assessment by Army and U.S. Department of Defense (DoD) leaders that the threat of large-scale ground combat with capable adversaries is increasingly likely and that the potential for future large-scale stability operations like those in Iraq and Afghanistan is dwindling.

Our results suggest that it is unlikely that the era of large U.S. stability operations has passed for good. The U.S. Army has used the experience in Iraq and Afghanistan to build strong expertise in how to execute complex stability operations in urban and other environments. The Army has developed training simulations and exercises to prepare its forces to operate in these environments and has acquired the equipment needed to support these operations. Letting this institutional capacity and knowledge atrophy would undermine the Army's ability to conduct these types of operations in the future and reduce the Army's responsiveness should such a demand arise again. To ensure that it has sufficient capacity to support a future large-scale stability operation, the Army would likely need to retain sufficient expertise in this field and keep it updated as the global operating environment changes, develop a surge strategy to hedge against the risk of a future large-scale stabilization mission, and maintain investments in a few critical long lead-time capabilities required for stabilization operations.

Great Power Competition and Deterrence

As noted elsewhere, our force projections for future deterrent interventions predict only relatively modest changes (some increases, some decreases depending on the scenario) from current demands. A potential new deterrent intervention in the Philippines is, for example, a frequent projection in many of our model iterations, and the present deterrent mission in the Sinai is typically the most likely such mission to be projected to end, depending on the scenario. However, it is important to note that our models are, by design, silent on one of the largest possible drivers of an increase in deterrent force demands: an increase in the size of current deterrent deployments. As discussed above, our models assume that all interventions ongoing in 2016 retain their 2016 size throughout the remainder of the time when the model projects them to continue. This was a necessary modeling simplification, but it has significant implications, especially given the current focus on great power competition. One possible response to the perceived threat presented to U.S. interests and allies by rising adversaries would be to increase the size of deterrent interventions intended to safeguard these interests and to place a check on adversary ambitions. For example, there are ongoing discussions about the appropriate size of a U.S. deterrent force in Eastern Europe, with some arguing that more forces are needed. Increasing size of ongoing deterrent interventions may then be a major driver of force demands in the future. Our model does not capture this increasing demand, but we mention it here as a limitation and to highlight it as another consideration for Army force planners.

Combat Interventions and Force Demands

Just as our forecasts highlight the plausibility of a future large-scale U.S. ground stability operation, our projections also suggest that there is a good chance of a sizable U.S. ground combat intervention into an ongoing armed conflict over the next two decades, particularly later into that period. We see this risk clearly in the baseline scenario, and it is even more dramatic in some of our alternative scenarios. More importantly, however, the implications of such an intervention for force demands would be substantial. As noted elsewhere, even where we expect the numbers of U.S. ground interventions into armed conflict to decline, such as in the baseline scenario, we still see sharp increases in the number of ground troops that could plausibly be committed to such interventions. In 90th percentile projections in the baseline scenario, armed conflict interventions are expected to require more than 150,000 troops in the early 2030s, while the average demand across all model iterations is roughly 75,000 troops. The projected demand for combat forces is dramatically higher under some of the alternative scenarios. In the Global Depression scenario, for example, demand for combat intervention forces is predicted to exceed 500,000 in the 90th percentile projection by 2040, with the average projection exceeding 100,000. These results are in keeping with historical patterns, as well. It has been the large combat interventions that have most significantly increased force demands in the past.

The need to undertake such a massive combat intervention is by no means a certainty. The 10th percentile projection for forces devoted to armed combat interventions remains much lower, in the low tens of thousands, even in the highly conflictual Global Depression scenario. For Army planners, however, the continued, sizable risk of a large-scale combat intervention is important, for a few reasons. First, the possibility that Army forces will be called on to engage in major combat operations is consistent with the increasing focus on great power competition and the Army's growing focus on the risk of large-scale conventional warfare. Our projections reinforce the value of the Army's strategic focus on these areas.

Second, our projections—and the review of similar, historical operation sizes that inform them—provide some insight into the size of possible force demands for combat interventions and can inform the decisions of Army leaders about training, force structure, recruiting, and retention. Notably, the average demand for combat intervention forces under the baseline scenario is small enough that the Army likely has sufficient capacity already. On the other hand, sustained combat force demands of 100,000, or perhaps many more, in an alternative strategic environment such as the Global Depression would place significantly more strain on the Army. Policymakers will need to assess the likelihood of such a dramatic worsening in the strategic environment, and the risk they are willing to accept, or not, in preparing for it.

Future Demand for Heavy and Light Forces

In addition to projecting overall force demands, we considered the likely future demand for both heavy and light forces. Generally speaking, we found a relatively consistent ratio of heavy to light forces, where heavy forces made up roughly one-fifth of the total number of forces required. This result is driven both by our assessment of the historical utilization of heavy forces in interventions of different activity types and characteristics and the frequency with which those different types of interventions are projected to occur in the future. In most of our results, including the baseline scenario, we project that future interventions are likely to have higher demands for heavy forces than the current mix of deployed forces today. In the baseline scenario, the 90th percentile projection suggests a demand heavy forces equal to the peak number of heavy forces deployed in Iraq and Afghanistan, with the average projection roughly two-thirds of that number. This average projection would represent an increase of roughly 10,000 additional heavy forces employed in U.S. ground interventions compared with today. It should be emphasized that these increases would occur on top of any potential increase in the size of heavy forces in existing deterrent interventions, which, as discussed previously, our model does not reflect. The 10th percentile projection, meanwhile, is roughly consistent with or slightly below the number of heavy forces already committed to interventions today.

The question of whether the Army could properly resource such an increased demand with the current mix of heavy and light forces in the continental United States was beyond the scope of this study. However, it is worth noting that during the peak of the wars in Iraq and Afghanistan, the United States struggled to meet the demand for heavy forces. Given the potential for an increase in demand for heavy troops to that level, it may make sense for the Army to explore increasing its heavy force capability, at a minimum, as a hedge against the risk that it might face that level of heavy force demand in the future. This could mean ensuring that sufficient numbers of personnel are trained in necessary occupations and might also suggest the need for careful and strategic investment to fill any equipment gaps or to support innovations and upgrades. The Army's fiscal year 2019 budget suggests that it may already be moving in this direction.

Risks of U.S. Isolationism

While the results of each of our scenarios provide useful insights, our U.S. Isolationism scenario highlights a potentially important dynamic given contemporary policy debates. In the wake of recent wars in Iraq and Afghanistan, it is not surprising that many in the policy community have expressed an interest in pulling back and reducing U.S. military activity overseas. There are also concerns among some policymakers and segments of the public about the constraints and entanglements of alliances, multinational institutions, and multilateral trade agreements, and some leaders have suggested that the United States would be better off without these commitments and relationships. Our projections for this scenario, however, provide a note of caution regarding the potential effects of such changes. Our models suggest that a broad-based reduction in U.S. engagement internationally may ultimately increase the risk of interstate war.

While some of these wars may occur in regions where U.S. policymakers would no longer feel compelled to intervene, such as the Middle East, this may not necessarily be the case everywhere. Our scenario generally projects a modest decline in U.S. troop commitments overseas. However, it also projects that in East/Southeast Asia a decline in U.S. troop commitments to deterrence missions would be more than offset by an increase in the average projected number of U.S. troops committed to combat missions in that region, resulting in an overall increase in U.S. troop presence there. While it is difficult to isolate whether the increased risk of and commitment to combat missions in the region was driven by the reduction in deterrence commitments specifically or to other aspects of the scenario, such as reductions in U.S. alliance commitments or support for international institutions and norms, this correlation is still worth bearing in mind, and it highlights that not all efforts to reduce U.S. commitments overseas may prove to be cost-effective in the long run.

Acknowledgments

The authors are grateful to the Deputy Chief of Staff, G-3/5/7 (Operations, Plans, and Training), Headquarters, Department of the Army (HQDA) for sponsoring this study. We are especially appreciative of support from MG Christopher McPadden and MG William Hix. We also thank Tony Vanderbeek and Mark Calvo for monitoring the study and providing constructive feedback during the course of research. Jonathan Moyer (University of Denver) and RAND colleagues Trevor Johnston and Maria DeYoreo provided invaluable reviews that greatly improved the final report.

The authors are grateful for the contributions of several RAND colleagues, including Bryan Rooney, Joe Cheravitch, Abby Doll, Dan Elinoff, Ben Harris, Mark Toukan, and Sean Zeigler. We also thank Tim Bonds and Sally Sleeper for their consistent support since the start of this project.

Abbreviations

AUC	area under the curve
BUR	Bottom-Up Review
CAPE	Cost Assessment and Program Evaluation
CINC	Composite Index of National Capability
CoW	Correlates of War
DMDC	Defense Manpower Data Center
DoD	U.S. Department of Defense
DoDD	Department of Defense Directive
DSCA	Defense Support to Civilian Authorities
GDP	gross domestic product
MTW	major theater war
ORBAT	order of battle
OSD	Office of the Secretary of Defense
PAE	Program Analysis and Evaluation
PITF	Political Instability Task Force
PRIO	Peace Research Institute Oslo
ROC	receiver operating characteristic
RUGID	RAND U.S. Ground Intervention Dataset
QDR	Quadrennial Defense Review
TO&E	table of organization and equipment
UCDP	Uppsala Conflict Data Program

CHAPTER ONE

Introduction

Assessments about current and emerging threats inform a range of different military decisions that are central to the military's ability to prepare for future operations: decisions about what types of equipment to invest in, how to train military personnel, where to station U.S. Army forces abroad, and how to manage units' deployment cycles. These assessments are informed by historical data, analyses of current trends, simulations, scenarios, and expert opinion.

To defend against potential threats, the U.S. Army devotes significant resources to strategic and operational planning. But U.S. Army planning is an exercise in risk management across a wide array of potential threats facing the United States. U.S. Army plans support the larger U.S. Joint Military Force, and Army resources are allocated to each threat based on policymakers' priorities about which threats pose the greatest danger to U.S. national security. This resourcing necessarily comes with knowledge that there are gaps in how U.S. ground forces prepare for and respond to individual threats as policymakers prepare U.S. ground forces for the contingencies that pose the greatest strategic risk to the United States while consciously accepting risk in deemphasizing preparations for other missions.

Past experience clearly reveals the limitations of planning tools that do not adequately account for the full spectrum of missions potentially undertaken by U.S. ground forces, and U.S. ground forces have not always been optimally positioned and prepared to respond to emerging conflicts and crises around the globe. U.S. ground forces, for instance, were far from fully prepared to conduct the type of counterinsurgency and stabilization missions they were required to undertake in Iraq and Afghanistan following the initial U.S. invasion and combat phases against the Iraqi military and Taliban, respectively. In the early years of those counterinsurgencies, U.S. ground forces lacked appropriate training and equipment. The military overall was too small in 2003 to handle the demands created by those simultaneous conflicts, and shortages in certain types of occupations and billets were severe throughout the U.S. military. Doctrine and processes were also insufficient for the evolving demands of the conflict.[1]

[1] See for example, Steven Metz, *Learning from Iraq: Counterinsurgency in American Strategy*, Carlisle, Pa.: U.S. Army War College, Strategic Studies Institute, January 2007; R. Jeffrey Smith, "Military Admits Major Mistakes

But by 2012, when U.S. ground forces had left Iraq and were drawing down in Afghanistan, the U.S. military had adapted to conditions on the ground, with training programs designed to mimic the counterinsurgency environment, better equipment, and new guidance to shape operational planning and strategy. But because future operations are unpredictable and unlikely to be exactly like the past, this cycle is likely to repeat itself in the future, compromising military efficacy and placing additional strain and risk on U.S. ground forces.

The U.S. Army experience in Iraq and Afghanistan highlights the dangers of force-planning tools that are misaligned with a changing strategic landscape. Perhaps more importantly, those counterinsurgency experiences, which appeared relatively unlikely prior to the terrorist attacks of September 11, 2001, highlight that policymakers' plans need to be both robust and flexible enough to accommodate multiple strategic futures, and that even unlikely events can significantly shape Army force requirements. To avoid those kinds of pitfalls, military planners need better tools to forecast future operating environments to determine future force requirements and preemptively shape and prepare U.S. Army forces for the kinds of missions they are most likely to encounter in future environments. Along those lines, to assist policymakers with the prioritization of future threats and risk management, we provide in this report a modeling approach that allows policymakers to dynamically develop robust forecasts of future trends in armed conflict and U.S. ground interventions and to systematically assess U.S. ground force requirements across a wide array of potential strategic environments

The Evolution of DoD Efforts to Anticipate Future Force Requirements

How much of and which kinds of military forces should the United States build and maintain for future missions? This question has been at the center of U.S. defense strategy debates for decades, and the processes the United States uses to answer this question have been continuously evolving.

During the late Cold War, U.S. military force planning was relatively straightforward. The Soviet Union (and, to a lesser extent, China and other minor Communist powers) provided a clear "pacing threat" against which the United States measured its capabilities. As the Soviet Union first retrenched and then collapsed, the United States required other standards by which to estimate its capability requirements. Beginning with the Bottom-Up Review (BUR) in 1993, the United States adopted the two "major regional conflict" (later known as the two "major theater war," or 2-MTW) standard—that is, the standard that U.S. military forces be sized to win two major

in Iraq and Afghanistan," *The Atlantic*, June 11, 2012; Michael Peck, "Don't Hold Your Breath for 'Sim Afghanistan," *Wired*, October 1, 2009.

regional conflicts nearly simultaneously. Although these two wars could hypothetically occur anywhere around the world, in practice 2-MTW force requirements were based on near-simultaneous wars against Iraq and North Korea.[2]

The great advantage of the 2-MTW standard set by the BUR was its clarity—it provided an intuitively understandable standard to outside audiences while conveying precise requirements for force planners. This force-sizing construct, however, suffered from two limitations. First, by focusing on a clear, concrete, and specific threat, it did a poor job of reckoning with the inherent uncertainty of international relations and potential threats—a weakness that became even more evident after the September 11th attacks and subsequent large-scale stabilization operations in Afghanistan and Iraq.[3] Second, it failed to reckon explicitly with the United States' many small-scale force commitments in Bosnia, Kosovo, the Sinai Peninsula, and elsewhere.[4] These forces could be re-deployed for higher-priority contingencies, such as a major theater war, but these minor commitments nonetheless decreased many kinds of military readiness and could slow deployment times for forces that had to be extracted from ongoing operations.

High-level defense planning processes in the 1990s suffered from shortcomings in the bureaucratic process, as well. The Joint Staff J-8 was responsible for joint warfighting assessments with some support from the Program Analysis and Evaluation (PAE) office within the Office of the Secretary of Defense (OSD), but much of the more detailed work was done by each military service within the U.S. Department of Defense (DoD). These parallel analytic efforts were typically based on different data sources, and there was little transparency about the assumptions informing the analyses or the outputs' sensitivity to alternative assumptions, leading to a largely disparate series of analytical efforts and force-planning constructs.

These recognized shortfalls of U.S. defense planning in the 1990s combined with legislative requirements to lay the foundations for what was initially known as the Analytic Agenda and later as Support for Strategic Analysis. The National Defense Authorization Act for Fiscal Year 2000 mandated that DoD routinely undertake a Quadrennial Defense Review (QDR) similar to the one that had first been conducted in 1997. These QDRs were intended to "define sufficient force structure, force modernization

---

[2] For an overview of the various defense strategy reviews of the 1990s and early 2000s, see Raphael S. Cohen, *The History and Politics of Defense Reviews*, Santa Monica, Calif.: RAND Corporation, RR-2278, 2018.

[3] Paul K. Davis, *Capabilities for Joint Analysis in the Department of Defense: Rethinking Support for Strategic Analysis*, Santa Monica, Calif.: RAND Corporation, RR-1469, 2016.

[4] By the second half of the 1990s, the United States was involved in many such missions, but these missions were not represented in the scenarios used to approximate future force-structure requirements. Eric V. Larson, Derek Eaton, Michael E. Linick, John E. Peters, Agnes Gereben Schaefer, Keith Walters, Stephanie Young, H. G. Massey, and Michelle Darrah Ziegler, *Defense Planning in a Time of Conflict: A Comparative Analysis of the 2001–2014 Quadrennial Defense Reviews, and Implications for the Army*, Santa Monica, Calif.: RAND Corporation, RR-1309, 2018.

plans, infrastructure, budget plans, and other elements of the defense program of the United States associated with that national defense strategy that would be required to execute successfully the full range of missions called for in that national defense strategy."[5] To develop the evidentiary and analytic foundation for the judgments now required by law, in 2002 DoD issued DoD Directive (DoDD) 8260.1. This directive established policy and assigned responsibilities among DoD offices to "generate, collect, develop, maintain, and disseminate data on current and future U.S. and non-U.S. forces in support of strategic analysis conducted by the Department of Defense."[6] More broadly, it laid the groundwork for a system in which DoD planners would develop scenarios for a range of potential future contingencies and concepts of operations given certain assumptions about U.S. force size and mix, posture, readiness, and so on. This system became more comprehensive and institutionalized throughout the 2000s, eventually culminating in DoDD 8260.05, *Support for Strategic Analysis*, issued in 2011.[7] The evolution in DoD's thinking about future force requirements is illustrated in Figure 1.1.[8]

Figure 1.1
Evolution in DoD Efforts to Forecast Demand

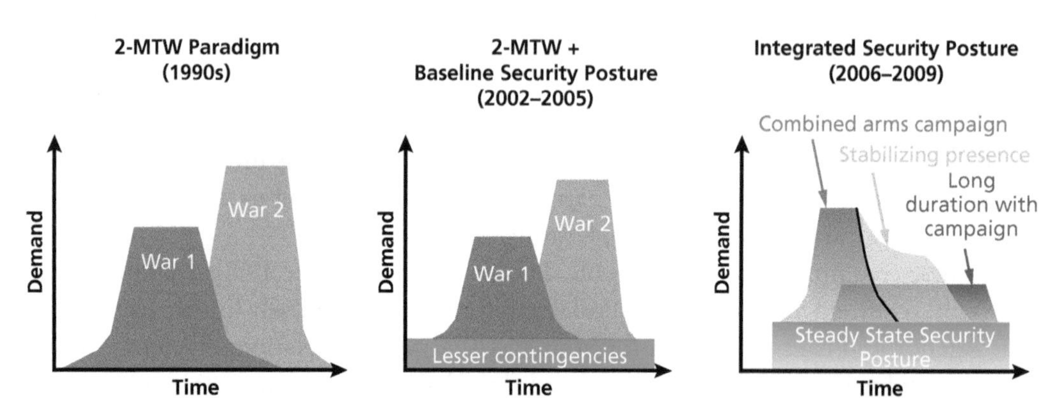

SOURCE: James R. Mitre, *Force Planning and Scenario Development*, Washington, D.C.: Office of the Secretary of Defense (Policy), May 5, 2015, not available to the general public.
NOTE: The contingencies represented in these figures are illustrative; they do not reflect actual DoD scenarios or force planning constructs.

---

5 Larson et al., 2018.

6 DoDD 8260.1, *Data Collection, Development, and Management in Support of Strategic Analysis*, Washington, D.C.: U.S. Department of Defense, December 6, 2002.

7 DoDD 8260.05, *Support for Strategic Analysis (SSA)*, Washington, D.C.: U.S. Department of Defense, July 7, 2011.

8 The third graphic in Figure 1.1 represented a much longer period of time than the others—years instead of months. It also represents only one of the possible force-sizing constructs; a 2-MTW construct was also in use at this time.

The Support for Strategic Analysis process made several additional improvements over old methods, including especially integrating a wide range of actors from the OSD, the Joint Staff, and the military services. It also developed common data and analysis into analytical baselines that could be used as jumping-off points for further, service-specific analyses. However, the system also had its critics. One report summarized complaints, noting that it was time-intensive, complex, and difficult to discuss with non-DoD and nontechnical audiences.[9]

Regardless of its specifics, however, DoD requires a process to estimate demands across a wide range of contingencies and over long periods of time, but one that is much less labor-intensive than the Support to Strategic Analysis process. Ideally, the process to estimate future force requirements would also be transparent (unclassified and understandable to policy audiences beyond those that are narrowly specialized in defense planning) and based in rigorous analysis of data.

A New Approach to Forecasting Demand?

In the case of military planning, where uncertainty about the future is both high and potentially costly, the value of better forecasting tools that can more accurately predict the timing, location, and nature of future demands on U.S. forces is clear. This is especially the case as the Army works to implement the National Defense Strategy and its call for Dynamic Force Employment that will enable a global response through a scalable, flexible, and deployable force.[10] Better forecasts can lead to a better-prepared and more efficient military by enabling forces that are more appropriately trained for future conflicts and a force structure and posture that increases both responsiveness and necessary capabilities. Current forecasting methods, however, fall short of this ideal. As a result, there is space for new, more flexible forecasting methods that can provide more sophisticated and detailed inputs into the Army planning process.

Advances in forecasting tools and methodologies provide a starting point for the development of improved approaches to forecasting demand for U.S. ground forces. The study of forecasting tools in areas related to conflict projection and early warning is not new, having roots in the 1960s and the first efforts to systematically study the causes and consequences of war.[11] Early work focused largely on event-based forecast-

---

[9] Davis, 2016.

[10] U.S. Department of Defense, *Summary of the 2018 National Defense Strategy of the United States of America*, Washington, D.C., 2018.

[11] See for example, Edward E. Azar, R. D. McLaurin, Thomas Havener, Craig Murphy, Thomas Sloan, and Charles H. Wagner, "A System for Forecasting Strategic Crises: Findings and Speculations About Conflict in the Middle East," *International Interactions*, Vol. 3, No. 3, 1977; Edward E. Azar, "The Conflict and Peace Databank (COPDAB) Project," *Journal of Conflict Resolution*, Vol. 24, No. 1, 1979; Charles A. McClelland and Gary D. Hoggard, *Conflict Patterns in the Interactions Among Nations*, Los Angeles, Calif.: University of Southern Califor-

ing and providing early warnings to policymakers with short-term projections.[12] Later approaches began incorporating algorithms that were used to build larger datasets that could be used for independent and dependent variables in forecasting models. These models could produce more fine-grained forecasts, often at the weekly or daily level, and could be applied to conflict events and domestic policy problems.[13]

More recent conflict forecasting efforts have varied in their methodology and approach, with many new efforts relying on advanced and complex statistical models. Most have focused on developing models that predict conflict onset or escalation in the near and medium term (up to two years in the future). One of the better-known efforts was the Political Instability Task Force (PITF), which attempted to predict different types of political instability, ranging from coups to conflict to revolutions, two years before they occurred. The PITF developed functioning statistical models that used only a small number of variables, as well as those with many more covariates.[14] PITF models were able to predict 80 percent of conflicts over a two-year window.

Developments over the past several years, however, have taken the sophistication of conflict forecasts far beyond the simple PITF models. In terms of time frames, some ongoing efforts continue to rely on daily or monthly data to predict changes in conflict levels in the short term, while others utilize country-year data to produce forecasts that look out several years and even decades. There have been relatively fewer attempts at

---

nia Press, 1968; J. David Singer and Michael D. Wallace, eds., *To Augur Well: Early Warning Indicators in World Politics*, Beverly Hills, Calif.: Sage, 1979.

[12] See for example, Benjamin E. Bagozzi, "Forecasting Civil Conflict with Zero-Inflated Count Models," *Civil Wars*, Vol. 17, No. 1, 2015; Patrick T. Brandt, John R. Freeman, and Philip A. Schrodt, "Real Time, Time Series Forecasting of Inter- and Intra-State Political Conflict," *Conflict Management and Peace Science*, Vol. 28, No. 1, 2011; Bruce Bueno de Mesquita, *Predicting Politics*, Columbus, Ohio: Ohio State University Press, 2002; Thomas Chadefaux, "Early Warning Signals for War in the News," *Journal of Peace Research*, Vol. 51, No. 1, 2014; David Muchlinski, David Siroky, Jingrui He, and Matthew Kocher, "Comparing Random Forest with Logistic Regression for Predicting Class-Imbalanced Civil War Onset Data," *Political Analysis*, Vol. 24, No. 1, 2016; Sean P. O'Brien, "Crisis Early Warning and Decision Support: Contemporary Approaches and Thoughts on Future Research," *International Studies Review*, Vol. 12, No. 1, 2010; Michael D. Ward, Nils W. Metternich, Cassy L. Dorff, Max Gallop, Florian M. Hollenbach, Anna Schultz, and Simon Weschle, "Learning from the Past and Stepping into the Future: Toward a New Generation of Conflict Prediction," *International Studies Review*, Vol. 15, No. 4, 2013.

[13] Brandt, Freeman, and Schrodt, 2011; Andy Doyle, Graham Katz, Kristen Summers, Chris Ackermann, Ilya Zavorin, Zunsik Lim, Sathappan Muthiah, Patrick Butler, Nathan Self, Liang Zhao, Chang-Tien Lu, Rupinder Paul Khandpur, Youssef Fayed, and Naren Ramakrishnan, "Forecasting Significant Societal Events Using the Embers Streaming Predictive Analytics System," *Big Data*, Vol. 2, No. 4, 2014; Philip A. Schrodt, Shannon G. Davis, and Judith L. Weddle, "Political Science: KEDS—A Program for the Machine Coding of Event Data," *Social Science Computer Review*, Vol. 12, No. 4, 1994; Philip A. Schrodt and Deborah J. Gerner, "Cluster-Based Early Warning Indicators for Political Change in the Contemporary Levant," *American Political Science Review*, Vol. 94, No. 4, 2000.

[14] Jack A. Goldstone, Robert H. Bates, David L. Epstein, Ted Robert Gurr, Michael B. Lustik, Monty G. Marshall, Jay Ulfelder, and Mark Woodward, "A Global Model for Forecasting Political Instability," *American Journal of Political Science*, Vol. 54, No. 1, 2010.

the latter, largely because of the complexity of forecasting complex social and political interactions and accounting for future changes in underlying relationships or key parameters over an extended period.[15] Previous RAND Arroyo Center research tackled this challenge of predicting long-term conflict trends at the global and regional level under various scenarios using projected data from the International Futures Project to estimate key factor trends.[16] This work illustrated that developing reasonable estimates of conflict several decades in the future is possible and even useful for thinking about regions that may be at risk for higher levels of conflict under various scenarios.[17]

Another more recent example of long-term conflict forecasting by Hegre et al. (2013) also used country-year data to predict changes in armed conflict from 2010 through 2050. That analysis relied on a dynamic estimation technique that proceeds in several stages. The first stage involves modeling the likelihood of conflict and identifying key predictors. The second stage generates forecasts using projected values for key covariates. The model is iterative, in the sense that it cycles through these two stages repeatedly, as it generates the conflict forecasts. The dynamic and iterative nature of this simulation allows not only for the prediction of conflict onsets but also overall incidence, as it is able to capture the duration of each predicted conflict.[18] Dynamic models can be particularly useful in generating longer-term estimates of conflict incidence, which depends on a constantly changing set of interactions between variables.[19]

Another important innovation in the area of conflict forecasting involves efforts to disaggregate temporal and spatial dynamics to provide more accurate and targeted forecasts—for example, identifying specific districts or "grids" where conflict is most likely.[20] Weidmann and Ward (2010), for instance, show clearly that incorporating geography and spatial mechanisms into conflict models greatly improves their predic-

---

[15] Stephen Watts, Jennifer Kavanagh, Bryan Frederick, Tova C. Norlen, Angela O'Mahony, Phoenix Voorhies, and Thomas S. Szayna, *Understanding Conflict Trends: A Review of the Social Science Literature on the Causes of Conflict,* Santa Monica, Calif.: RAND Corporation, RR-1063/1-A, 2017c.

[16] Frederick S. Pardee Center for International Futures, website, undated; Thomas S. Szayna, Angela O'Mahony, Jennifer Kavanagh, Stephen Watts, Bryan Frederick, Tova C. Norlen, and Phoenix Voorhies, *Conflict Trends and Conflict Drivers: An Empirical Assessment of Historical Conflict Patterns and Future Conflict Projections*, Santa Monica, Calif.: RAND Corporation, RR-1063-A, 2017; Stephen Watts, Bryan Frederick, Jennifer Kavanagh, Angela O'Mahony, Thomas S. Szayna, Matthew Lane, Alexander Stephenson, Colin P. Clarke, *A More Peaceful World? Regional Conflict Trends and U.S. Defense Planning*, Santa Monica, Calif.: RAND Corporation, RR-1177-A, 2017a.

[17] Watts et al., 2017a.

[18] Håvard Hegre, Joakim Karlsen, Håvard Mokleiv Nygård, Håvard Strand, and Henrik Urdal, "Predicting Armed Conflict, 2010–2050," *International Studies Quarterly*, Vol. 57, No. 2, 2013.

[19] Håvard Hegre, Halvard Buhaug, Katherine V. Calvin, Jonas Nordkvelle, Stephanie T. Waldhoff, and Elisabeth Gilmore, "Forecasting Civil Conflict Along the Shared Socioeconomic Pathways," *Environmental Research Letters*, Vol. 11, No. 5, 2016.

[20] Lars-Erik Cederman and Nils B. Weidmann, "Predicting Armed Conflict: Time to Adjust Our Expectations?" *Science*, Vol. 355, No. 6324, 2017.

tive accuracy.[21] Incorporating geo-located data opens doors to building forecasting models that appear more realistic and are able to capture a larger number of factors that may influence predictive accuracy. For example, Witmer et al. (2017) used geolocation to build a climate-sensitive approach to forecasting models of conflict in sub-Saharan Africa that are able to assess how changing temperatures and their effects on population and other variables will affect regional conflict.[22]

The forecasting models described thus far rely on theoretically driven statistical approaches, but there is also growing attention to forecasting conflict using machine learning tools to predict conflict.[23] Models that rely on machine learning respond, in part, to frustration with the low accuracy of traditional theoretically driven statistical models for prediction, given the rarity of conflicts. In the simplest terms, machine learning tools, such as the Random Forest approach, use algorithms and the specified dependent and independent variables to determine relationships between variables and to generate forecasts.[24] The advantages of this approach are that it can lead to more accurate forecasts, is robust to outliers, and deals well with missing data. On the other hand, it does not provide the same information about the causal relationships and interactions between factors that a statistical model would. These relationships may be very important to understanding the drivers of conflict and interventions.[25] Muchlinski et al. (2016), for example, used a Random Forests approach to project civil war onset and found that this approach has better predictive accuracy than several different logit specifications. Their analysis was able to identify the most important covariates, as well, and, for the most part, the results are similar to the covariates identified in the statistical literature as drivers of conflict onset.[26] The ViEWS project underway at Uppsala University also includes machine learning methods in its conflict estimates, but the researchers use these as a supplement to dynamic simulations like those described above. The ViEWS models also incorporate spatial and temporal data, making them potentially "state-of-the-art" in terms of applying all recent advances in forecasting methodologies. The ViEWS researchers argue that this "one step ahead" forecasting allows them to combine the strengths of both machine learning and classi-

---

21 Cederman and Weidmann, 2017.

22 Frank D. W. Witmer, Andrew M. Linke, John O'Loughlin, Andrew Gettelman, and Arlene Laing, "Subnational Violent Conflict Forecasts for Sub-Saharan Africa, 2015–65, Using Climate-Sensitive Models," *Journal of Peace Research*, Vol. 54, No. 2, 2017.

23 Cederman and Weidmann, 2017.

24 David S. Siroky, "Navigating Random Forests and Related Advances in Algorithmic Modeling," *Statistics Surveys*, Vol. 3, 2009.

25 Muchlinski et al., 2016.

26 Muchlinski et al., 2016.

cal statistical approaches.[27] The models forecast conflict at the monthly level and initial diagnostics show strong performance. Machine learning approaches can also prove useful for forecasting other types of events such as leadership changes,[28] and they can even be applied to help inform and improve model development.[29]

Looking to the future, two general trends characterize current conflict-oriented forecasting. The first is a trend toward a greater focus on prediction of new conflicts and out-of-sample forecasting, rather than explanations of the causes of conflict. An emphasis on understanding which factors are most important to predictive accuracy reflects the desire to understand not only which factors explain underlying trends, but also which are best at identifying future ones. Out-of-sample projections are also necessary to prevent model overfitting, which can limit the utility of forecasting models. In practice, however, out-of-sample projections can be challenging in the field of conflict research, which must forecast rare events that oftentimes occur for idiosyncratic reasons. Second, researchers are working to develop more transparent ways to report their results, including a focus on reporting the extent to which a model produces accurate out-of-sample forecasts (rather than just using a measure based on "area under the curve"), including a reporting of false positives and false negatives. From the perspective of policymakers and military planners, this is an important development, because even a highly accurate forecasting model that has lots of false positives or negatives will have little utility for policy decisions and force-planning purposes.[30]

The methodology used in this report very much moves in the direction of the trends outlined above, particularly that of the dynamic simulations just described. Despite these advancements, it is important to emphasize that forecasting conflict is necessarily complicated and difficult, given the many different actors and factors all interacting simultaneously in a system in which the fundamental dynamics are constantly changing.[31] However, from the perspective of forecasting force requirements, the methodology and predictions offered here represent an improvement over existing empirical approaches. That is, by relying on publicly available data and transparent methods, our forecasting approach provides both a flexible and transparent input

---

[27] Håvard Hegre, Marie Allansson, Matthias Basedau, Michael Colaresi, Mihai Croicu, Hanne Fjelde, Frederick Hoyles, Lisa Hultman, Stina Högbladh, Remco Jansen, Naima Mouhleb, Sayyed Auwn Muhammad, Desirée Nilsson, Håvard Mokliev Nygård, Gudlaug Olafsdottir, Kristina Petrova, David Randahl, Espen Geelmuyden Rød, Gerald Schneider, Nina von Uexkull, and Jonas Vestby, "ViEWS: A Political Violence Early-Warning System," *Journal of Peace Research*, Vol. 56, No. 2, 2019.

[28] Michael D. Ward and Andreas Beger, "Lessons from Near Real-Time Forecasting of Irregular Leadership Changes," *Journal of Peace Research*, Vol. 54, No. 2, 2017.

[29] Michael Colaresi, and Zuhaib Mahmood, "Do the Robot: Lessons from Machine Learning to Improve Conflict Forecasting," *Journal of Peace Research*, Vol. 54, No. 2, 2017.

[30] Håvard Hegre, Nils W. Metternich, Håvard Mokleiv Nygård, and Julian Wucherpfennig, "Introduction: Forecasting in Peace Research," *Journal of Peace Research*, Vol. 54, No. 2, 2017.

[31] Cederman and Weidmann, 2017.

to future generations of defense planning scenarios and integrated security postures based on key factors that can be parameterized and systematically varied to produce alternative projections. Still, all forecasts should be interpreted with caution. They can provide valuable information to structure thinking about the future, but even the best predictions have substantial associated uncertainty.

Objective of This Report

This report seeks to address the demand for more robust and systematic assessments of future demands for U.S. ground forces by presenting a dynamic forecasting model that projects trends in future conflict and the resulting demands on U.S. forces for combat, stabilization, and deterrent operations out to 2040. We present both a baseline, "no surprises" future that assumes most current trends in global structural conditions continue on their current path, as well as four alternative futures that make different assumptions about the future to help assess how changes in those global conditions could affect demands for U.S. forces abroad. To do so, we build on several years of RAND Arroyo Center work focused on identifying and forecasting conflict trends at the global and regional level and work that explored where and at what scale the United States is mostly likely to send ground troops overseas. Our model improves on, updates, and integrates these models to generate forecasts of future U.S. ground interventions, and we use historical data and trends to assess the approximate scale and force mix required for those forecasted interventions.

We aim to provide several different types of forecasts in this study. First, we project trends in future interstate and intrastate conflicts through 2040 at the global and regional levels. Second, we forecast trends in future U.S. ground interventions using forecasted future conflicts as a "demand signal" and our statistical models of U.S. ground interventions to determine where and when the United States is mostly likely to intervene abroad. We distinguish between deterrence, combat, and stabilization interventions and provide projections again at the regional level. Finally, we use historical data to develop heuristic rules of thumb for broad categories of military interventions (by context and operating environment) that provide baseline estimates for the average force requirements for the interventions predicted by our analyses. Our estimates, therefore, are more precise when discussing the regional location of conflicts and U.S. ground interventions but use general guidelines on the likely size and capabilities of those interventions to assess the characteristics of those forecasted interventions.

The forecasts presented in this report will be useful to the U.S. military generally and the U.S. Army in particular in several ways. First, our analysis identifies key factors that can serve as early warning indicators of future conflicts, U.S. ground interventions, or surges in demand for U.S. forces to enable early planning. Second, our analysis provides an improved empirical basis for estimating the frequency, magnitude, duration, and overlap of future contingencies to help military personnel plan for a scalable and tailorable force. Third, our forecasts may help increase the responsiveness of U.S. forces in the event of an overseas crisis by informing force posture decisions that more accurately position U.S. forces around the world. Finally, the methodology presented in this report also provides an empirical process for replacing the Support for Strategic Planning process to support steady state Army planning, filling an emerging gap with a more flexible and dynamic process.

Limitations

Our approach to forecasting future conflict trends and demands for U.S. ground forces comes with a number of limitations worth noting at the outset. First, any set of forecasts has predictive error associated with it. Empirical models are only approximations of the world; they omit variables and rely on particular assumptions about the future. Because the models presented in this report use one set of forecasts (conflict) and build another set of forecasts (interventions) on top, and then apply qualitative assessments of ground force characteristics using historical data, the error at each stage of our forecasting process will compound. There will, therefore, be some degree of uncertainty associated with our results, even under the best of circumstances.

A related limitation, however—the process of matching forces to predicted interventions—is more difficult to address. Although we use dynamic models to forecast conflicts and interventions, our models of intervention size are ill-suited to a dynamic forecasting process. As a result, we use historical patterns to identify how intervention size and the relative use of heavy and light forces vary based on key criteria (taken from empirical models) and then assign a "force package" to each projected intervention.[32] While this will give the Army a general estimate of expected size for each projected intervention, it will not provide fully precise estimates. This may, however, be enough for military planners, who need to consider deployment rules, available forces, force flow plans, and other considerations when making plans and decisions about intervention size.

The realities of existing data may also be a source of error that limits the accuracy of our estimates. For some factors that may affect the likelihood of conflict or the potential for an intervention, we may use a proxy that only imperfectly captures the underlying factor, or data may be inconsistently available across our historical series. For example, gross domestic product (GDP) per capita is often used to measure state capacity. This is likely reasonable, as states with higher GDP per capita likely also have higher state capacity, but it is not exact, and GDP per capita datasets are also more likely to have missing observations for countries and years that are poorer and more prone to conflict.

---

[32] This method and others are described in more detail in later chapters and in full detail in Appendixes A and B.

We have taken several steps to account for and minimize, to the extent possible, these statistical limitations. First, we implemented 500 repeated simulations of each model in an effort to gradually reduce the uncertainty in any individual projection through a larger number of forecasts. Second, in our discussion of model results, we are clear and explicit regarding the degree of confidence attached to each individual result, so that readers can interpret the trends projected in the appropriate context.

Another limitation concerns the difficulty of using statistical models to realistically model the demand for Army forces, given the existence of Army-specific constraints and processes and the fact that intervention decisions are actually surprisingly idiosyncratic. We have tried to navigate this challenge by developing empirical models that are as realistic as possible and by carefully considering which factors to include in our empirical models based on surveys of the existing literature and on previous RAND Arroyo Center empirical research on armed conflict and interventions.

Finally, the future strategic environment will quite likely be different than expectations and projections built on historical trends. In addition to our baseline projections, we consider a range of potential alternative futures developed by modifying model parameters based on experience and expectations. This approach helps us identify which trends in our projections are particularly sensitive—or not—to the nature of the anticipated future environment. One of our scenarios, for example, considers the implications of an isolationist United States, while another considers the impact of a global pandemic. In addition to providing insights into the robustness of different projected trends, this alternative futures approach is one of the greatest advantages and contributions of the models provided here: They can be easily modified to take into account new factors and new parameters to providing updated forecasts as threats evolve, emerge, or disappear. Because our alternative futures all exhibit characteristics that have historical precedents, our approach provides planners with archetypes that can be further modified to create and test new assumptions about the future, providing a way to flexibly and responsively adapt planning tools to a wide range of plausible futures.

Organization of This Report

The remainder of this report proceeds as follows. In the next chapter, we discuss historical trends in armed conflict and U.S. ground interventions that help to shape our expectations about the future and inform our modeling strategy. Chapter Three discusses the methodology used for the forecasts presented in this report. Chapter Four presents the results, and Chapter Five summarizes key results and insights, as well as implications for the Army. We also include three technical appendixes that provide additional details on our conceptual approach and overall forecasting methodology.

CHAPTER TWO

Historical Trends in Armed Conflict and U.S. Ground Interventions

The starting point for any forecasting effort is a careful assessment of historical trends. Even if the future is likely to be very different than the past, an analysis of historical data provides insight into past patterns, outliers, and possibly into key underlying factors or contextual influences. Historical patterns also provide an important foundation from which to compare and judge the forecasts presented elsewhere in this report. This chapter briefly describes the historical trends in armed conflicts and U.S. ground interventions—two of the key building blocks for assessing future demands for U.S. forces that will be utilized in our forecasting models discussed in subsequent chapters. Trends in armed conflicts provide important background because they identify potential opportunities for many types of U.S. ground interventions, either directly in those conflicts or in stability operations later on. Trends in U.S. ground interventions are helpful, not only for grounding our baseline expectations regarding the frequency and resources with which the U.S. Army may intervene in the future, but also for identifying the types of interventions that are most likely in particular circumstances.

Historical Trends in Armed Conflict

Armed conflict, both in terms of conflicts within and between states, has generally declined in frequency and intensity in recent decades. Intrastate armed conflicts declined in frequency and intensity beginning after the end of the Cold War, although they have seen a notable increase since 2012, as shown in Figure 2.1.[1] While the annual

[1] We assessed trends in intrastate armed conflict in the post–World War II period using data from the Uppsala Conflict Data Program (UCDP)/Peace Research Institute Oslo (PRIO) Armed Conflict Dataset, focusing on both lower-intensity intrastate armed conflicts and higher-intensity intrastate wars. The UCDP/PRIO Armed Conflict Dataset defines conflict as "a contested incompatibility that concerns government and/or territory where the use of armed force between two parties, of which at least one is the government of a state, results in at least 25 battle-related deaths." The UCDP/PRIO dataset also identifies intrastate wars as a conflict with more than 1,000 battle-related deaths. The UCDP/PRIO dataset defines battle-related deaths as any death "caused by the warring parties that can be directly related to combat." This means that traditional battlefield fighting, guerilla

number of new intrastate conflicts remained relatively stable, the annual number of ongoing intrastate armed conflicts rose precipitously between the end of World War II and the end of the Cold War, with numbers of ongoing intrastate conflicts peaking in the early 1990s after the collapse of the Soviet Union. Though the number of states in the international system, and therefore the number of potential venues for intrastate conflict, increased in the postcolonial period, making it difficult to directly compare trends from the early Cold War to the present, the consistent increase in levels of conflict throughout the Cold War suggests an increasingly conflict-prone international system and a potential shift toward intrastate conflict in the post–World War II world.

Even with a sharp increase in new intrastate conflicts through the 1990s, levels of intrastate conflict declined over the two decades following the Cold War and gradually returned to a level of armed activity well below the peak of the early 1990s. This lower level of armed conflict, both in terms of new and ongoing conflicts, remained relatively stable until about 2012, when both numbers of new and ongoing intrastate

Figure 2.1
Trends in Intrastate Conflict and Conflict Onsets, 1946–2016

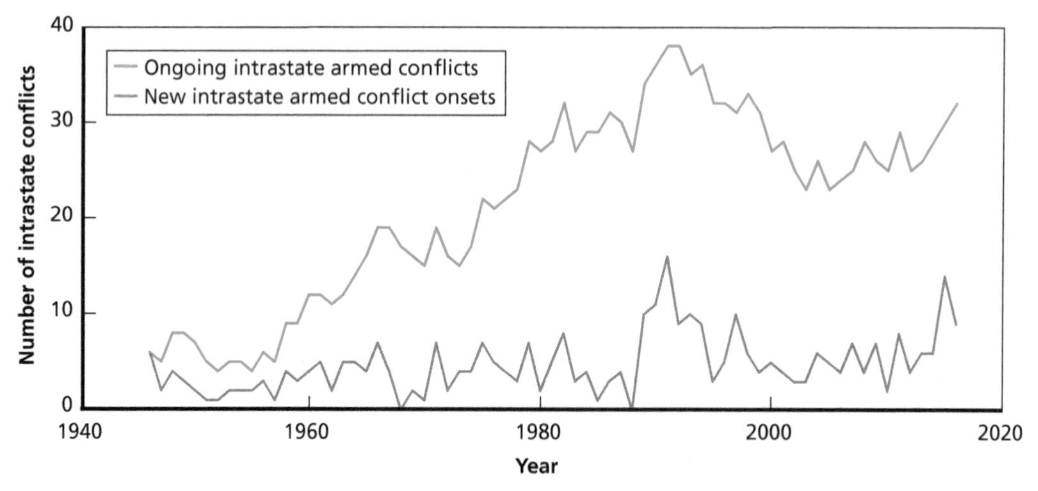

SOURCE: RAND Arroyo Center analyses using the UCDP/PRIO Armed Conflict Dataset (Gleditsch et al., 2002).
NOTES: The red line shows the total number of ongoing intrastate armed conflicts in the international system for each year between 1946 and 2016. The blue line shows the number of new intrastate armed conflict onsets for each year between 1946 and 2016.

warfare, urban warfare, and attacks on military bases are included, as are civilians killed in attacked on military targets. Civilians who die of starvation or other conditions are not considered in these totals. Injured persons who later die due to those injuries are included. For more information, see Marie Allansson, Erik Melander, and Lotta Themnér, "Organized Violence, 1989–2016," *Journal of Peace Research*, Vol. 54, No. 4, 2017; Nils Petter Gleditsch, Peter Wallensteen, Mikael Eriksson, Margareta Sollenberg, and Håvard Strand, "Armed Conflict, 1946–2001: A New Dataset," *Journal of Peace Research*, Vol. 39, No. 5, 2002.

conflicts precipitously increased again. At least some of this increase may reflect recent or increasing instability in certain regions, such as the Middle East and North Africa, which has facilitated the rise of new insurgent groups, such as ISIS, and the regional spread of transnational armed groups throughout areas of weakened governmental capacity.[2] Though current levels of intrastate conflict remain below the peak levels of the early 1990s, it is still unclear whether this recent trend is a temporary increase in conflict or the beginning of a long-term trend that will support further increases in intrastate conflict over the next several years.

In contrast to the rise and fall of intrastate armed conflicts, the decline in interstate wars has been considerably more pronounced and longer-lasting since the end of World War II. Interstate war has been comparatively rare throughout the post–World War II period, and particularly rare since the mid-1970s, as shown in Figure 2.2.[3] This overall trend culminated in the early 2000s; there have been no new interstate wars since 2003. This general decline in interstate war is particularly notable for its impact

Figure 2.2
Trends in Interstate War and War Onsets, 1900–2016

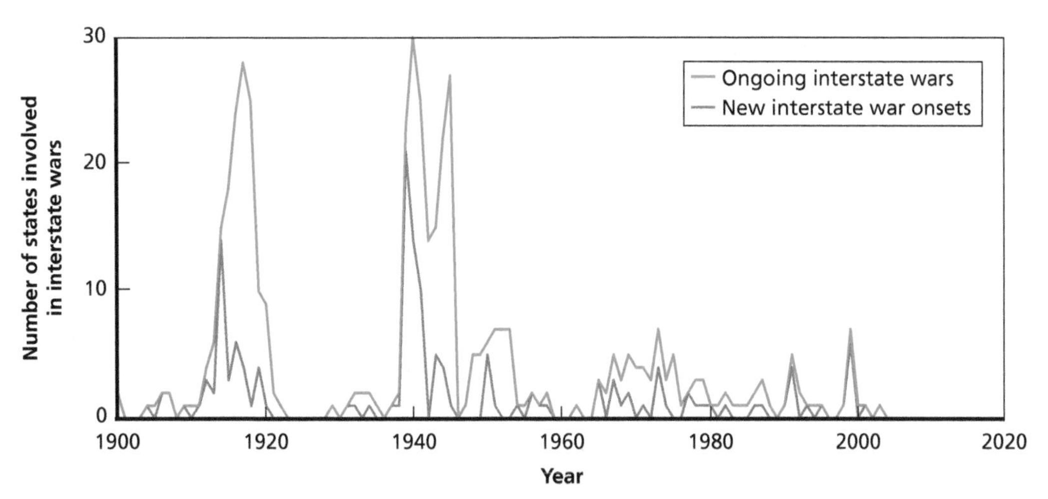

SOURCE: RAND Arroyo Center analyses using the UCDP/PRIO Armed Conflict Dataset (Gleditsch et al., 2002) and the CoW interstate war data (Correlates of War Project, undated).
NOTES: The red line shows the total number of states involved in ongoing interstate wars in the international system for each year between 1900 and 2016. The blue line shows the number of states involved in interstate war onsets each year between 1900 and 2015.

[2] Mark Mazzetti, *The Way of the Knife: The CIA, a Secret Army, and a War at the Ends of the Earth*, New York, N.Y.: Penguin Books, 2013; Joby Warrick, *Black Flags: The Rise of Isis*, New York, N.Y.: Penguin Random House, 2015; Michael Weiss and Hassan Hassan, *ISIS: Inside the Army of Terror*, New York, N.Y.: Reagan Arts, 2015.

[3] We focus, both in this chapter and in the remainder of the report, on interstate wars rather than lower-intensity interstate conflicts. Both interstate wars and conflicts have been relatively rare since 1945, meaning that including data prior to 1945 is particularly important for a longer-term understanding of historical trends. However, data on interstate conflicts are not as readily available prior to 1945. The data source that is available for this

on the aggregate intensity of overall conflict in the international system—in terms of battle-related deaths, interstate wars have typically been significantly more deadly than intrastate conflicts. As such, the overall severity of conflict has generally declined throughout the post–World War II period, even as levels of intrastate conflict have risen.

Regional Trends

While global-level trends are important, regional trends in armed conflict, particularly those that may diverge from global patterns, are also worthy of careful consideration. Figures 2.3 and 2.4 show regional trends in intrastate conflict and interstate war, respectively, in each of the geographic regions used in this analysis.

These figures suggest a number of key observations. Interstate war, while less common than intrastate conflict across the board, has been most frequent since 1946 in the Middle East, South Asia, and East and Southeast Asia. Intrastate conflict has been concentrated in four main regions: East and Southern Africa, East and Southeast Asia, the Middle East, and South Asia. No region has avoided intrastate conflict completely, and some regions have, as alluded to above, experienced some rise in the frequency of conflict as of late. This is especially true of West Africa, East and Southern Africa, and the Middle East.

Summary

This presentation of historical conflict trends highlights the long-term decline in both interstate war and intrastate conflict at both the global and the regional levels. Although levels of interstate war have sharply declined in recent decades, the decline in intrastate conflict has been much more moderate, and certain regions continue to frequently experience steady or even increasing levels of such violence.

There are several ways to interpret these trends. First, it is possible that conflict is simply becoming less frequent and less deadly over time.[4] Conversely, it is possible that we are merely in a temporary lull of high-intensity conflict, with a possible increase in conflict incidence approaching or beginning. Somewhat similarly, it is possible that we are now returning to a more normal level of armed conflict—that in fact it was the high levels of conflict during the Cold War that are the anomaly. Finally, it is possible

---

earlier period, from the Correlates of War Project (CoW), identifies only interstate wars that cause at least 1,000 battle-related deaths (Correlates of War Project, website, undated). To be consistent across the full historical period, we therefore focus only on interstate wars. This is in contrast to intrastate conflict, which is sufficiently frequent after 1945 when data on these lower-intensity events are available. For information on the CoW interstate war data, see Meredith Reid Sarkees and Frank Whelon Wayman, *Resort to War: 1816–2007*, Washington, D.C.: CQ Press, 2010.

[4] Tanisha M. Fazal, "Dead Wrong? Battle Deaths, Military Medicine, and Exaggerated Reports of War's Demise," *International Security*, Vol. 39, No. 1, 2014; Steven Pinker, *The Better Angels of Our Nature: The Decline of Violence in History and Its Causes*, London, UK: Penguin Books, 2011.

Figure 2.3
Regional Trends in Intrastate Conflict, 1946–2016

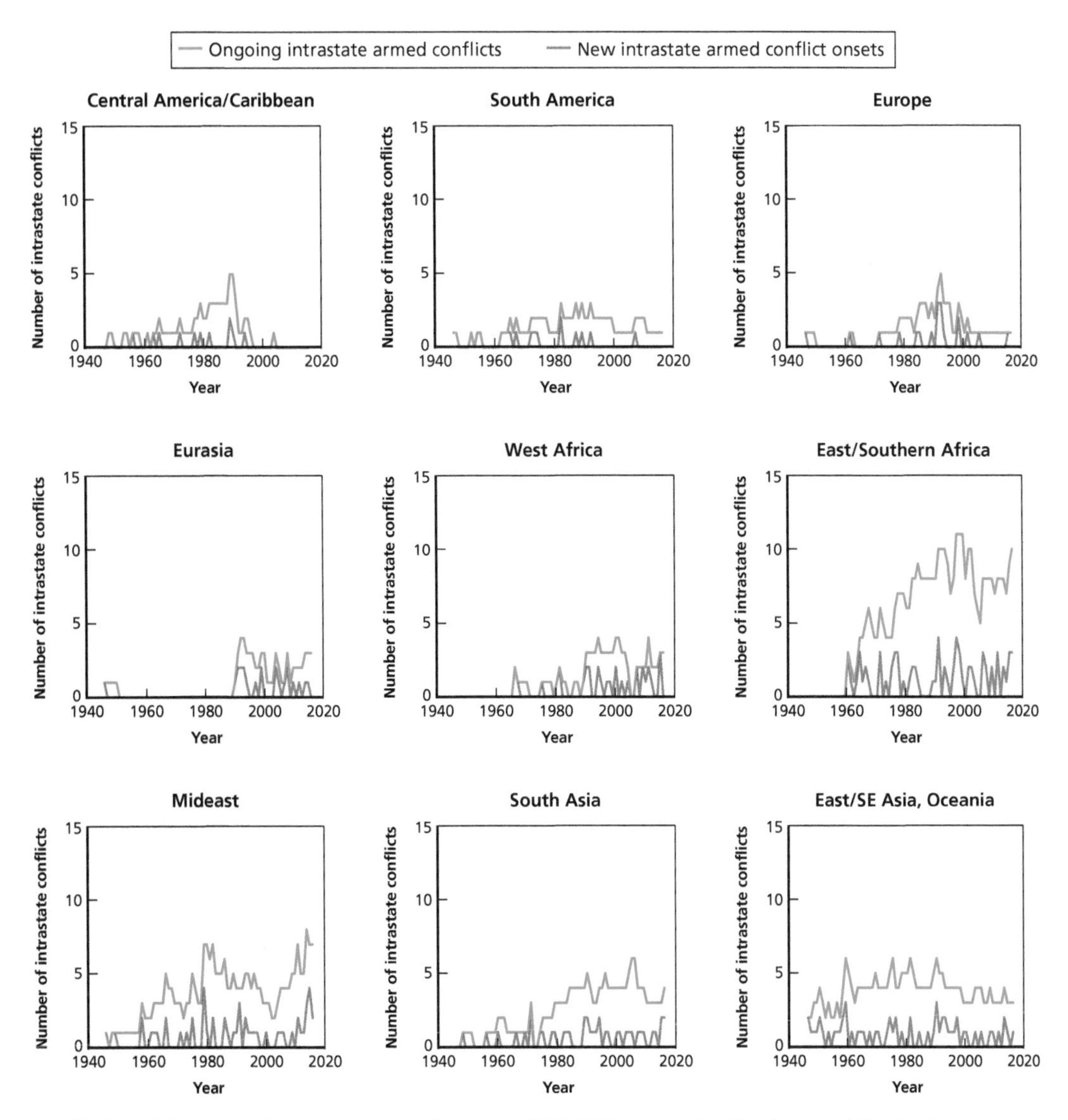

SOURCE: RAND Arroyo Center analyses using the UCDP/PRIO Armed Conflict Dataset (Gleditsch et al., 2002).
NOTES: The red line shows the total number of ongoing intrastate armed conflicts in the international system for each year between 1946 and 2016. The blue line shows the number of new intrastate armed conflict onsets for each year between 1946 and 2016.

that conflict is merely changing form, shifting from prolonged interstate war to more numerous, lower-intensity intrastate conflicts.[5]

---

5 Szayna et al., 2017.

Figure 2.4
Regional Trends in Interstate War, 1946–2016

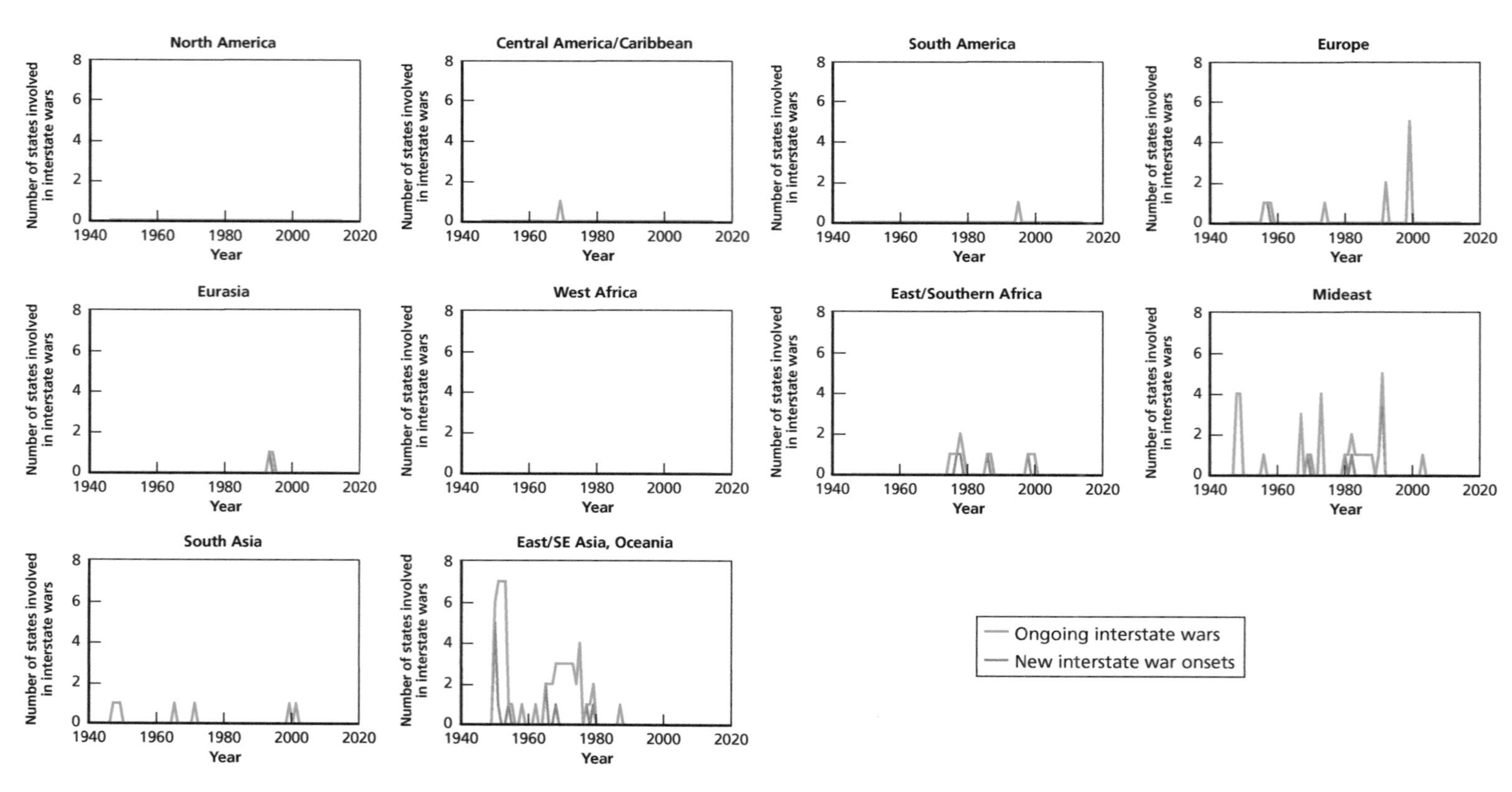

SOURCE: RAND Arroyo Center analyses using the UCDP/PRIO Armed Conflict Dataset (Gleditsch et al., 2002) and CoW interstate war data (Correlates of War Project, undated).
NOTES: The red line shows the total number of states in ongoing interstate wars in the international system for each year between 1900 and 2016. The blue line shows the number of states involved in interstate war onsets each year between 1900 and 2016.

In this report, the trends described here inform both our future conflict projections and our projections of future U.S. ground interventions, as will be discussed in detail in Chapter Three. In general, the more conflicts that occur, the more opportunities we expect there to be for the United States to intervene in them, or to stabilize their aftermath. The United States is by no means equally likely to intervene in all conflicts, however. It is therefore important to first survey historical patterns in U.S. ground interventions to provide an improved understanding of how, where, and when the United States has tended to employ its forces abroad.

Historical Trends in U.S. Ground Interventions

In many ways, we can consider the incidence of conflict in the international system to be a "demand signal" for U.S. forces. The United States, however, does not intervene every time it has the opportunity. Instead, the timing, geographic location, and types of interventions conducted by the United States have varied considerably in ways that are not linearly related to historical trends in armed conflict. This section describes the historical trends in U.S. ground interventions and highlights key patterns that informed the development of our intervention projection models, as discussed in Chapter Three.

Historical Trends

Prior RAND Arroyo Center data collection efforts identified 145 U.S. ground, air, and naval interventions since 1945, which we use to identify historical patterns of U.S. ground interventions.[6] To ensure consistency across time and avoid collecting interventions that are irrelevant for most policy purposes, ground interventions had to pass a size threshold to be included in this dataset—100 person-years of activity for each included intervention.

Figure 2.5 shows the total number of ongoing U.S. ground interventions in each year from 1946 to 2016. While the number of ongoing ground interventions has fluctuated over time, the United States experienced a marked increase in ground interventions in the period after the end of World War II, which reached a peak in 1960 before declining through the mid-1970s. The number of interventions increased again in the 1980s, fell at the end of the Cold War, and then increased again, reaching a level that has been more or less sustained in recent years.

We explored the regional distribution of U.S. ground interventions as well, as shown in Figure 2.6. U.S. ground interventions have historically been concentrated in four key regions. First, U.S. ground interventions in Central America and the Carib-

---

[6] All information included in this section is drawn from Jennifer Kavanagh, Bryan Frederick, Matthew Povlock, Stacie L. Pettyjohn, Angela O'Mahony, Stephen Watts, Nathan Chandler, John Speed Meyers, and Eugeniu Han, *The Past, Present and Future of U.S. Ground Interventions: Identifying Trends, Characteristics, and Signposts*, Santa Monica, Calif.: RAND Corporation, RR-1831-A, 2017.

Figure 2.5
Number of Ongoing U.S. Ground Interventions, 1946–2016

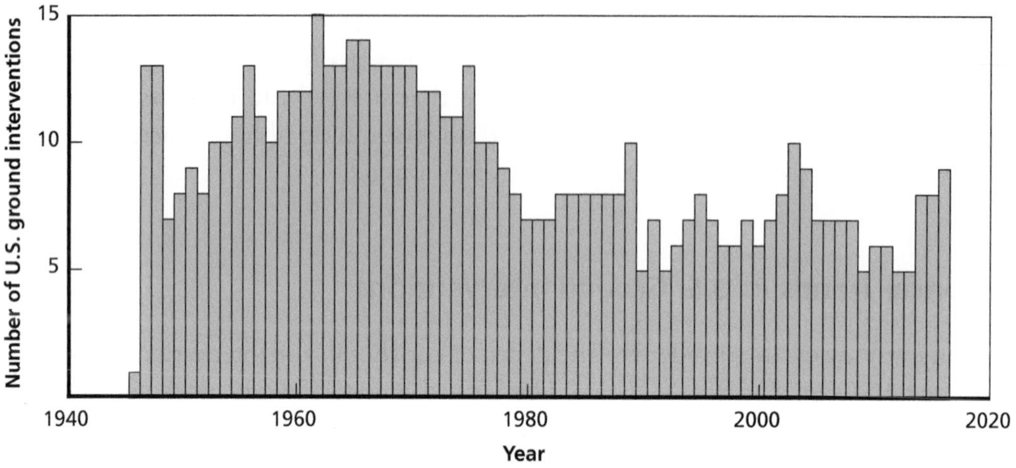

SOURCE: RAND Arroyo Center analyses using U.S. Ground Intervention Database (RUGID) data (Kavanagh et al., 2017).
NOTE: Bars show the total number of ongoing U.S. ground interventions for each year between 1946 and 2016.

bean were most common in the period leading up to World War II. Since then, interventions in Europe, the Middle East, and East and Southeast Asia have been most common. U.S. ground interventions in Africa, on the other hand, have been rare, and although there has been a slight increase in their frequency in recent years, those that have occurred have remained relatively small in size and limited in scope.

We also consider U.S. ground interventions by activity type. Since World War II, the majority of U.S. military operations have haven fallen into one of three primary types, as shown in Figure 2.7:[7]

1. **Armed conflict interventions:** Combat operations that occur in armed conflicts, including conventional warfare and counterinsurgency. This category includes major combat operations, such as the Vietnam war and the two World Wars; the wars in Iraq and Afghanistan; and also smaller operations such as in Panama.
2. **Stabilization interventions:** Stability operations that include efforts to maintain and institutionalize peace and stability either during a conflict or immediately following it, with activities such as peacekeeping and institution building. Examples include operations in the later parts of the Iraq and Afghanistan conflict, post–World War II rebuilding in Europe and Asia, and postconflict operations in Bosnia.

[7] Kavanagh et al., 2017.

Figure 2.6
Regional Distribution of U.S. Ground Interventions, 1946–2016

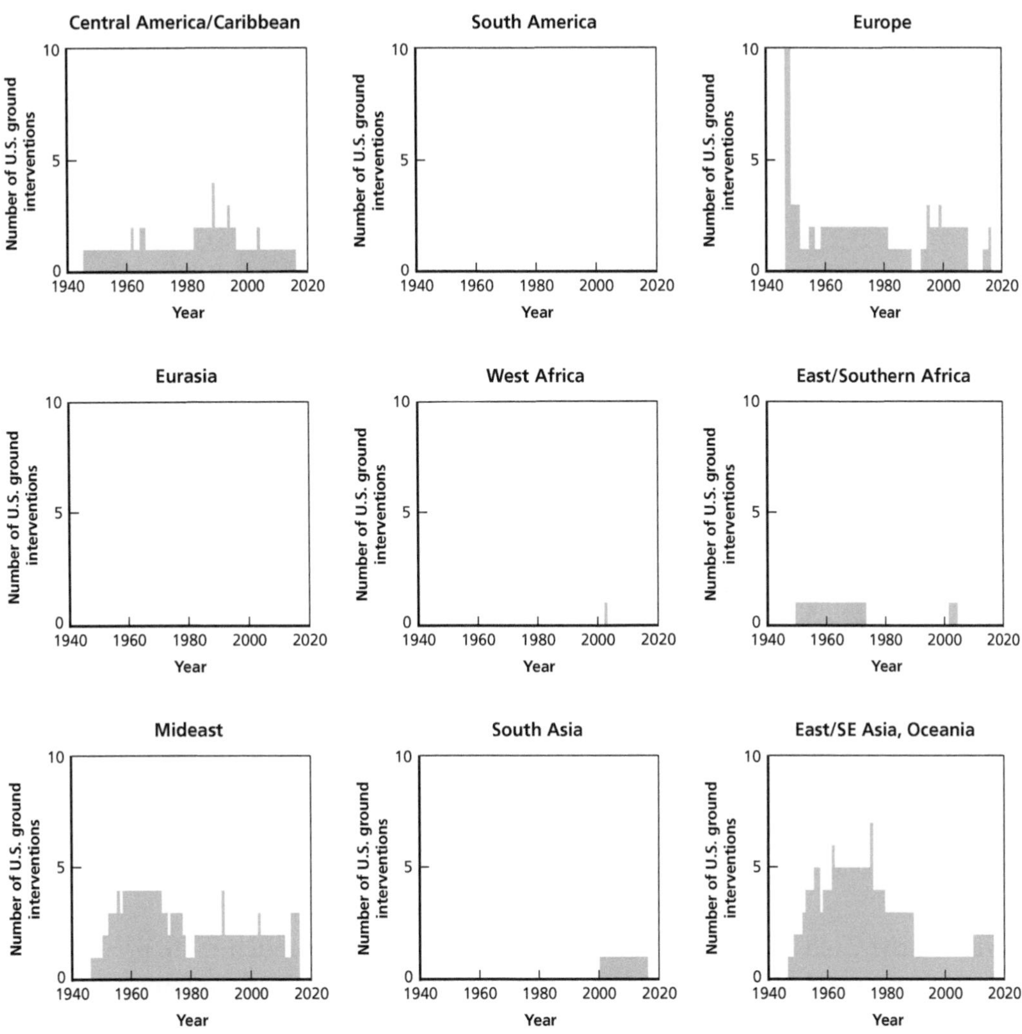

SOURCE: RAND Arroyo Center analyses using RUGID data (Kavanagh et al., 2017).
NOTE: Bars show the total number of ongoing U.S. ground Interventions for each year between 1946 and 2016.

3. **Deterrence Interventions**—deterrence operations intended to shape the behaviors of allies and adversaries, to protect allies and major U.S. interests, and to prevent allies from taking steps that might compromise these U.S. interests.

The frequency of these different activities has changed over time. In the years prior to and including World War II, armed conflict interventions into conventional wars was the largest single activity for U.S. forces. This seems to have shifted with the

Figure 2.7
Number of Ongoing U.S. Ground Interventions by Intervention Type, 1946–2016

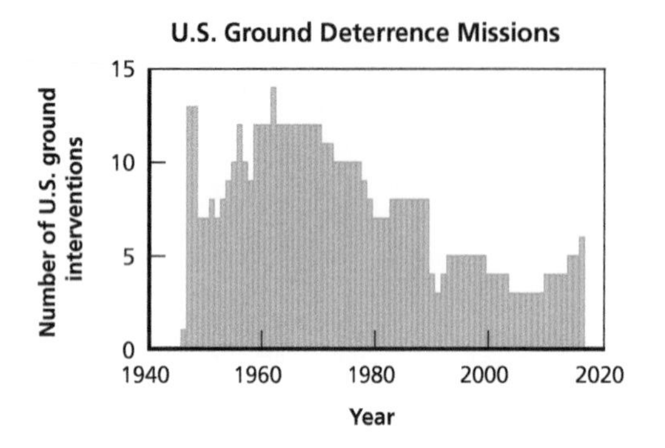

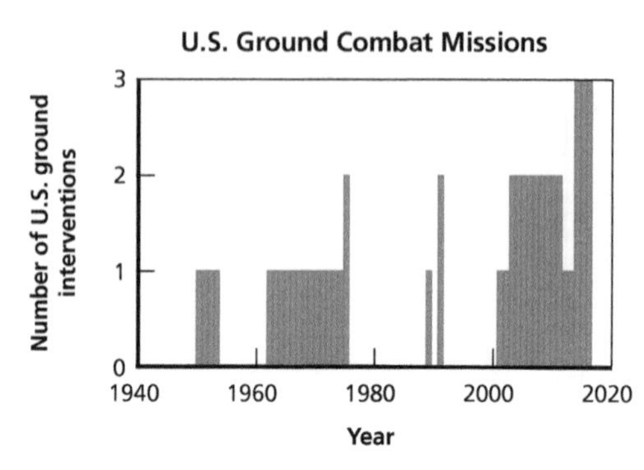

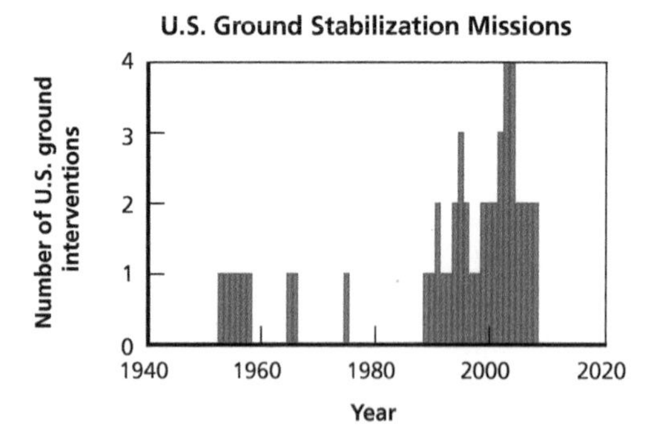

SOURCE: RAND Arroyo Center analyses using RUGID data (Kavanagh et al., 2017).
NOTES: The blue bars in the top left graph show the total number of ongoing U.S. ground force deterrent interventions for each year between 1946 and 2016. The red bars in the top right graph show the total number of ongoing U.S. ground force armed conflict interventions for each year between 1946 and 2016. The green bars in the bottom left graph show the total number of ongoing U.S. ground force stabilization interventions for each year between 1946 and 2016.

advent of the Cold War, when deterrence operations became much more prevalent. In recent years, meanwhile, forces committed to stabilization (and counterinsurgency) have become more numerous. This has been especially true since 2001.

Interventions of different types also vary in their size and duration—key aspects of the overall demand that interventions place on U.S. ground forces. Among U.S. ground interventions, interventions into armed conflicts tend to be larger; historically, almost 70 percent of combat operations have involved more than 20,000 troops, whereas only slightly more than half of stability operations and deterrence missions combined have reached this threshold. While armed conflict interventions have often been the largest missions undertaken, however, stability operations and deterrence missions have been significantly more likely to last longer than armed conflict interventions; 70 percent of deterrence missions and 60 percent of stability operations have lasted at least three years, respectively.

These general trends also hold among large U.S. ground interventions, and further reveal an additional interesting pattern. While only a small number of large combat operations have lasted for longer than three years, large deterrent and stability operations have been more likely to last longer than three years. Thus, although combat operations have been more likely to require large force commitments, large deterrent and stability operations may ultimately place more strain on U.S. ground forces due to their typically prolonged length.

These three activity types—combat, deterrence, and stability operations—have also accounted for the vast majority of U.S. ground forces employed in interventions abroad, as shown in Figure 2.8. Armed conflict interventions have historically driven massive, but short-lived, increases in demand for ground forces, as shown in the notable spikes around the world wars, the Vietnam War, the Korean War, and the 1991 Gulf War. Deterrent interventions, meanwhile, have tended to involve longer-lasting, more stable commitments of troops, constituting the dominant use of ground troops

Figure 2.8
Number of U.S. Ground Troops by Intervention Type, 1946–2016

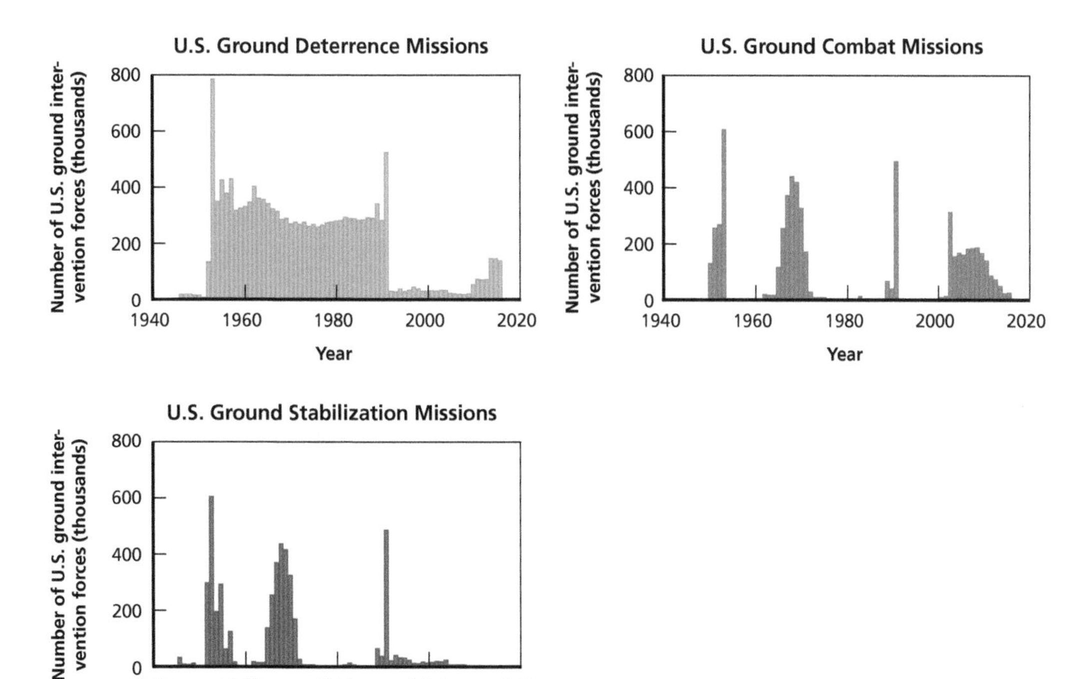

SOURCE: RAND Arroyo Center analyses using RUGID data (Kavanagh et al., 2017).
NOTES: The blue bars in the top left graph show the total number of ongoing U.S. ground force deterrent interventions for each year between 1946 and 2016. The red bars in the top right graph show the total number of ongoing U.S. ground force armed conflict interventions for each year between 1946 and 2016. The green bars in the bottom left graph show the total number of ongoing U.S. ground force stabilization interventions for each year between 1946 and 2016.

throughout most of the Cold War period. High levels of troop commitments to stabilization missions has been a more recent development, but this type of intervention has been the primary driver of force demands since 2001.

While the troops demand for these three types of interventions follow different patterns, taken together they constitute the overwhelming share of troops in U.S. ground interventions, with other activities, such as assistance missions, security, and humanitarian missions, making up a much smaller fraction. For this reason, this report focuses on developing models to project future demand for U.S. forces only across these three activity types, as will be discussed in detail in Chapter Three.

CHAPTER THREE

Methodology for Forecasting Future Armed Conflicts and U.S. Ground Interventions

In this chapter, we describe how we built our model for projecting future trends in armed conflict and future demands for U.S. ground forces in military interventions. The model combines a series of statistical projections of future opportunities for intervention—primarily the future incidence of armed conflict and its aftermath—with projections of the likelihood that the United States will choose to intervene in those opportunities. It then relies on historical data to estimate the forces the United States would be likely to commit to those projected interventions.

This discussion was written for an audience not versed in statistical modeling, and we attempt to use nontechnical language wherever possible. Readers interested in the technical details of our analysis should consult Appendix A and Appendix B, in which we provide more insights into our coding rules for assessing historical intervention force requirements and extensive technical details on the construction of our models and the results of the statistical regressions built into our forecasting architecture, respectively.

Forecasting Model Architecture

Broadly, our forecasting model works by sequentially developing annual predictions of armed conflict and U.S. ground interventions for each year in the 2017–2040 time frame.[1] The model then uses those predictions to inform the subsequent year's predictions of conflict and interventions. For example, for 2017, our model first predicts levels and locations of intrastate and interstate armed conflicts, followed by predictions about the types and locations of U.S. ground interventions, followed by estimates of

[1] 2016 is the final year for which we have a complete set of historical data across all of our input variables, so our projection models necessarily begin in 2017.

the forces required by those interventions. Considering the results of those 2017 forecasts, the model then follows the same process for the year 2018, and so forth.[2]

More specifically, the overall architecture of our forecasting model is built around four main components, summarized in Figure 3.1. These four components work interdependently to predict future trends in armed conflict, U.S. ground interventions, and U.S. ground force requirements across a broad range of future strategic environments. A single iteration of our model involves the full simulation of each model component annually for each year from 2017 through 2040. Each subsequent iteration then re-simulates the entire 2017–2040 period using a different random seed.

Each of these four components is discussed individually in detail in the sections that follow. We first discuss how we project different future strategic environments and operationalize key differences between those alternate futures that likely affect the future incidence of either armed conflict or U.S. ground interventions. We then detail the creation of our statistical models of intrastate and interstate armed conflict. Following that, we describe how we model the likelihood of future U.S. ground interventions using statistical models, before proceeding to a discussion of how we use historical data to estimate the U.S. ground forces likely to be required in these interventions. A concluding section provides additional details about how we tie these different components together to produce our forecasts of future armed conflicts and U.S. ground interventions.

Figure 3.1
Forecasting Architecture Conceptual Diagram

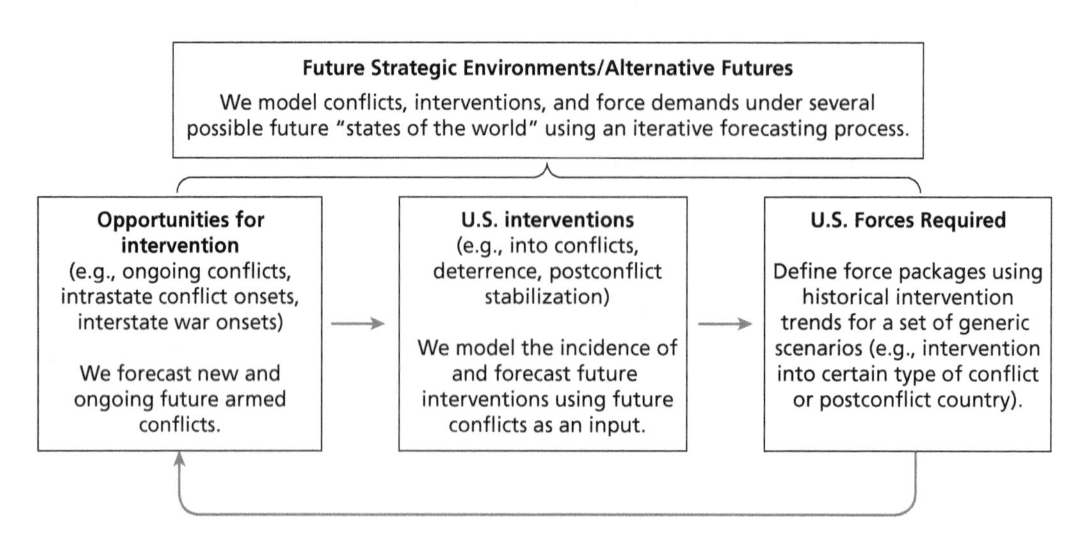

2 Appendix B provides a step-by-step discussion of how the forecasting model works, along with a discussion of the assumptions and coding decisions built into the model at various points in the forecasting process.

Projecting the Future Strategic Environment

To project future trends in armed conflict and U.S. ground interventions, we need to project the future strategic environment in which these events may take place. We developed multiple future strategic environments, including a baseline, most likely future and a series of alternative future scenarios designed to assess how demands for U.S. ground forces might vary in response to a number of dramatic events or structural changes in the international system.[3]

As a precursor to projecting the future strategic environment, we first defined what characteristics of that environment might affect our projections of future armed conflicts and U.S. ground interventions. We then developed statistical models that assess which key factors are most likely to be associated with armed conflict and U.S. ground interventions, respectively. Building on prior RAND Arroyo Center research, we developed separate tailored statistical models for each of our outcomes of interest.[4] For all of the historical data for the key factors used in our statistical models, we rely on data from social science datasets that are publicly available, transparent in their construction, and widely used by academics and analysts in the policy community. The key factors used in each model represent those factors that are most strongly associated with each outcome of interest, as based on prior RAND Arroyo Center research.[5]

These statistical models tell us the structural factors of the international systems that have historically affected the risk of armed conflicts and U.S. ground interventions, as well as whether those factors significantly increase or decrease that risk of conflict or interventions. These factors also therefore define the strategic environment that we needed to project into the future in order to utilize these statistical models to project the future incidence of conflict and intervention.

To project the key factors needed for these future strategic environments, and particularly the baseline scenario, we largely relied on the University of Denver Pardee Center's International Futures tool.[6] Drawing on more than 40 years of historical data, the International Futures tool projects future values for a range of structural vari-

---

[3] The design of these alternative future scenarios, along with their results, are discussed in greater detail in Chapter Four. The alternative future scenarios presented here are meant to be illustrative examples of the kinds of alternative strategic environments that can be crafted to analyze alternate trends in conflict and U.S. ground interventions. Through our flexible modeling approach, other alternative futures could be crafted to suit policymakers' specific analytic needs.

[4] Kavanagh et al., 2017; Watts et al., 2017a.

[5] As will be detailed in subsequent sections, however, the key factors that we were able to include in our statistical models were somewhat limited to those factors that could be reasonably projected forward into the 2017–2040 time frame. As such, while our models include many broad and often-utilized structural characteristics of the international system, such as population size, GDP, and balance of military capabilities, they do not include other likely important factors that cannot be reasonably projected, such as refugee flows from ongoing armed conflicts and future battle-related deaths.

[6] Frederick S. Pardee Center for International Futures, undated.

ables across 186 countries through sophisticated techniques that dynamically model the interaction of multiple economic, demographic, sociopolitical, education, health, agricultural, energy, and infrastructure variables. Future projections for our key factors that are used as inputs to our conflict and intervention models, such as GDP per capita and population demographics, were drawn directly from the International Futures tool projections or were calculated using component projections drawn from the International Futures tool.

Projecting Future Opportunities for Interventions—Intrastate and Interstate Armed Conflicts

To project the future opportunities for the United States to intervene, we went through three main stages. First, we built statistical models of the historical likelihood of the onset and ending of both intrastate and interstate armed conflicts, as will be discussed in detail in this section. Second, we utilized these models and data from the projected future strategic environment to project the likelihood of conflicts occurring in specific countries, or between pairs of countries, in the future year being projected. This process produced two of the three sets of opportunities for future intervention—interventions directly into the armed conflicts being projected and stability operations that occurred in the aftermath of historical or projected conflicts ending. The third set of opportunities for intervention, deterrent interventions, were assumed to be able to occur in any country in a given year, provided that the country was not in the midst of an interstate war, although of course the likelihood of such a deterrent intervention varies dramatically depending on the characteristics of the country in question.

Building Our Statistical Models of Intrastate Armed Conflict

We developed two statistical models for forecasting future levels of intrastate armed conflict—one model examining the key factors affecting the onset of intrastate armed conflicts and another model examining the cessation of ongoing intrastate armed conflicts. The universe of cases for our statistical model of intrastate armed conflict onset includes every country-year from 1960 to 2016, providing a full universe of 9,409 country-years for our analyses.[7] The universe of cases for our models of intrastate conflict cessation includes all years in which countries experienced an ongoing intrastate armed conflict from 1960 to 2016, providing a universe of 1,432 country-years for our analyses.

Data on intrastate armed conflicts are drawn from the UCDP/PRIO Armed Conflict Dataset, which records whether each state is involved in a civil war in a given

[7] In the 1960–2016 period, there were 7,977 years of peace, 1,432 years of intrastate armed conflict, and 302 intrastate armed conflict onsets.

year.[8] Using these data, we measured an intrastate armed conflict onset as whether a new period of civil war broke out in a given country and year.[9] Using these same data, we measured the cessation of an intrastate armed conflict as the last year of each intrastate armed conflict.

Our statistical models utilize several key factors repeatedly highlighted in the academic and policy literatures as significantly impacting the risk of intrastate armed conflict. We discuss the underlying conceptual logic of those key factors' effects on intrastate conflict here:

- **Economic development:** Economic deprivation may foster resentment against incumbent governments, and dim economic outlooks may provide greater opportunities for would-be insurgent leaders to mobilize fighters seeking change against incumbent governments. Conversely, wealthier states can more effectively provide welfare, security, and social services, and state wealth can proxy for more capable governmental institutions, which should decrease states' risk of conflict.[10]
- **Political representation:** Broadly representative and inclusive institutions, such as democracies, can limit the risk of armed conflict by resolving ideological differences and political disagreements through peaceful institutional means. Alternatively, authoritarian regimes may effectively utilize institutionalized repression to preemptively crush opposition before conflict or rely on force to degrade domestic challengers. In contrast, anocracies, which exist between consolidated democracies and authoritarian regimes and which often lack both the institutional inclusivity of democracies and the repressive force of authoritarian regimes, are at a greater risk for armed conflict.[11] Similarly, political regimes undergoing

[8] Allansson, Melander, and Themner, 2017; Gleditsch et al., 2002.

[9] In our statistical models, we use a 25 battle-related deaths threshold to measure the occurrence of and intrastate armed conflict. Relatedly, at least two years of peace must occur between conflicts for a new conflict to be measured as the start of a new period of civil war, rather than as a continuation of a previous conflict.

[10] Halvard Buhaug, Lars-Erik Cederman, and Kristian Skrede Gleditsch, "Square Pegs in Round Holes: Inequalities, Grievances, and Civil War," *International Studies Quarterly*, Vol. 58, No. 2, 2014; Lars-Erik Cederman, Nils B. Weidmann, and Kristian Skrede Gleditsch, "Horizontal Inequalities and Ethnonationalist Civil War: A Global Comparison," *American Political Science Review*, Vol. 105, No. 3, 2011; Paul Collier and Anke Hoeffler, "Greed and Grievance in Civil War," *Oxford Economic Papers*, Vol. 56, No. 4, 2004; James D. Fearon and David D. Laitin, "Ethnicity, Insurgency, and Civil War," *American Political Science Review*, Vol. 97, No. 1, 2003.

[11] Hanne Fjelde, "Generals, Dictators, and Kings: Authoritarian Regimes and Civil Conflict, 1973–2004," *Conflict Management and Peace Science*, Vol. 27, No. 3, 2010; Håvard Hegre, Tanja Ellingson, Scott Gates, and Nils Petter Gledtisch, "Toward a Democratic Civil Peace? Democracy, Political Change, and Civil War, 1816–1992," *American Political Science Review*, Vol. 95, No. 1, 2001; James Raymond Vreeland, "The Effect of Political Regime on Civil War: Unpacking Anocracy," *Journal of Conflict Resolution*, Vol. 52, No. 3, 2008.

significant political transitions, which often entail associated changes in incumbent political institutions, are often at a heightened risk of intrastate conflict.[12]

- **Ethnic discrimination:** States may also be at an increased risk of intrastate conflict if large portions of their populations, particularly among ethnic minorities, face discrimination from incumbent political systems. Political discrimination along ethnic lines enflames grievances against incumbent regimes. In many cases, such ethnic discrimination is coupled with exclusion from political power, leaving minority groups few avenues other than armed resistance to redress their grievances against the state.[13]
- **Societal opportunity and population pressures:** Large youth populations, or youth bulges, increase the risk that states experience intrastate conflicts, because younger populations often depress extant economic opportunities and can provide a robust supply of potential militants to insurgent groups.[14] Relatedly, states with large populations have been repeatedly shown to be at a higher risk of intrastate armed conflict due to a congruence of economic and social opportunities and a larger pool of potential militants.
- **Ongoing and recent intrastate conflicts:** Ongoing conflicts within a region can subject states to a heightened risk of intrastate conflict, because ongoing conflicts in neighboring states may spill across borders, either through a direct expansion of conflict zones or indirectly through the transmission of weapons and revolutionary ideas. Relatedly, ongoing armed conflicts within the state may open opportunities for additional conflicts, either by increasing grievances among the populace or military or by making it relatively easier to take up arms against the state. Similarly, recent conflicts may leave economic and social conditions in such a poor state that opportunities for new conflicts persist even after fighting ends.[15]
- **Geostrategic environment:** Intrastate armed conflicts are by definition driven by internal threats to incumbent political regimes, but external relations can sig-

---

[12] Fearon and Laitin, 2003.

[13] Halvard Buhaug, Lars-Erik Cederman, and Jan Ketil Rød, "Disaggregating Ethno-Nationalist Civil Wars: A Dyadic Test of Exclusion Theory," *International Organization*, Vol. 62, No. 3, 2008; Lars-Erik Cederman, Andreas Wimmer, and Brian Min, "Why Do Ethnic Groups Rebel? New Data and Analysis," *World Politics*, Vol. 62, No. 1, 2010; Andreas Wimmer, Lars-Erik Cederman, and Brian Min, "Ethnic Politics and Armed Conflict: A Configurational Analysis of a New Global Data Set," *American Sociological Review*, Vol. 74, No. 2, 2009.

[14] Jack A. Goldstone, "Population and Security: How Demographic Change Can Lead to Violent Conflict," *Journal of International Affairs*, Vol. 56, No. 1, 2002; Henrik Urdal, "A Clash of Generations? Youth Bulges and Political Violence," *International Studies Quarterly*, Vol. 50, No. 3, 2006; Omer Yair and Dan Miodownik, "Youth Bulge and Civil War: Why a Country's Share of Young Adults Explains Only Non-Ethnic Wars," *Conflict Management and Peace Science*, Vol. 33, No. 1, 2016.

[15] Alex Braithwaite, "Resisting Infection: How State Capacity Conditions Conflict Contagion," *Journal of Peace Research*, Vol. 47, No. 3, 2010; Halvard Buhaug, and Kristian Skrede Gleditsch, "Contagion or Confusion? Why Conflicts Cluster in Space," *International Studies Quarterly*, Vol. 52, No. 2, 2008; Barbara F. Walter, "Does Conflict Beget Conflict? Explaining Civil War Recurrence," *Journal of Peace Research*, Vol. 41, No. 3, 2004.

nificantly affect the risk of such internal threats. Perhaps most significantly, the superpower competition of the Cold War era notably increased the risk that states experienced intrastate armed conflicts as great powers stoked the flames of rebellion through proxy wars and peripheral competition in much of the international system.

These key factors appear as needed in our statistical models of intrastate armed conflict onset and cessation. To gain a better sense of how we operationalize each concept in our statistical models, and of which concepts are used in which statistical model, Table 3.1 summarizes the metrics used in each of our two models of intrastate conflict and how those metrics relate to our key factors.[16]

Results of Our Statistical Models of Intrastate Armed Conflict

Table 3.2 summarizes the relationship between our key factors and states' risk of experiencing the onset or cessation of an intrastate armed conflict. For intrastate conflict onset, green-colored cells indicate factors that are associated with a statistically significant decrease in states' risk of intrastate armed conflict, and red-colored cells indicate factors that are associated with a statistically significant increase in states' risk of intrastate armed conflict. For intrastate conflict cessation, green-colored cells indicate factors that are associated with a statistically significant increase in the chance that ongoing intrastate armed conflicts will end in a given year, and red-colored cells indicate factors that are associated with a statistically significant decrease in the chance that ongoing armed conflicts will end in a given year, meaning that those conflicts are more likely to continue.

Only a few key factors in our statistical model of intrastate armed conflict decrease states' risk of experiencing an intrastate conflict. Not surprisingly, wealthier states are significantly less likely to experience an intrastate armed conflict, as they likely face fewer economic grievances among their populations, one of the main drivers of intrastate conflict. Relatedly, states that have been at peace for long periods are more likely to remain at peace in any given year. Finally, states in the East/Southeast Asia region are significantly less likely to experience an intrastate conflict. This is not necessarily surprising; looking through the historical record, only about 3.5 percent of country-years in East/Southeast Asia experienced an intrastate conflict, compared with over 12 percent in South Asia, 8 percent in Eurasia, and 6 percent in the Middle East.

---

[16] Table 3.1 also includes several other variables that are necessary as statistical controls for spatial and temporal interdependence in our statistical models. Nathaniel Beck, Jonathan N. Katz, and Richard Tucker, "Taking Time Seriously: Time-Series-Cross-Section Analysis with a Binary Dependent Variable," *American Journal of Political Science*, Vol. 42, No. 4, 1998; Nathaniel Beck and Jonathan N. Katz, "Modeling Dynamics in Time-Series-Cross-Section Political Economy Data," *Annual Review of Political Science*, Vol. 14, 2011; David B. Carter and Curtis S. Signorino, "Back to the Future: Modeling Time Dependence in Binary Data," *Political Analysis*, Vol. 18, No. 3, 2010.

Table 3.1
Key Factor Concepts and Metrics Affecting Intrastate Armed Conflict Onset and Cessation

Key Factor Name	Key Factor Metric	Intrastate Conflict Onset Model	Intrastate Conflict Cessation Model
Economic development	The state's GDP per capita (per 1,000 people)	X	X
Political representation	Whether the state is an anocracy	X	X
	Whether the state has experienced a significant regime transition in the prior five years	X	
Ethnic discrimination	Whether the state political apparatus discriminates against a significant portion of the population	X	X
Societal opportunity	Whether the percentage of the population between the ages of 15 and 29 exceeds 45% of the total population	X	X
	The state's population size	X	X
Ongoing and recent intrastate conflicts	The number of ongoing intrastate armed conflicts among each state's regional neighbors	X	
	The number of previous intrastate armed conflicts experienced by the state	X	
Geostrategic environment	Whether the Cold War is ongoing	X	X
Regional and temporal interdependencies	Whether the state is in the Middle East, Eurasia, or East Asia	X	X
	The number of years since the previous intrastate armed conflict in the state	X	
	The number of years that an active intrastate armed conflict in the state remains ongoing		X

NOTE: Data sources for our key factor metrics are available in Appendix B.

In contrast, most key factor metrics in our statistical model of intrastate armed conflict onset significantly increase states' risk of experiencing an intrastate conflict. Societal pressures stoked by increasing population sizes, the presence of youth bulges within the state, and high levels of ethnic discrimination within the state significantly increase the risk that states fall victim to intrastate conflict. Inconsistent or transient political institutions also favor intrastate conflicts, as anocracies and states having recently experienced significant political regime changes are at a significantly heightened risk of intrastate conflict. Geo-strategically, states in regions beset by ongoing

Table 3.2
Effects of Key Factors on Intrastate Armed Conflict Onset and Cessation

Key Factor Metric	Effect on Intrastate Armed Conflict Onsets	Effect on Intrastate Armed Conflict Cessation
GDP per capita (per 1,000 people)	Less Likely	More Likely
Anocracy	More Likely	
Significant regime transition	More Likely	
Ethnic discrimination	More Likely	Less Likely
Youth bulge	More Likely	Less Likely
State population size	More Likely	Less Likely
Neighborhood/regional intrastate conflicts	More Likely	
Number of previous intrastate conflicts	More Likely	
Cold War	More Likely	Less Likely
Middle East	More Likely	Less Likely
Eurasia	More Likely	
East/Southeast Asia	Less Likely	
Number of years since last intrastate armed conflict	Less Likely	
Number of years of ongoing intrastate armed conflict		Less Likely
Number of observations	6,235	1,330
Model pseudo R^2	0.1524	0.1495

NOTES: Models use a threshold of 25 battle-related deaths to mark the onset of intrastate armed conflicts. For intrastate armed conflict onset, green-colored cells indicate factors that are associated with a statistically significant decrease in states' risk of intrastate conflict, and red-colored cells indicate factors that are associated with a statistically significant increase in states' risk of intrastate conflict. For intrastate armed conflict cessation, green-colored cells indicate factors associated with a statistically significant increase in the likelihood that an ongoing conflict ends, and red-colored cells indicate factors associated with a statistically significant increase in the risk that ongoing conflicts remain ongoing. Shading indicates the degree of statistical significance: The dark green and dark red cells (which have white text) indicate a higher level of statistical significance ($p < 0.05$) than light green and light red cells (which have black text) ($p < 0.10$). Gray-colored cells indicate variables included in our models that are not associated with statistically significant changes in intrastate armed conflict. Cells without any color-coding indicate variables that were not included in a particular statistical model. Models also include squared and cubic polynomials of nonconflict and ongoing conflict years, respectively (not shown).

intrastate conflicts are also significantly more likely to experience conflict themselves, as ongoing conflicts in neighboring states spill across international borders and fan the flames of rebellion in nearby states. Similarly, states may be at an increased risk of intrastate conflict if they fall between two competing superpowers, as states were more likely to experience intrastate conflicts in the Cold War period. In line with notions that states can fall victim to "conflict traps," in which states repeatedly fall victim to intrastate conflicts, states become increasingly likely to experience new intrastate conflict as the number of intrastate conflicts previously fought by the regime increases.

GDP per capita was the only key factor metric in our model of intrastate conflict cessation to significantly increase the chances that an ongoing intrastate conflict would end in a given year, suggesting that wealthier states do a better job not only of deterring would-be rebels but also of resolving conflicts quickly when they arise. Like our model of intrastate conflict onset, societal pressures from increasing population sizes, the presence of youth bulges within the state, and high levels of ethnic discrimination within the state significantly increase the risk that ongoing intrastate conflicts persist. Further suggesting that geo-strategic elements and proxy wars impact intrastate conflict dynamics, conflicts during the Cold War were significantly less likely to end than conflicts in the post–Cold War period. Finally, similar to the prior notion that states at peace are more likely to remain at peace, intrastate conflicts are more likely to continue as they persist, suggesting that long-standing conflicts within states are increasingly likely to continue significantly longer than new conflicts.

Building Our Statistical Models of Interstate War

Like our models of intrastate armed conflict, we developed two statistical models for forecasting future levels of interstate war—one model examining the key factors affecting the onset of interstate wars and another model examining the cessation of ongoing interstate wars.[17]

Our models of interstate war focus on the set of relevant pairs of states that could realistically engage in war against each other in the post–World War II period.[18] Within that period, our universe of cases includes all state pairs, or dyads, that could project military power and initiate military conflicts against each other. We include two categories of dyads that meet these criteria. Assuming that all states have the ability to project power against their immediate neighbors, we first include dyads for all pairs of contiguous states.[19] Because more powerful states can project military power

---

[17] As noted in Chapter Two, we focus here on interstate wars, those exceeding a threshold of 1,000 battle deaths, rather than lower-intensity intrastate conflicts, which exceeded a threshold of 25 battle deaths, for reasons of historical data availability.

[18] Zeev Maoz and Bruce M. Russett, "Normative and Structural Causes of Democratic Peace, 1946–1986," *American Political Science Review*, Vol. 87, No. 3, 1993; Douglas Lemke, "The Tyranny of Distance: Redefining Relevant Dyads," *International Interactions*, Vol. 21, No. 1, 1995.

[19] For measurement purposes, we include in our analyses contiguous pairs of states that either shared a land border or were separated by no more than 150 miles of open water. The contiguity data used to define this set of dyads were drawn from the Correlates of War Direct Contiguity Data (Correlates of War Project, "Direct Contiguity Data, 1816–2006," Version 3.1, 2016; Douglas M. Stinnett, Jaroslav Tir, Philip Schafer, Paul F. Diehl, and Charles Gochman, "The Correlates of War Project Direct Contiguity Data, Version 3," *Conflict Management and Peace Science*, Vol. 19, No. 2, 2002).

well beyond their borders, we also include state pairs between regional powers and all other states in relevant geographic regions. For instance, because of its colonial history and continuing interests in the region, we include state pairs between France and every state in West Africa, even though France is physically located in Western Europe. Drawing on prior RAND Arroyo Center research, we include regional power dyads in geographic regions in which those states possessed at least 10 percent of the total military capabilities in the region.[20] Table 3.3 defines the list of great or regional powers for each geographic region identified by this measure.

These criteria provide us with 1,103 relevant state pairs and provide a full universe of 42,630 dyad-years for our analyses.[21] We define the onset of an interstate war as whether each state pair began a new interstate war in a given year, using the Correlates of War (CoW) data on interstate wars and the UCDP/PRIO Armed Conflict Dataset.[22] Using these same data, we measure the cessation of an interstate war as the last year of an interstate war in the state pair.

Our statistical models utilize several key factors repeatedly highlighted in the academic and policy literatures as significantly influencing the risk of interstate war. We discuss the underlying conceptual logic of those key factors' effects on interstate war here:

- **Degree of regional hegemony:** Regional hegemons, which build a preponderance of power in a region through a combination of military capabilities, economic resources, or the enforcement of regional norms, may deter conflict within the region by extending protection to weaker states. Conversely, the erosion of an incumbent hegemon's power and a significant transition of military capabilities within a region may spark increased conflict between rising and falling powers.[23]
- **Balance of military capabilities:** Power parity between states increases the uncertainty of either side's victory in armed conflict, increasing the risk of mis-

---

[20] Watts et al., 2017a.

[21] In the 1946–2015 time frame, there were 42,496 dyad-years of peace, 134 dyad-years of interstate conflict, and 51 interstate armed conflict onsets.

[22] In our statistical analyses, we use a threshold of 1,000 battle-related deaths to measure the occurrence of an interstate war in a state pair. Sarkees and Wayman, 2010; Allansson, Melander, and Themner, 2017; Gleditsch et al., 2002.

[23] A. F. K. Organski and Jacek Kugler, *The War Ledger*, Chicago, Ill.: University of Chicago Press, 1981; Paul K. Huth, "Extended Deterrence and the Outbreak of War," *American Political Science Review*, Vol. 82, No. 2, 1988; Robert Gilpin, *War and Change in World Politics*, Cambridge, UK: Cambridge University Press, 1981; Robert O. Keohane, *After Hegemony: Cooperation and Discord in the World Political Economy*, Princeton, N.J.: Princeton University Press, 1984; Douglas Lemke, "The Continuation of History: Power Transition Theory and the End of the Cold War," *Journal of Peace Research*, Vol. 34, No. 1, 1997; Angela O'Mahony, Miranda Priebe, Bryan Frederick, Jennifer Kavanagh, Matthew Lane, Trevor Johnston, Thomas S. Szayna, Jakub P. Hlavka, Stephen Watts, and Matthew Povlock, *U.S. Presence and the Incidence of Conflict*, Santa Monica, Calif.: RAND Corporation, RR-1906-A, 2018.

Table 3.3
Regional Powers in the Post–World War II Period

Region	Regional Powers
Central America/Caribbean	• United States • Mexico (1975–)
South America	• United States • Argentina (1946–1990) • Brazil
Western Europe	• United States • France • Italy • Russia/Soviet Union • United Kingdom • West Germany/Germany
Eurasia	• United States • China (2005–) • Russia/Soviet Union
West Africa	• United States • France • Nigeria
East and Southern Africa	• United States • South Africa (1965–) • Russia/Soviet Union (1960–1991)
Middle East	• United States • Egypt (1991–) • France (1946–1980) • Iran (1965–) • Iraq (1975–1991) • Russia/Soviet Union (1946–1991) • Saudi Arabia (1975–) • Turkey • United Kingdom (1946–1975)
South Asia	• United States • China • India
East and Southeast Asia	• United States • China • Japan • Russia/Soviet Union (1946–1991)

NOTE: No dates are provided for states that were regional powers for the entire period of analysis. Although Mexico is geographically part of North America, we include it as a regional power that can project military power into the Central American/Caribbean region.

calculation and leading states to be more likely to believe they can prevail in a conflict, incentivizing them to take that risk. In contrast, when tensions arise between states with a greater difference in military capabilities, the likely outcome of a direct conflict is clear to both sides, and they are more likely to be able to resolve their dispute without conflict.[24]

- **Territorial contestation:** Territorial claims are often flashpoints of militarized conflict and significantly increase the risk of conflicts between states escalating to war. This is especially true when the territory being contested is highly valued by the belligerents involved.[25]
- **Extent of economic interdependence:** Increased economic activity and trade between states should make conflict less likely, because conflict between trading partners can significantly harm both sides' economies, even following military victory in conflict.[26]
- **Political congruence:** Democratic regimes are significantly less likely to fight one another, given the normative and structural similarities between democratic regimes in the international system.[27]
- **Strength of international norms:** Strong international norms set expectations about proper behaviors of states, which in the post–World War II era have consistently promoted sustained peace. Strong norms can avert the escalation of conflict by punishing states that violate these norms of peace without sufficient cause. Strong international organizations have proven key to sustaining peace in the post–World War II era by helping to shape and enforce international norms, by helping to align states' preferences, and by offering states venues to mediate conflicts without the use of force.

---

[24] Geoffrey Blainey, *The Causes of War*, London, Macmillan, 1973; Paul K. Huth, D. Scott Bennett, and Christopher Gelpi, "System Uncertainty, Risk Propensity, and International Conflict Among the Great Powers," *Journal of Conflict Resolution*, Vol. 36, No. 3, 1992; Paul K. Huth, Christopher Gelpi, and D. Scott Bennett, "The Escalation of Great Power Militarized Disputes: Testing Rational Deterrence Theory and Structural Realism," *American Political Science Review*, Vol. 87, No. 3, 1993.

[25] Douglas M. Gibler, *The Territorial Peace: Borders, State Development, and International Conflict*, Cambridge, UK: Cambridge University Press, 2012; Paul R. Hensel, "Charting a Course to Conflict: Territorial Issues and Interstate Conflict, 1816–1992," *Conflict Management and Peace Science*, Vol. 15, No. 1, 1996; Paul K. Huth, *Standing Your Ground: Territorial Disputes and International Conflict*, Ann Arbor, Mich.: University of Michigan Press, 1998; John A. Vasquez and Marie T. Henehan, *Territory, War, and Peace*, New York, N.Y.: Routledge, 2010.

[26] John R. Oneal and Bruce Russett, "The Classical Liberals Were Right: Democracy, Interdependence, and Conflict, 1950–1985," *International Studies Quarterly*, Vol. 41, 1997; John R. Oneal and Bruce Russett, *Triangulating Peace: Democracy, Interdependence, and International Organizations*, New York, N.Y.: W.W. Norton and Company, 2001.

[27] Bruce Bueno de Mesquita, James D. Morrow, Randolph M. Siverson, and Alastair Smith, "An Institutional Explanation of the Democratic Peace," *American Political Science Review*, Vol. 93, No. 4, 1999; Erik Gartzke, "Kant We All Just Get Along? Opportunity, Willingness, and the Origins of the Democratic Peace," *American Journal of Political Science*, Vol. 42, No. 1, 1998; Maoz and Russett, 1993.

These key factors appear as needed in our statistical models of interstate war onset and cessation. To gain a better sense of how we operationalize each concept in our statistical models, and of which concepts are used in which statistical model, Table 3.4 summarizes the metrics used in each of our two models of interstate war and how those metrics relate to our key factors.[28]

Table 3.4
Key Factor Concepts and Metrics Affecting Interstate War Onset and Cessation

Key Factor Name	Key Factor Metric	Interstate War Onset Model	Interstate War Cessation Model
Degree of regional hegemony	Ratio of capabilities between first- and second-most powerful states in each region	X	X
	Power transition: whether the regional capabilities ratio crossed a 2:1 threshold in the previous five years	X	X
	Number of U.S. heavy ground forces forward deployed in each region	X	X
Balance of military capabilities	Ratio of military capabilities between both states in a dyad	X	X
	Whether both states in a dyad fall under a nuclear umbrella	X	
Territorial contestation	Whether the states in a dyad contest a territorial claim of medium or high salience	X	X
	Whether the states in a dyad are contiguous by a land border	X	X
Economic interdependence	The minimum ratio of bilateral trade to GDP in the dyad	X	X
	Whether both states in a dyad belong to the same or different trading blocs	X	
Political congruence	Whether both states in a dyad are established democracies	X	
Strength of international norms	Percentage of states in each region that have ratified multiple multilateral treaties requiring the pacific settlement of international disputes	X	X
Temporal interdependencies	The number of years since the previous interstate armed conflict in the state	X	
	The number of years that an active interstate armed conflict in the state remains ongoing		X

NOTE: Data sources for our key factor metrics are available in Appendix B.

[28] Table 3.4 also includes several other variables that are necessary as statistical controls for spatial and temporal interdependence in our statistical models. Beck, Katz, and Tucker, 1998; Beck and Katz, 2011; Carter and Signorino, 2010.

Results of Our Statistical Models of Interstate War

Table 3.5 summarizes the relationship between our key factors and dyads' risk of experiencing the onset or cessation of an interstate war. For interstate war onset, green-colored cells indicate factors that are associated with a statistically significant decrease in dyads' risk of interstate armed conflict, and red-colored cells indicate factors that are associated with a statistically significant increase in dyads' risk of interstate armed conflict. For interstate war cessation, green-colored cells indicate factors that are asso-

Table 3.5
Effects of Key Factors on Interstate War Onset and Cessation

Key Factor Metric	Effect on Interstate War Onsets	Effect on Interstate War Cessation
Regional hegemony ratio	Less Likely	More Likely
Power transition	More Likely	
U.S. heavy forces forward presence	Less Likely	
Dyadic balance of capabilities		
Nuclear umbrella	Less Likely	
Medium or high-salience territorial claim	More Likely	
Land border		More Likely
Bilateral trade/GDP in dyad		
States in different trading blocs	More Likely	
Dyadic democracy	Less Likely	
Prevalence of regional norms	Less Likely	
Number of years since last interstate armed conflict		
Number of years of ongoing interstate armed conflict		
Number of observations	43,313	72
Model pseudo R^2	0.2132	0.2644

NOTES: Models use a threshold of 1,000 battle-related deaths to mark the onset of interstate war. For interstate war onset, green-colored cells indicate factors that are associated with a statistically significant decrease in dyads' risk of interstate war, and red-colored cells indicate factors that are associated with a statistically significant increase in dyads' risk of interstate war. For interstate war cessation, green-colored cells indicate factors associated with a statistically significant increase in the likelihood that an ongoing war ends, and red-colored cells indicate factors associated with a statistically significant increase in the risk that ongoing wars remain ongoing. Shading indicates the degree of statistical significance: Dark green and dark red cells (with white text) indicate a higher level of statistical significance ($p < 0.05$) than light green and light red cells (with black text) ($p < 0.10$). Gray-colored cells indicate variables included in our models that are not associated with statistically significant changes in interstate war. Cells without any color-coding indicate variables that were not included in a particular statistical model. Models also include squared and cubic polynomials of nonconflict and ongoing conflict years, respectively (not shown).

ciated with a statistically significant increase in the chance that ongoing interstate wars will end in a given year, and red-colored cells indicate factors that are associated with a statistically significant decrease in the chance that ongoing wars will end in a given year, meaning that those wars are more likely to continue.

Not surprisingly, dyads that experience more hostile territorial claims over valuable territory are significantly more likely to fight, as are pairs of states in different trading blocs that do not share little direct economic interdependence. In contrast, dyads in which both states are democratic regimes are at less risk of militarized conflict, often because democracies pursue peaceful means of resolving their disputes.

State pairs in regions dominated by a strong hegemon are significantly less likely to engage in militarized conflict. Often, this hegemonic pacifying is brought about by U.S. presence in the region, with an increase in regional U.S. heavy forces also associated with a significant decrease in the risk of interstate war. However, when the balance of power begins shifting in the region, giving rise to a new hegemon, the risk of militarized conflict between states significantly increases.

From a strategic perspective, there appears to be a pacifying effect from extending the nuclear umbrella, as pairs of states that either possess nuclear weapons or fall under the protection of nuclear-armed allies face a lower risk of militarized interstate conflict. International norms also have a significant pacifying effect on interstate war, as increases in the number of signatories to pacifying organizational treaties within a region lower the risk of interstate war.

Fewer of our key factors significantly affect the cessation of interstate hostilities. Although regional hegemons are associated with a lower risk of interstate war, the presence of a hegemon within a region also entails shorter conflicts, as interstate wars are more likely to end in a given year. Somewhat surprisingly, pairs of states sharing a contiguous land border also fight shorter conflicts, as the presence of land border in a dyad significantly increases the likelihood that an ongoing interstate war ends in a given year.

Projecting Future U.S. Ground Interventions

Like we did with our statistical models of intrastate conflict and interstate war, we developed separate statistical models to examine the key factors associated with decisions to undertake or end U.S. ground interventions. Because we are interested in three types of U.S. ground interventions—deterrence interventions, interventions into ongoing armed conflicts, and postconflict stabilization interventions—we developed statistical models tailored around the key factors associated with each type of intervention.

We identified historical U.S. ground interventions using the RAND U.S. Ground Intervention Database (RUGID), which codes all major U.S. ground interventions from 1898 to 2016.[29] Again, our models are concerned with three types of U.S. ground interventions: deterrent interventions, interventions into ongoing armed conflicts, and postconflict stabilization interventions. In all three cases, we measured the onset of an intervention as the first year that the intervention was undertaken. Similarly, we measured the cessation of each type of intervention as the last year that the intervention was conducted by U.S. ground forces.[30]

As noted above, the opportunities for intervention, or the set of country-years that could potentially experience a U.S. ground intervention, varies by each type of intervention. The universe of cases for possible U.S. ground deterrent interventions is all country-years not experiencing an interstate war.[31] Conversely, the universe of cases for possible U.S. ground interventions into ongoing armed conflicts includes all years in which states experienced an ongoing intrastate or interstate armed conflict.[32] The universe of cases for possible U.S. ground postconflict stabilization interventions includes those country-years that occur in a five-year window after the end of an intrastate or interstate armed conflict.[33]

Our statistical models of U.S. ground interventions utilize several key factors previously used in our statistical models of intrastate and interstate armed conflict that were found to significantly affect U.S. ground interventions in previous RAND Arroyo Center studies.[34] These key factors include domestic characteristics of the United States, the characteristics of states targeted for interventions, characteristics of the U.S.-target state relationship and U.S. strategic interests, and characteristics of the broader geo-strategic environment.[35] It is also important to reiterate that we treat

---

[29] Kavanagh et al., 2017.

[30] We include both combat and counterinsurgency missions as interventions into ongoing armed conflicts. As noted more extensively in Appendix B, we make a simplifying assumption in our forecasting model that once U.S. ground forces undertake an intervention into an ongoing armed conflict, it continues that intervention until the conflict ends. This assumption is mostly in keeping with the historical record of U.S. interventions into armed conflicts. Therefore, we do not develop a separate model for the cessation of U.S. Army interventions into ongoing armed conflicts, because none is needed for our forecasting process.

[31] Of the 10,317 nonconflict years in the post–World War II period, there were 1,150 country-years of ongoing deterrent interventions by the U.S. Army, and 9,167 country-years without an ongoing U.S. ground deterrent intervention.

[32] Of the 1,648 country-years of ongoing intrastate or interstate armed conflicts in the post–World War II period, 53 involved U.S. ground interventions, while 1,595 country-years did not include an intervention into an ongoing armed conflict. To clarify, if a future armed conflict is projected in which the United States is a direct party (e.g., an interstate war between the United States and another state), then a U.S. armed conflict intervention is automatically identified for that event, as well.

[33] Of the 1,287 postconflict country-years since World War II, 37 included U.S. ground stabilization interventions, while 1,250 country-years did not include a U.S. ground stabilization intervention.

[34] Kavanagh et al., 2017.

[35] As noted previously, although characteristics of ongoing or recently ended conflicts, such as battle-related deaths and refugee flows, have been shown to be significantly associated with the likelihood of U.S. ground interventions, our larger forecasting process precludes us from including such factors in our statistical models.

armed conflicts that directly involve the United States as U.S. armed conflict interventions, so the factors discussed above that affect the likelihood of such a U.S.-involved conflict may be particularly important. Here, we discuss the underlying conceptual logic of the effects on intrastate conflict of the factors employed in the separate intervention models:

- **U.S. economic outlook:** The United States has generally been more willing to undertake interventions abroad under favorable economic outlooks, because devoting resources to ongoing interventions presents comparatively less risk and appears less costly under more favorable economic conditions.[36] Conversely, declining budgets may force the United States to be more selective about when and where it intervenes abroad.
- **U.S. military capabilities:** States in general may also be less inclined to undertake interventions when they lack the necessary military resources to sustain an ongoing intervention, and the United States has historically been less likely to undertake new interventions abroad when its military resources are already committed elsewhere. That said, the United States has historically increased its interventionist tendencies in the post–World War II era as its relative military capabilities have increased.[37]
- **Target state economic development and resources:** A significant body of literature suggests that the United States is most likely to intervene in regions or states that possess significant economic resources or in which the United States seeks to improve its political and economic influence.[38] The United States may be especially prone to undertaking deterrence or stabilization missions in wealthier states in an effort to prevent the occurrence or recurrence of conflict. Similarly, significant oil resources, a major driver of global economic wealth, may prompt the United States to intervene on behalf of partner states to avoid conflict.[39]
- **Target state political system:** Partner states' levels of democracy may significantly influence U.S. decisions to undertake interventions in support of partner

---

[36] Kavanagh et al., 2017.

[37] Paul K. Huth, "Major Power Intervention in International Crises, 1918–1988," *Journal of Conflict Resolution*, Vol. 42, No. 6, 1998; Nicolas Rost and J. Michael Greig, "Taking Matters into Their Own Hands: An Analysis of the Determinants of State-Conducted Peacekeeping in Civil Wars," *Journal of Peace Research*, Vol. 48, No. 2, 2011.

[38] Michael T. Klare, *Beyond the "Vietnam Syndrome": U.S. Interventionism in the 1980s*, Washington, D.C.: Institute for Policy Studies, 1981; Frederic S. Pearson and Robert A. Baumann, "Foreign Military Intervention and Changes in United States Business Activity," *Journal of Political and Military Sociology*, Vol. 5, No. 1, 1977; Mi Yung Yoon, "Explaining U.S. Intervention in Third World Internal Wars, 1945–1989," *Journal of Conflict Resolution*, Vol. 41, No. 4, 1997.

[39] Aysegul Aydin, "Where Do States Go? Strategy in Civil War Intervention," *Conflict Management and Peace Science*, Vol. 27, No. 1, 2010.

regimes. Some research suggests that the United States may be more likely to undertake interventions into armed conflicts when it can promote democracy abroad.[40] Similarly, the United States may be more likely to undertake interventions to stabilize anocratic partner regimes to protect them from backsliding into a recurrence of conflict or toward greater authoritarianism in an effort to deter intrastate conflict.

- **Partner states under threat:** The United States may be especially likely to come to the aid of partners and undertake deterrence missions when partner states face severe threats from adversaries, especially if the adversary is also a U.S. rival.[41]
- **Strategic relationship with the United States:** Strong military ties are significant predictors of military interventions. States sharing formal alliances are significantly more likely to intervene in support of their partners.[42] States close to the United States that fall victim to armed conflicts are significantly more likely to experience U.S. interventions in a bid by U.S. forces to prevent the spread of conflict close to the homeland.[43] Interventions into armed conflicts abroad may also tie the United States to partner states over an extended period, as U.S. forces are significantly more likely to intervene to stabilize postconflict states if the United States was previously involved in the prior conflict.
- **Geostrategic environment:** Ongoing conflicts often spill across interstate borders, and states surrounded by conflicts in neighboring states are particularly susceptible to armed conflict. In an effort to prevent the further spread of conflict within particularly conflict-prone regions, the United States may be more likely to undertake interventions into ongoing armed conflicts when they occur in particularly conflict-prone regions.[44]

These key factors appear as needed in our statistical models of U.S. ground intervention onset and cessation. To gain a better sense of how we operationalize each concept in our statistical models, and of which concepts are used in which statistical model, Table 3.6 summarizes the metrics used in each of our models of U.S. ground

---

[40] Andrew J. Enterline and J. Michael Greig, "Beacons of Hope? The Impact of Imposed Democracy on Regional Peace, Democracy, and Prosperity," *Journal of Politics*, Vol. 67, No. 4, 2005; James Meernik, "United States Military Intervention and the Promotion of Democracy," *Journal of Peace Research*, Vol. 33, No. 4, 1996.

[41] H. W. Brands Jr., "Decisions on American Armed Intervention: Lebanon, Dominican Republic, and Grenada," *Political Science Quarterly*, Vol. 102, No. 4, Winter 1987–1988; Patrick James and John O'Neal, "The Influence of Domestic and International Politics on the President's Use of Force," *Journal of Conflict Resolution*, Vol. 35, No. 2, 1991; Mark P. Lagon, "The International System and the Reagan Doctrine: Can Realism Explain Aid to 'Freedom Fighters'?" *British Journal of Political Science*, Vol. 22, No. 1, 1992; Yoon, 1997.

[42] Michael G. Findley and Tze Kwang Teo, "Rethinking Third-Party Interventions into Civil Wars: An Actor-Centric Approach," *Journal of Politics*, Vol. 68, No. 4, 2006; Huth, 1998.

[43] Kavanagh et al., 2017.

[44] Braithwaite, 2010; Buhaug and Gleditsch, 2008.

Table 3.6
Key Factor Concepts and Metrics Affecting U.S. Ground Intervention Onset and Cessation

Key Factor Name	Key Factor Metric	Deterrence Intervention Onset	Deterrence Intervention Cessation	Armed Conflict Intervention Onset	Stabilization Intervention Onset	Stabilization Intervention Cessation
U.S. Economic Outlook	U.S. GDP growth	X	X			
U.S. Military Capabilities	U.S. aggregate military capabilities	X	X			
	Number of ongoing U.S. ground interventions			X		
Partner state economic and strategic resources	Partner state GDP per capita (per 1,000 people)	X	X	X	X	X
	Partner state oil production	X	X	X		
	Partner state population size				X	
Partner state political system	Whether the state is an anocracy				X	
	Level of partner state democracy			X		
Partner states under threat	Whether the partner state is the target of a high-value territorial claim by an adversary state	X	X			
U.S.-partner state strategic relationship	U.S. alliance with partner state	X	X		X	X
	Distance between a partner state and the United States				X	
	Whether the United States was involved in a prior armed conflict intervention in the state				X	
Geostrategic environment	The number of ongoing armed conflicts among each state's regional neighbors			X		

Table 3.6—continued

Key Factor Name	Key Factor Metric	Deterrence Intervention Onset	Deterrence Intervention Cessation	Armed Conflict Intervention Onset	Stabilization Intervention Onset	Stabilization Intervention Cessation
Regional and temporal interdependencies	Whether the state is in Europe	X	X			
	Whether the state is in sub-Saharan Africa				X	
	The number of years since the previous intervention (of each type) in the state	X		X	X	
	The number of years that an intervention in the state remains ongoing		X			

NOTE: Data sources for our key factor metrics are available in Appendix B.

interventions and how those metrics relate to our key factors.[45] The inclusion of particular metrics in specific models builds on research conducted by Kavanagh et al., who found that the drivers of different types of interventions varied substantially from one intervention activity type to another.[46]

Results of Our Statistical Models of U.S. Ground Interventions

Table 3.7 summarizes the relationship between our key factors and the likelihood that U.S. ground forces undertake a deterrence intervention, intervention into an ongoing armed conflict, or a stabilization intervention in a postconflict state. For intervention onset, green-colored cells indicate factors that are associated with a statistically significant increase in the likelihood that U.S. ground forces undertake an intervention, and red-colored cells indicate factors that are associated with a statistically significant decrease in the likelihood of U.S. ground interventions. For the cessation of interventions, green-colored cells indicate factors that are associated with a statistically significant increase in the chance that ongoing interventions will end in a given year, and red-colored cells indicate factors that are associated with a statistically significant decrease in the chance that ongoing interventions will end in a given year, meaning that those interventions are more likely to continue.

---

[45] Table 3.6 also includes several other variables that are necessary as statistical controls for spatial and temporal interdependence in our statistical models (Beck, Katz, and Tucker, 1998; Beck and Katz, 2011; Carter and Signorino, 2010).

[46] Kavanagh et al., 2017.

Table 3.7
Effects of Key Factors on U.S. Ground Intervention Onset and Cessation

Key Factor Metric	Deterrence Intervention Onset	Deterrence Intervention Cessation	Armed Conflict Intervention Onset	Stabilization Intervention Onset	Stabilization Intervention Cessation
U.S. GDP growth	Less Likely	More Likely			
U.S. aggregate military capabilities	More Likely				
Number of ongoing U.S. ground interventions					
Partner state GDP per capita (per 1,000 people)			More Likely	Less Likely	
Partner state oil production	More Likely				
Partner state population size				Less Likely	
Whether the state is an anocracy					
Level of partner state democracy			More Likely		
Whether the partner state is the target of a high-value territorial claim by an adversary state					
U.S. alliance with partner state	More Likely				More Likely
Distance between a partner state and the United States				Less Likely	
Whether the United States was involved in a prior armed conflict intervention in the state				More Likely	Less Likely
The number of ongoing armed conflicts among each state's regional neighbors					
Whether the state is in Europe	More Likely				
Whether the state is in sub-Saharan Africa					
The number of years since the previous intervention (of each type) in the state	Less Likely				
The number of years that an intervention in the state remains ongoing					

Table 3.7—continued

Key Factor Metric	Deterrence Intervention Onset	Deterrence Intervention Cessation	Armed Conflict Intervention Onset	Stabilization Intervention Onset	Stabilization Intervention Cessation
Number of observations	8,138	917	1,264	1,218	36
Pseudo R^2	0.3843	0.1133	0.2114	0.3418	0.3269

NOTES: For models of intervention onset, green-colored cells indicate factors that are associated with a statistically significant increase in the likelihood of U.S. ground interventions, and red-colored cells indicate factors that are associated with a statistically significant decrease in the likelihood of U.S. ground interventions. For models of intervention cessation, green-colored cells indicate factors associated with a statistically significant increase in the likelihood that an ongoing intervention ends, and red-colored cells indicate factors associated with a statistically significant decrease in the likelihood that an ongoing intervention ends. Shading indicates the degree of statistical significance: Dark green and dark red cells (with white text) indicate a higher level of statistical significance ($p < 0.05$) than light green and light red cells (with black text) ($p < 0.10$). Gray-colored cells indicate variables included in our models that are not associated with statistically significant changes in the likelihood of U.S. ground interventions. Cells without any color-coding indicate variables that were not included in a particular statistical model. Models also include squared and cubic polynomials of nonconflict and ongoing conflict years, respectively (not shown).

In contrast to our statistical models of intrastate conflict and interstate war, the effects of our key factors on U.S. ground interventions are more varied. Contrary to our expectations, the United States is significantly less likely to undertake new deterrence missions when U.S. GDP growth is high. Similarly, the United States is significantly more likely to end ongoing deterrence missions in periods when U.S. GDP growth is high. While a strong U.S. economy has an adverse effect on U.S. deterrence missions, however, increasing U.S. military capabilities significantly increases the likelihood that the United States undertakes a deterrence mission abroad. Such military capabilities, however, have no statistical effect on whether the United States ends or continues an ongoing deterrence mission in a given year.

While partner-state economic development has no significant effect on U.S. decisions concerning deterrence missions, oil-producing states are significantly more likely to host a U.S. deterrence mission than states with smaller oil reserves, lending credence to our expectations about the relationship between partner-state strategic resources and U.S. willingness to defend those partners. Such strategic resources, however, have no discernible effect on U.S. decisions about ending ongoing deterrence missions in our analyses. In line with our expectations concerning strategic relationships between the United States and partner states, U.S. allies are significantly more likely to host U.S. deterrence missions. Somewhat surprisingly, however, states facing severe threats from high-value territorial claims by adversary states are no more likely to host U.S. ground force deterrence interventions than states facing less severe threats.

Not surprisingly, partner states in Europe, home to NATO and host to many long-lasting deterrence missions, are significantly more likely to host U.S. ground deterrence interventions than states in other regions. Like the findings of our conflict models, states become increasingly less likely to host U.S. ground deterrence missions over time; the longer a state goes without hosting a deterrence mission, the less it is to host a new deterrence mission in any given year. However, the same cannot be said of ongoing deterrence missions: In our statistical models, deterrence missions are no more or less likely to end the longer they last.

Our statistical models of U.S. ground interventions into ongoing armed conflicts support many expectations about the effects of partner-state characteristics. The United States is more likely to undertake interventions in support of wealthier and more democratic states that experience armed conflict. Many strategic factors, however, do not necessarily entail a significant relationship with U.S. ground interventions. Partner-state oil production, the number of ongoing U.S. ground interventions, and the level of ongoing conflict in a state's region do not significantly affect the likelihood of U.S. ground interventions into ongoing armed conflicts.

Characteristics of partner states appear to similarly drive U.S. ground stabilization interventions in postconflict environments. Wealthier partners are less likely to host U.S. stabilization missions, probably because such partners are more capable of maintaining or building stability without U.S. assistance. Similarly, states with large populations are significantly less likely to host U.S. stabilization missions. Strategically, although U.S. allies in postconflict environments are not any more likely to host U.S. stabilization missions, postconflict states that hosted U.S. ground interventions during the previous conflict are significantly more likely to continue hosting U.S. forces as the U.S. mission transitions from combat to stabilization at the cessation of hostilities. Such ongoing stabilization missions are likely to be prolonged missions, as stabilization missions following U.S. involvement in prior conflicts are significantly less likely to end than stabilization missions undertaken by U.S. forces when the United States was not previously involved in hostilities in a partner state.

Estimating Force Requirements for Projected U.S. Ground Interventions

The fourth component of our model involves estimating the forces the United States is likely to commit to projected future interventions. The forces required for military interventions have historically depended on the goals the United States seeks to achieve and the contexts in which U.S. ground forces are deployed. As discussed earlier, the goal of this report is not to develop precise estimates of the forces required for specific U.S. ground interventions. Rather, we seek to provide "rules of thumb" for forecasting broad categories of military intervention forces in a manageable number of typical or abstracted geostrategic contexts and local operating environments. Although this approach cannot provide accurate estimates of the force requirements for specific contingencies, it can provide a useful baseline estimate of average requirements that can be expected, based on historical trends, for multiple interventions occurring over extended periods of time.

Analytic Approach

To develop our approximations for U.S. ground force requirements in forecasted interventions, we used historical data on U.S. ground interventions, the numbers and types of forces deployed in those interventions, and the contexts—both local and geostrategic—in which those forces operated. This section summarizes our data and approach. In Appendix A, we provide an extensive discussion of the underlying data, assumptions used, and other details of our analyses.

Even with an expansive definition of military interventions that encompasses several different types of U.S. ground missions, such operations are relatively rare; according to the definitions used in this study, there have only been 145 U.S. ground interventions since 1898. This number of cases makes it possible to use statistical analyses to assess certain broad trends, such as the circumstances under which the United States is likely to intervene abroad. It is much harder, however, to use sophisticated statistical techniques for narrower purposes, including estimating the number and types of forces required for certain types of military interventions.[47] For example, in the RUGID data used in this study, there are only 20 cases of combat interventions, which is too small a sample size for complex statistical analyses with multiple key factors.[48]

Fortunately, it is possible to develop empirically grounded, transparent estimates of broad force requirements using simpler methods. Given the small number of cases involved in our analyses, we created a typology of different operational environments for each category of intervention based on a few key contextual factors relevant to each. We then used this typology to estimate differences on three important characteristics of U.S. ground interventions: the duration of an intervention, the average number of forces deployed in each year of the intervention, and the proportion of those forces that could be considered "heavy" combat forces. The rest of this section briefly reviews the definitions, rationale, and data for our typologies and the typical forces deployed in each environment, and the following section summarizes the "rules of thumb" for U.S. ground force requirements that emerge from this analysis.[49]

---

[47] Other studies have circumvented this problem by combining interventions by the United States and other countries, thus obtaining a sufficiently large number of cases to analyze through statistical models. This approach, of course, encounters different challenges—specifically, differences in the ways that various countries conduct military interventions. See, for instance, Kyle Beardsley, "Peacekeeping and the Contagion of Armed Conflict," *Journal of Politics*, Vol. 73, No. 4, 2011; Reed M. Wood, Jacob D. Kathman, and Stephen E. Gent, "Armed Intervention and Civilian Victimization in Intrastate Conflicts," *Journal of Peace Research*, Vol. 49, No. 5, 2012.

[48] For more details on the RUGID data, see Kavanagh et al., 2017.

[49] In constructing these typologies, we roughly adopted the "typological" approach to qualitative research suggested by Alexander George and Andrew Bennett. The main difference is that we did not seek to confirm (or disconfirm) specific hypotheses for each conjunction of independent variables. Rather, our hypothesis was simply

Categories of Interventions

As discussed above, the United States deploys forces abroad for a wide variety of missions. Most of these types of missions, however, are not major drivers of U.S. Army force structure. They are typically very small, involving a few hundred or at most a few thousand personnel. Most advisory missions, for instance, never involved more than 1,000 uniformed military personnel (U.S. military support to Plan Colombia, for example, one of the larger advisory missions in U.S. Army history, involved only 900 military personnel at its height). In contrast, humanitarian assistance and disaster response missions can be much larger—often involving a few thousand personnel and substantial heavy equipment. But these missions tend to be very short-duration, lasting only a few months. Moreover, many of these types of missions represent lower-priority commitments for the United States and are rarely drivers of decisions about military force structure and force employment. Were a major crisis to erupt, the United States could pull forces away from these operations.[50]

Consequently, we focus our analyses on only the three types of interventions that represent substantial military commitments (as described and defined above) and are significant drivers of U.S. military force management decisions:

- combat missions (including conventional warfare and counterinsurgency missions)
- deterrence missions
- stabilization missions.

These types of interventions represent significant resource demands on U.S. forces, often because they require a significant number of forces or because they have historically been long-duration missions that draw on U.S. forces for prolonged periods. As such, estimating the expected demands of these missions in conjunction with forecasted levels of future interventions should provide a significant proportion of the overall expected demands facing future U.S. ground forces. Additionally, these missions represent the majority of strategic planning considerations facing military planners concerned with force allocation and force management strategies.

Characterizing Force Requirements

As described in the previous chapter, DoD and the U.S. Army have extensive processes for estimating future force requirements—in particular, Support to Strategic Analysis

---

that different conjunctions of variables suggested from the existing social science literature on military interventions would define operational environment "types" with relatively consistent force requirements (i.e., the scale, composition, and duration of force deployments would be relatively similar for each conjunction of independent variables). This hypothesis largely proved correct, as the following section shows. On typological approaches, see Alexander L. George and Andrew Bennett, *Case Studies and Theory Development in the Social Sciences*, Cambridge, Mass.: MIT Press, 2005.

[50] Though, admittedly, re-allocating forces from existing missions in such situations would take some time.

and the Army's service-specific process, Total Army Analysis.[51] To support more quick-turn, long-term force planning, we seek to provide a much simpler characterization of force requirements in future scenarios involving U.S. ground interventions. More specifically, we estimate three characteristics of force requirements for different types of intervention:[52]

- **Size:** We estimate the number of ground-force personnel (both Army and Marine Corps) deployed in an intervention.[53] This number is expressed as the *average* annual number of personnel deployed over the course of an intervention (the maximum number deployed at the height of an intervention can be substantially higher).
- **Duration:** We estimate the duration of an intervention from the month of first deployment to the month of total withdrawal.[54]
- **Proportion of heavy forces:** To characterize the types of combat forces required, we used historical data on orders of battle (ORBAT) and tables of organization and equipment (TO&E) to determine which deployed forces could be considered "heavy" forces. This analysis was conducted primarily at the battalion level. For the purposes of this study, heavy forces include armored battalions, armored cavalry squadrons, mechanized battalions (i.e., those with a large proportion of infantry fighting vehicles and/or armored personnel carriers), fires (artillery) battalions, attack aviation (e.g., attack reconnaissance battalions), and any units that combine any of the previous types of forces (e.g., U.S. Army combined arms battalions or Marine Expeditionary Units).[55]

---

[51] For a brief overview of Total Army Analysis and how it supports broader Army force management requirements, see U.S. Army War College, *How the Army Runs: A Senior Leader Reference Handbook 2015–2016*, Carlisle, Pa., 2015, Chapter Three, "Force Management."

[52] The selection of these specific characteristics was informed (but not constrained) by the intervention size models described above.

[53] These data have been collected from numerous sources such as official histories published by the U.S. Army's Center of Military History; academic journals and studies; statistical reference publications, such as *The Military Balance* (International Institute for Strategic Studies, *The Military Balance*, Washington, D.C., multiple years); military graduate theses published by the various U.S. war colleges; and other existing databases, namely the Defense Manpower Data Center's (DMDC's) historical time-series publications titled *Worldwide Manpower Distribution by Geographical Area*.

[54] Kavanagh et al., 2017.

[55] Unfortunately, the historical record is often incomplete, and orders of battle for historical U.S. ground interventions are not always readily available at this level, necessitating the use of imperfect data and simplifying assumptions. Consequently, our estimates should be understood as rough approximations rather than as precise reckonings of history.

Geostrategic and Local Operating Environments

The size, composition, and duration of military deployments obviously vary depending on the category of mission being conducted in the intervention (combat, deterrence, or stabilization) and also on the context in which they occur. For each category of intervention, we define context of the intervention through the conjunction of three variables at the geostrategic and local levels.

Historical Geostrategic Environments

The United States has conceived of its national security requirements and how to best deal with national security threats in very different terms over the course of its history. These varying conceptions of the national interest have important implications for the goals the United States pursues in its military interventions and the military commitments it is willing to make to achieve those goals.

For most of its history, the United States sought to avoid "foreign entanglements" and limit its military adventurism abroad. Its primary security goal, as expressed in the Monroe Doctrine, was to prevent other great powers from exercising influence in the Western Hemisphere. That said, the United States was never truly isolationist; it was willing to use military force to pursue its continental expansion and to protect and expand its commercial interests (primarily in the Western Hemisphere, but ultimately as far afield as China during the Boxer Rebellion). But until World War II, uses of military force outside of North America and the Caribbean were infrequent, usually of short duration, and generally for very limited aims.

The United States' ambitions gradually increased in tandem with its growing power, especially in the post–World War II era. These ambitions first became clear in the Spanish-American War of 1898. Soon after, World War I demonstrated that the United States had clearly entered the ranks of the world's great powers. But even in these cases, the United States' commitments remained fairly narrow and short-lived. It was not until World War II that the United States clearly began to adopt a grand strategy premised on enduring international leadership—an approach known as "globalism" or "primacy."[56]

The differences in these two periods are stark. Before World War II, the United States had military forces permanently stationed abroad in only a single country, Panama. Even on the eve of World War II, its standing Army was relatively small, numbering 185,000 personnel. In the decades that followed, however, the United States maintained standing military forces numbering over one million personnel, forged standing alliances with dozens of countries, and maintained permanent military bases across Europe and Asia. Not only did its capabilities increase; its ambitions similarly expanded. In the latter period, the United States often pursued expansive

[56] See, for instance, Stephen E. Ambrose, *Rise to Globalism: American Foreign Policy Since 1938*, sixth edition, New York, N.Y.: Penguin Books, 1991; Christopher Layne, "From Preponderance to Offshore Balancing: America's Future Grand Strategy," *International Security*, Vol. 22, No. 1, 1997.

goals such as maintaining a "liberal world order" or reshaping the domestic politics of states in which it intervened.[57]

There is no clear line demarcating precisely when the United States transitioned from one grand strategy to the other. The beginnings of this transition can clearly be seen before World War II, although the full institutionalization of the United States' sprawling alliance system and commitment to maintaining a large standing military did not occur until the Korean War. Most scholars, however, argue that the critical shift in American thinking took place over the course of World War II, beginning with the start of the war in 1939, strengthening with the attack on Pearl Harbor in 1941, and solidifying during the course of multiple summits during the war on the shape of the postwar order.[58] For simplicity's sake, our analysis uses 1940 as the dividing line between the earlier period of U.S. foreign policy and the later, "globalist" period.

Local Operating Environments

Force requirements, of course, are not determined solely by broad geopolitical factors but also by the local context of an intervention. Drawing on the social science literature, we derived two factors for each category of intervention that are particularly influential for determining force requirements.

Force requirements for combat interventions are shaped primarily by the type of war being fought and the strength of the adversary:

- **War type:** Most states are capable of mobilizing populations and resources on a scale that dwarfs the capabilities of all but the most capable nonstate belligerents. On the other hand, states have vulnerabilities that nonstate actors do not. Because states are defined as political units that fuse territorial control with population control, interstate wars can be brought to a decisive end when the more powerful state seizes control of critical enemy territory (such as its capital) or imposes such costs on the enemy that it threatens its opponents' ability to retain control over its population, territory, and other key assets. Interstate wars therefore tend to be larger in scale but shorter in duration than intrastate wars. They also often involve a higher proportion of heavy combat forces than wars against nonstate actors. For the purposes of our analysis, we characterized a U.S. intervention as an *interstate intervention* if the United States intervened in an ongoing interstate war fought against another sovereign state. We characterized an intervention as an *intrastate intervention* if the United States intervened in an ongoing intrastate conflict, or

---

[57] See, for instance, Kavanagh et al., 2017; Ambrose, 1991; Martha Finnemore, *The Purpose of Intervention: Changing Beliefs about the Use of Force*, Ithaca, N.Y.: Cornell University Press, 2003; Christopher Layne, *The Peace of Illusions: American Grand Strategy from 1940 to the Present*, Ithaca, N.Y.: Cornell University Press, 2006.

[58] Ambrose, 1991; Layne, 2006.

if an intrastate conflict developed during the course of an ongoing U.S. ground intervention.[59]

- **Adversary strength:** Because state and nonstate actors typically fight in different ways, their strength must be calculated differently. State strength can be estimated on the basis of its overall material capabilities, including the size of its standing military, the population from which it can conscript additional soldiers, its economic resource base, and its technological sophistication.[60] Nonstate adversaries represent a greater challenge. Whereas states are extremely durable (almost never disappearing in the era covered in our analysis), insurgencies come and go much more frequently. Trying to characterize nonstate adversaries' strength in a way that can be projected into the future is therefore extremely difficult. For the purposes of our analyses, we adopted a simplifying assumption: Insurgents' strength varies inversely to the strength of the state against which it is fighting, with state strength approximated by its level of economic development—or, more specifically, GDP per capita.[61] Although this simplifying assumption is not entirely satisfying, it is consistent with conventions sometimes used in social science research on the topic.[62]

Extended deterrence deployments do not need to be large enough to defeat an adversary; they only need to be large enough to convince an adversary that it is unlikely to be unable to easily achieve its goals through military force.[63] Consequently, force requirements for extended deterrence missions in support of partner states depend less on the capabilities of the adversary being deterred than they do on the perceived intentions of an adversary and the United States' degree of commitment to its allies:

- **Level of threat from adversaries:** We use the presence of a higher salience territorial claim against the host or target state to indicate a heightened threat from potential adversaries.[64]

---

59 In cases where a conflict shifted from one type to the other, such as Operation Iraqi Freedom, U.S. military operations are considered two separate interventions. Data on both inter- and intrastate conflicts was taken from the UCDP/PRIO Armed Conflict Dataset (Allansson, Melander, and Themner, 2017; Gleditsch et al., 2002).

60 Correlates of War National Material Capabilities (v5.0) dataset (J. David Singer, "Reconstructing the Correlates of War Dataset on Material Capabilities of States, 1816–1985," *International Interactions*, Vol. 14, No. 2, 1988).

61 Jutta Bolt, Robert Inklaar, Herman de Jong, and Jan Luiten van Zanden, "Rebasing 'Maddison': New Income Comparisons and the Shape of Long-Run Economic Development," GGDC Research Memorandum 174, 2018.

62 See in particular Fearon and Laitin (2003).

63 John J. Mearsheimer, *Conventional Deterrence*, Ithaca, N.Y.: Cornell University Press, 1983.

64 Frederick et al., 2017.

- **Level of commitment to partners:** We use the presence of a formal defensive alliance between the United States and another country as an indicator of the extent of the United States' commitment to partner states.[65]

Finally, force requirements for stabilization missions depend primarily on the level of local consent for U.S. forces and the size of the country to be stabilized:

- **Consent:** If the local government and population are largely supportive of a foreign military presence (as in cases of United Nations consensual peace operations), force requirements are often relatively small. If the government or a large portion of the population are hostile, or are believed to be hostile, to stabilization efforts, force requirements are typically much larger.[66] Accurately measuring level of consent, of course, is extremely difficult even for ongoing military operations; making future projections of consent with any precision is impossible. Consequently, we adopt a simple assumption to characterize the level of consent in the operating environments of stabilization missions: If the United States fought a war against the government or a major nonstate actor in the country immediately prior to the start of the stabilization intervention, then U.S. presence is considered to have a low level of consent. Conversely, if the United States fought a war against an occupying power, such as the Japanese in Korea in World War II, the subsequent stabilization mission is considered to have a higher degree of consent.
- **Population size:** Because stabilization operations are generally considered "population-centric" missions as much as or more than "enemy-centric," force requirements are typically based at least in substantial part on the population size of the target country.[67]

The variables we use to characterize operating environments for each category of intervention are summarized in Table 3.8.

---

[65] In this assumption, we follow Kavanagh et al., 2017. For an explanation of the alliance data, see Douglas M. Gibler, *International Military Alliances, 1648–2008*, Washington, D.C.: CQ Press, 2009.

[66] See, for instance, James Dobbins, Seth G. Jones, Keith Crane, and Beth Cole DeGrasse, *The Beginner's Guide to Nation-Building*, Santa Monica, Calif.: RAND Corporation, MG-557-SRF, 2007.

[67] There is an enormous literature that discusses—and criticizes—"force-to-population" ratios as a basis for calculating force requirements for stabilization. See for instance James T. Quinlivan, "Force Requirements in Stability Operations," *Parameters: U.S. Army War College Quarterly*, Vol. 25, No. 4, Winter 1995/96; Jeffrey A. Friedman, "Manpower and Counterinsurgency: Empirical Foundations for Theory and Doctrine," *Security Studies*, Vol. 20, No. 4, 2011; and Stephen Watts, Patrick B. Johnston, Jennifer Kavanagh, Sean M. Zeigler, Bryan Frederick, Trevor Johnston, Karl P. Mueller, Astrid Stuth Cevallos, Nathan Chandler, Meagan L. Smith, Alexander Stephenson, and Julia A. Thompson, *Limited Intervention: Evaluating the Effectiveness of Limited Stabilization, Limited Strike, and Containment Operations*, Santa Monica, Calif.: RAND Corporation, RR-2037-A, 2017b. In this analysis, we do not need a precise ratio between stabilizing forces and the local populations; for our purposes, it is sufficient to note that larger populations generally require much larger forces to stabilize.

Table 3.8
Summary of Variables Defining Operating Environments

Intervention Category	Factors Determining Operating Environment	Coding	Data Sources
Combat	Geostrategic era	Globalist era/ pre-globalist era	Pre/post-1940
	War type	Interstate war/ Intrastate war	UCDP/PRIO Armed Conflict Dataset
	Adversary strength	Major adversary/ minor adversary	CoW NMC data; Maddison Project estimates
Deterrence	Geostrategic era	Globalist era/ pre-globalist era	Pre/post-1940
	Level of threat	Higher threat/ lower threat	Index developed from multiple sources
	Level of commitment (alliance)	Treaty ally/ non-ally	CoW alliance data
Stabilization	Geostrategic era	Globalist era/ pre-globalist era	Pre/post-1940
	Level of consent	More consensual/ less consensual	UCDP/PRIO data on prior conflict
	Population size	Larger population/ smaller population	World Development Indicators

All of these ways of characterizing interventions' operating environments are only rough approximations. Readers will no doubt disagree with the assessments of some individual cases. In aggregate, however, these factors offer reasonable approximations of the degree of challenge the United States faces in a given category of intervention, and they can all be projected into the future using the models developed in this study. They thus can serve as a basis for estimating future force requirements in very broad terms.

Results

The combination of local factors and the overall geopolitical context together provide useful—albeit rough—indicators of the likely scale, composition, and duration of U.S. ground interventions. Table 3.9 provides a summary of the number of cases in each type, the average number of forces deployed in interventions of a given type, the average duration of interventions of that type, and the average ratio of the forces involved that were heavy troops, for post-1940 interventions where such data were available.

A few patterns are apparent across these summary data. As might be expected, the largest deployments occur for combat interventions—especially interstate wars against major adversaries. The lengthiest deployments are for deterrence missions, while interstate wars are typically comparatively short. Interventions also show clear differences

Table 3.9
Summary Information on Historical Force Requirements

Intervention Category	Historical Era	Local Operating Environment	Number of Cases (n)	Average Number of Deployed Forces	Average Duration	Heavy Troop Ratio (%, est.)
Combat	Globalist	Interstate war, major adversary	6	516,000	48	28.3
Combat	Globalist	Interstate war, minor adversary	3	12,000	2	8.1
Combat	Globalist	Intrastate war, major adversary	3	137,000	156*	27.5
Combat	Globalist	Intrastate war, minor adversary	1	5,000	48*	11.1
Combat	Pre-globalist	Interstate war, major adversary	3	223,000	11	–
Combat	Pre-globalist	Interstate war, minor adversary	0	N/A	N/A	–
Combat	Pre-globalist	Intrastate war, major adversary	1	13,000	9	–
Combat	Pre-globalist	Intrastate war, minor adversary	0	N/A	N/A	–
Deterrence	Globalist	Higher threat, treaty ally	6	46,000	440*	12.3
Deterrence	Globalist	Higher threat, non-ally	2	3,500	69	0.0
Deterrence	Globalist	Lower threat, treaty ally	3	18,000	84*	21.9
Deterrence	Globalist	Lower threat, non-ally	9	6,100	221*	1.7
Deterrence	Pre-globalist	Higher threat, treaty ally	0	N/A	N/A	–
Deterrence	Pre-globalist	Higher threat, non-ally	0	N/A	N/A	–
Deterrence	Pre-globalist	Lower threat, treaty ally	1	12,000	892	–
Deterrence	Pre-globalist	Lower threat, non-ally	0	N/A	N/A	–
Stabilization	Globalist	Less consensus, larger population	5	160,000	133*	21.4
Stabilization	Globalist	Less consensus, smaller population	6	11,000	95*	13.1
Stabilization	Globalist	More consensus, larger population	4	32,000	48*	10.2
Stabilization	Globalist	More consensus, smaller population	9	7,500	33*	6.9
Stabilization	Pre-globalist	Less consensus, larger population	1	67,000	49	–
Stabilization	Pre-globalist	Less consensus, smaller population	0	N/A	N/A	–
Stabilization	Pre-globalist	More consensus, larger population	3	3,600	63	–
Stabilization	Pre-globalist	More consensus, smaller population	7	3,000	106	–

NOTE: "N/A" designates categories for which there were no historical instances of U.S. intervention.
* One or more interventions in this category are ongoing.

between the globalist and pre-globalist eras. Deterrence missions in the pre-globalist era are almost nonexistent; with one exception (the U.S. military presence at the Panama Canal), the United States eschewed such open-ended support to other countries before World War II. Interventions before 1940 were also typically much smaller and shorter-lasting than those that occurred later.

A more detailed discussion of the individual intervention activity categories and the patterns of troops committed to such interventions is included in Appendix A. The average number of deployed forces and heavy troop ratios listed for each category in Table 3.9, however, are the crucial results for the purposes of our modeling effort, as it is these figures that are assigned to future projected interventions that fit into the relevant categories, in order to estimate the U.S. forces that are likely to be committed in the future.

Using Historical Data to Anticipate Future Requirements

Historical data on force deployments can help planners anticipate future requirements—but only if used with care. Two groups of problems prevent a straightforward extrapolation of past experience to potential future contingencies: problems associated with the historical data and problems associated with future requirements.

The historical data provide only an approximation of the "true" historical demand. Some problems arise from measurement error, including the limitations of Defense Manpower Data Center (DMDC) data on numbers of personnel deployed abroad and the imperfect record of unit types deployed.[68] Other problems arise from parameterizing these data—that is, reducing the constantly changing number and composition of forces into simple averages for entire missions, some of which last for decades. Yet other problems relate to historical force structure and force generation: In many cases, the United States deployed what units were available, not which ones were truly "required," and in other cases the United States relied on conscription to generate numbers of forces that would be nearly impossible under today's all-volunteer system.

Just as the past presents problems for estimating force requirements, so too does the future. The United States might approach future contingencies much differently than it has ones in the past. Perhaps, having grown weary of the enormous burdens associated with the wars in Iraq and Afghanistan, the United States will not again attempt to engage in large-scale stabilization operations, instead relying on much smaller deployments or avoiding such operations altogether. Technological innovations may give rise to military capabilities and concepts of operations that are much less manpower-intensive than previous operations.

These caveats notwithstanding, historical experience provides an important baseline for anticipating future needs. Problems of measurement error and the idiosyncrasies of individual cases prove less daunting as challenges if the typologies developed

---

[68] Issues of measurement error are described in greater detail in Appendix A.

in this chapter are understood to provide only rough "rules of thumb" for anticipating the broad outlines of force requirements. For instance, out of the five post–World War II ground stabilization interventions that took place in the most challenging type of operational environment, where local support for the U.S. military presence was low and the local population was large, only one required fewer than 100,000 U.S. personnel in an average year. On the other hand, of the 19 post–World War II stabilization operations that took place in all other environments, only four required more than 20,000 forces. These patterns suggest that planners can relatively reliably anticipate that stabilization missions will require no more than a division and usually substantially less—unless they occur in large countries that the United States recently fought in an interstate war, in which case they reliably require forces on the magnitude of two full corps deployed for several years.

The future, of course, might differ in important ways from the past. But using these historical patterns or "rules of thumb" serves as an important baseline estimate of future requirements. Technology might of course profoundly alter future force requirements, and the United States might avoid mistaken policies in the future that it has adopted in the past. But such claims have been made before (for instance, during the period of the so-called Revolution in Military Affairs), only to be proven incorrect. While planners should not uncritically engage in linear extrapolations from the past, they also would be unwise to ignore long-standing trends. Historical data should inform baseline estimates, which can be adjusted to reflect well-founded beliefs about future developments but should not be discarded altogether.[69]

Modeling Approach Summary

This chapter has summarized how each of the four main components of our projection model, as illustrated in Figure 3.1, work. In this section, we briefly emphasize and illustrate how these different components work together to produce our projections of future trends in armed conflict and the demands for U.S. ground forces in future military interventions.

Briefly, each subsequent component of the model relies on the preceding component to produce its own results, and they do so on an annual basis. First, leveraging the statistical models of armed conflict outlined above, our model predicts the risk that different states will experience an intrastate or interstate armed conflict in a given year. The model also predicts which, if any, ongoing armed conflicts will end in a given year.

Those forecasts of armed conflict provide opportunities for U.S. ground interventions. In a process that parallels our forecast of armed conflict, our model then

[69] Daniel Kahneman, *Thinking, Fast and Slow*, New York, N.Y.: Farrar, Straus and Giroux, 2011; Philip E. Tetlock, *Expert Political Judgment: How Good Is It? How Can We Know?* Princeton, N.J.: Princeton University Press, 2005.

predicts, again leveraging our statistical models of past trends in U.S. ground interventions, which states will experience an intervention in a given year. The model also predicts which, if any, ongoing U.S. ground interventions end in a given year. Deterrence interventions are forecast during future years of peace, interventions into armed conflicts in a state are forecast during years of ongoing intrastate or interstate conflicts, and postconflict stability interventions are only forecast during postconflict years that follow armed conflicts forecasted to end by our model.

The final component of our forecasting model concerns the characteristics of forecasted U.S. ground interventions. Using the characteristics of historical U.S. ground interventions detailed in the previous chapter, we assign each forecast intervention with an average number of troops deployed and an average number of heavy forces deployed while the intervention remains ongoing. Once the characteristics of ongoing interventions have been adjudicated, the model starts the process over and begins forecasting armed conflicts for the subsequent year.

As a hypothetical example of how this process works, consider that, in our baseline forecasts, an intrastate armed conflict is predicted to begin in Jordan in 2017. In 2017, our model predicts that U.S. ground forces do not undertake an intervention into the conflict. As such, no assessments of troops are adjudicated. Our model then projects that the armed conflict begun in 2017 continues into 2018. However, our model predicts that in 2018 U.S. ground forces undertake an intervention into the armed conflict. Based on characteristics of the Jordanian conflict, our model then adjudicates the average number of total and heavy forces to be deployed as part of the intervention. The model predicts that, in 2019, even with added U.S. involvement, the Jordanian conflict persists, along with the U.S. intervention. As such, our model adjudicates the same average number of troops for the intervention, based on our qualitative assessments of historical trends. However, our model predicts that the Jordanian conflict ends in 2020, which transitions the U.S. intervention to a stability mission. Based on the change in mission, our model then adjudicates a new average force size for the intervention. This process is repeated for each year through 2040.

In addition to helping explain how the components of our forecasting model work together, this example also highlights the semi-stochastic and path-dependent nature of our forecasting model. Because our forecasts are developed using predicted probabilities from historical trends, our model's predictions about the onset and cessation of armed conflicts and interventions are partly probabilistic, meaning that they slightly change every time the model is run. Additionally, because the components of our forecasting model are interconnected (e.g., the model can only forecast the start of an armed conflict intervention if the model has previously forecast the state of an armed conflict), then slight changes in the model's forecasts for different components can ultimately lead to significantly diverging forecasts between different runs of the model. To minimize these differences and make our forecasts more robust, we ran 500 iterations of our forecasting model for each future strategic environment and present the mean predicted trends in armed conflicts and U.S. ground interventions across those iterations. In addition, we also present the 10th and 90th percentile projections from those 500 iterations of our forecasting model to show the range of our forecasts. These results are discussed in Chapter Four.

CHAPTER FOUR

Future Demand for U.S. Ground Forces: Forecasts of Armed Conflicts and U.S. Military Interventions

This chapter describes the results from our forecasting models—forecasts of future armed conflicts and future U.S. ground interventions in response to those conflicts. We first describe our forecasts in a baseline, "no surprises," future strategic environment, and then move on to detail and analyze four alternative future scenarios. In each, we first discuss the scenario itself, the changes made to key model parameters to implement the scenario, if any, and then describe the resulting forecasts, drawing comparisons with the baseline case as appropriate. We provide information on interstate war and intrastate conflict and include projections at the regional level where possible. In projecting U.S. ground interventions, we consider interventions into armed conflict, stability operations, and deterrence operations. We also include forecasts for the total number of U.S. ground interventions in each year and the number of troops involved, both overall and in each type of intervention.

In all scenarios, we provide two summary measures of our forecasts. First, we provide the mean estimates of our forecasts from 500 iterations of our forecasting model, which provides our model's average forecasted trend for armed conflict and U.S. ground interventions. Second, we provide the 10th and 90th percentiles of those 500 iterations of our forecasting model. These percentiles provide insights into the degree of uncertainty and bounds of our forecasts. Effectively, these percentiles provide the range in which the true trend is most likely to fall.[1] The area between the 10th and 90th percentiles is demarcated by a gray shaded area in the graphs in this chapter, and the mean projection appears as a red line.[2] In most cases, the red line will fall within the gray shaded area. However, in a handful of cases, it does not. In these cases, there are large outlier estimates influencing the average value, of which we still felt it was

---

[1] Specifically, there is only a 20 percent overall chance that the actual value will lie outside of these bounds in any case.

[2] We use the *mean value* rather than the *median* because we want our summary metric to account for the full range of values produced by the 500 iterations of our model. In other words, if we have a handful of outliers, we want the summary metric to be influenced by these outliers to give us an accurate summary of the full range of the data.

important for planners to be aware. Providing both types of summary information about our projections allows us to be clearer about the level of uncertainty associated with our estimates.

Baseline Forecasts

As discussed in Chapter Three, the baseline scenario reflects the most likely anticipated future strategic environment. More specifically, our baseline scenario largely relies on projections of our key factors from the International Futures model base run, which anticipates mostly gradual changes in our key factors in the 2017–2040 period.[3] Globally, levels of state wealth continue to gradually increase, leading to higher overall GDP and levels of GDP per capita. Global population levels also continue to increase, but with the populations of many developing countries rising much faster than those of developed countries, and youth bulges largely disappearing outside of sub-Saharan Africa. Global levels of democracy also continue to gradually increase, with more states transitioning from autocracies to anocracies and some even becoming consolidated democracies.

Forecasts of Trends in Armed Conflict

Overall, our baseline projections suggest a long-term decline in the incidence of intrastate conflict, but a modestly increasing risk of interstate war. The decline in intrastate conflict is projected to take place across geographic regions, while the increased risk of interstate war is largely confined to the Middle East and Eurasia in our projections.

Interstate War

Figures 4.1 and 4.2 show our projections for future trends in the number of states involved in new interstate wars and the number of states involved in ongoing interstate wars, respectively, while Figure 4.3 shows our projections for the numbers of states involved in interstate wars at the regional level.

As shown in Figure 4.1, our baseline projections reflect a mean of about one new state involved in an interstate war each year between 2017 and 2040, a trend that remains relatively stable through 2040. Because interstate wars, of course, require at

---

[3] In many cases, the variable projections in our models are drawn directly from the International Futures tool, including for GDP growth, population changes, and regime type (Frederick S. Pardee Center for International Futures, undated). In other instances, including for interstate territorial claims, nuclear weapon capability, state military power or capabilities, and international normative strength, sufficient proxies were not directly available from International Futures, so we instead created new projection models using inputs that were available. This process is discussed in greater detail in Appendix B. For still other variables, we were unable to develop plausible projection models, and so made the simplifying assumption that their status in 2017 would continue into the future, including treaty alliance relationships, ethnic discrimination, and World Trade Organization membership, although these assumptions were modified in some alternative scenarios, as discussed later in this chapter.

Figure 4.1
Baseline Forecasts of Interstate War Onsets, 2017–2040

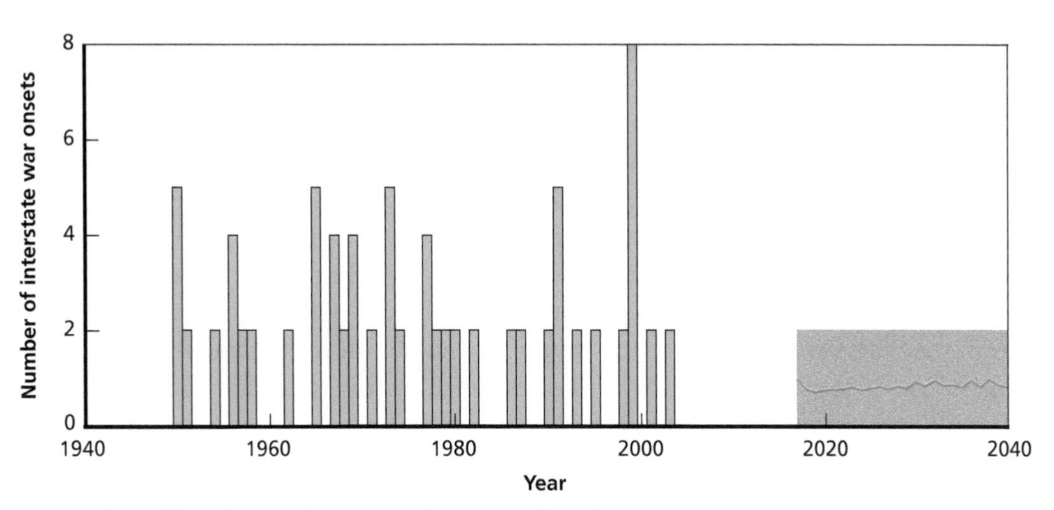

NOTES: The red line denotes the projected mean number of states involved in new interstate wars each year, based on 500 iterations of our forecasting model. The gray shaded area represents the range of forecasts bounded by the 10th and 90th percentiles of state involvement in interstate war onsets each year, based on 500 iterations of our forecasting model.

Figure 4.2
Baseline Forecasts of Interstate War Occurrence, 2017–2040

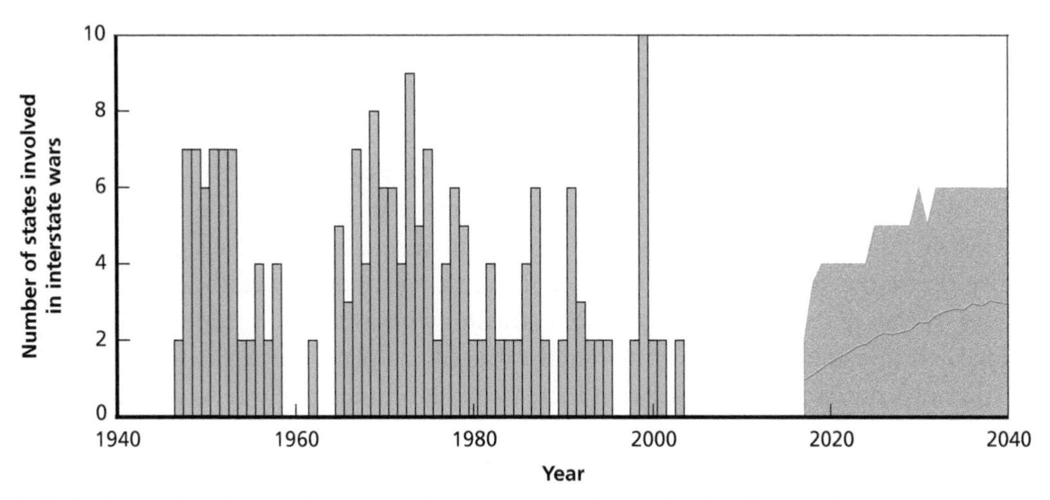

NOTES: The red line denotes the projected mean number of states involved in interstate wars each year, based on 500 iterations of our forecasting model. The gray shaded area represents the range of forecasts bounded by the 10th and 90th percentiles of state involvement in interstate war each year, based on 500 iterations of our forecasting model.

least two states to be fought, this mean projection can be interpreted as reflecting a roughly equal chance of an interstate war beginning, or not, each year. Even this average number of states involved in interstate war onsets is a notable divergence from recent trends, as there have been no new interstate wars since 2003. However, it is also relatively in line with longer historical trends from the pre-2003 period. The percentile projections from our baseline forecast provide additional context to this assessment. Throughout our forecasting period, our baseline forecasts suggest that there may be as many as two new states (or one war, in which the two states fight one another) and as few as zero new states involved in new interstate wars in each year. Again, when compared against the historical record, this range is fairly realistic. Given the recent rarity of interstate wars, however, a steady occurrence of one new interstate war each year in the 90th percentile projection would be a notable development, and our models suggest that it is plausible. Much of the relevance of this trend for U.S. policymakers would depend on the states likely to be involved, however, as will be discussed in greater detail below.

This trend in interstate war onsets is naturally also reflected in our projections of the number of states involved in ongoing interstate wars each year. In Figure 4.2, the average number of states involved in ongoing interstate wars gradually increases throughout the 2017–2040 period, increasing from an average of one state involved in interstate war in 2017 to an average of three states involved in ongoing interstate wars in 2040. Taken together with the projections in Figure 4.1, this increase in the number of states involved in ongoing interstate wars implies that at least some of those new interstate wars continue for multiple years. As this mean number of states involved in ongoing interstate wars increases over time, so to do our 90th percentile projections of interstate war. While our forecasts for 2017 predict between zero and two states involved in active interstate wars, this range quickly increases to as many as four states involved in active interstate wars, and eventually increases further to as many as six states involved in interstate wars. The increasing range of these 90th percentile projections, while the 10th percentile projections remain at zero, reflects the significant variability across our forecasts. Although our forecasting model generally expects global rates of interstate war to increase over time, futures with as many as three pairs of states engaged in interstate war each year remain roughly as plausible as futures with zero. So, although our models anticipate an increased risk of interstate war to 2040, such an increase is by no means certain.[4]

It is also worth putting this forecasted change in perspective. Our mean projection line culminates at around about three states involved in interstate wars by 2040. This would suggest only one additional state compared with levels in the early 2000s.

---

[4] One important factor contributing to this relative projected increase in the risk of interstate war is the reduction, in many regions, of the degree of hegemony exercised by the most powerful state (in many cases, the United States). Historically, such declines in regional hegemony and shifts toward more multipolar systems have been associated with an increased risk of interstate war.

Figure 4.3
Baseline Regional Forecasts of Interstate War Occurrence, 2017–2040

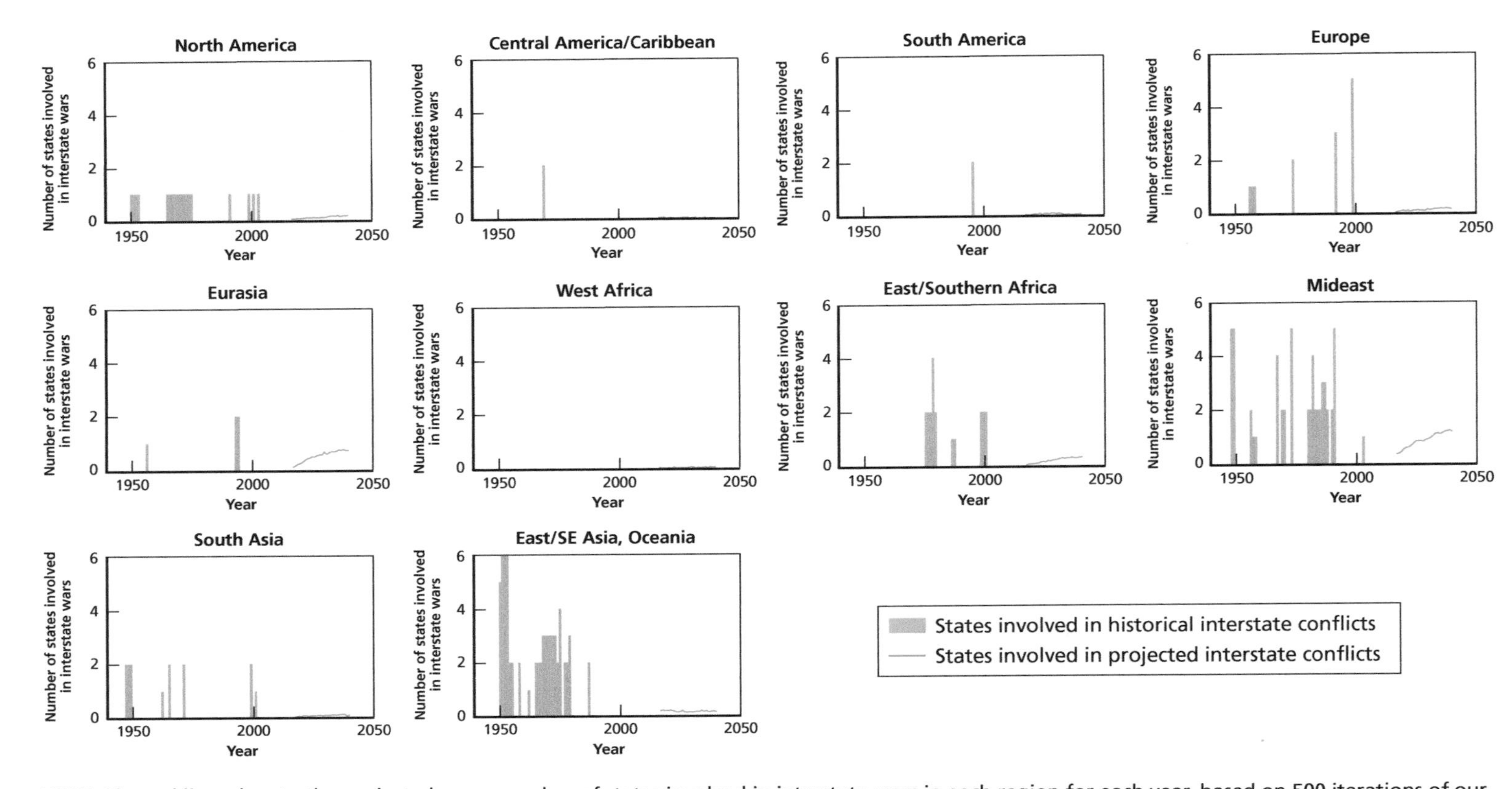

NOTE: The red lines denote the projected mean number of states involved in interstate wars in each region for each year, based on 500 iterations of our forecasting model.

Even at the upper end of our forecasting range, we would expect only about six states involved in ongoing interstate wars, which, while a sizable increase given the small historical number of interstate wars in recent years, would still be somewhat lower than peak levels of interstate war even in the post-1945 period.

Our regional forecasts of the number of states involved in interstate wars are shown in Figure 4.3 and suggest rather significant divergence in levels of interstate war between regions. Most regions, namely North America, Central America, South America, Europe, West Africa, South Asia, and East/Southeast Asia, show essentially no change over the 2017–2040 period, with state involvement in interstate wars expected to remain a rarity. Other regions show marked change over our forecasting period. Our forecasts expect notable increases in state involvement in interstate war in Eurasia and in the Middle East compared with recent historical trends, and a much more modest potential for increase in East and Southern Africa. Even in the most dramatic case, however, this expected increase does not amount to more than one state per year, on average, involved in interstate war, or a roughly even chance of two states involved in an interstate war in a given year. While this would be a particularly notable development in Eurasia, given the region's historical lack of interstate war, it is still likely below historical rates of involvement in interstate war in the Middle East in the pre-2003 period. However, this expected increase in the risk of interstate war in Eurasia and the Middle East may have important implications for U.S. policymakers. While not as concerning as forecasts of interstate war increases in Europe or East Asia would be, Eurasia and the Middle East are areas of important strategic interest for the

Figure 4.4
Baseline Forecasts of Intrastate Conflict Onsets, 2017–2040

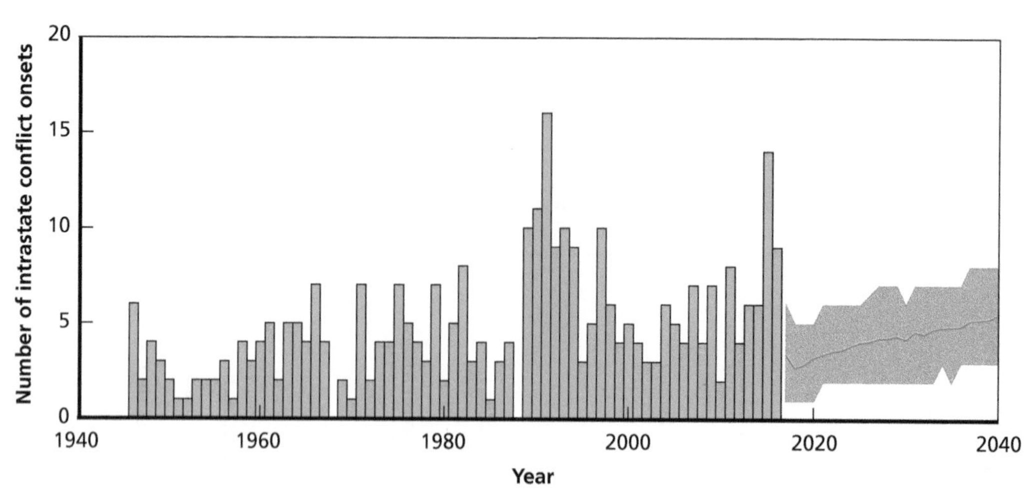

NOTES: The red line denotes the projected mean number of new intrastate conflict onsets each year, based on 500 iterations of our forecasting model. The gray shaded area represents the range of forecasts bounded by the 10th and 90th percentiles of intrastate conflict onsets each year, based on 500 iterations of our forecasting model.

Figure 4.5.
Baseline Forecasts of Intrastate Conflict Occurrence, 2017–2040

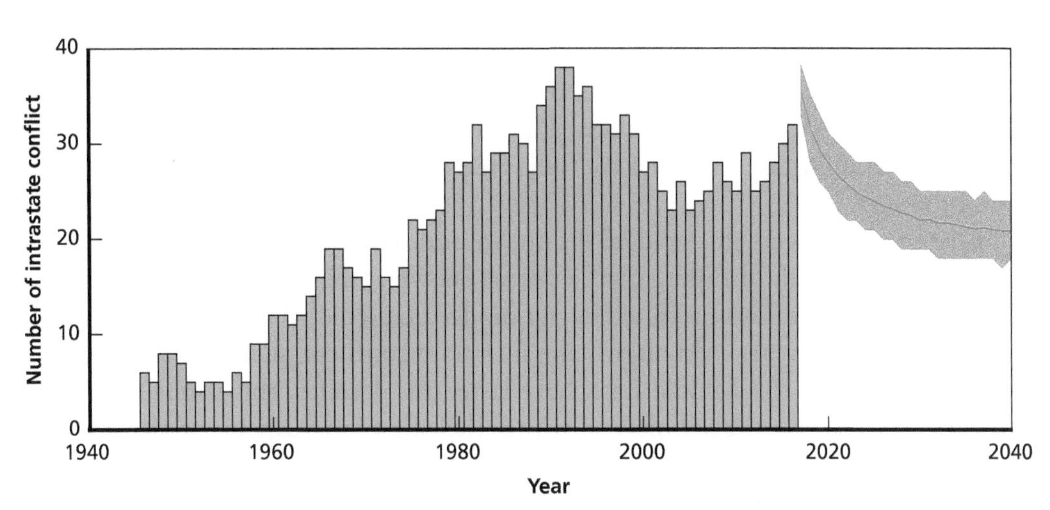

NOTES: The red line denotes the projected mean number of intrastate conflicts each year, based on 500 iterations of our forecasting model. The gray shaded area represents the range of forecasts bounded by the 10th and 90th percentiles of intrastate conflicts each year, based on 500 iterations of our forecasting model.

United States and may be accompanied by a heightened risk of intervention in such wars.

Intrastate Conflict

Our projections suggest that intrastate conflicts overall are likely to decline in incidence through 2040, with a modest increase in new conflict onsets being offset by a more dramatic increase in the rate at which intrastate conflicts are expected to end. Figures 4.4 and 4.5 show our forecasts of intrastate conflict onset and intrastate conflict incidence, respectively. Figure 4.6 shows our projections of intrastate conflict incidence at the regional level.

As noted, our baseline projections suggest a gradual increase in the onset of new intrastate conflicts. Over the 2017–2040 period, the annual number of new intrastate conflicts almost doubles, increasing from about three in 2017 to more than five new intrastate conflicts in 2040. In contrast with our interstate war projections, both the 10th and 90th percentile projections also increase, suggesting a total number of intrastate conflict onsets of about two more or less than this mean projection. Therefore, by 2040, there could be as many as seven or as few as three new intrastate armed conflicts beginning each year. While these rates of intrastate conflict onset are markedly lower than the significant numbers of new intrastate conflicts beginning in 2015 and 2016, or the even higher levels near the end of the Cold War, they still represent a heightened level over the broader historical average.

Our forecasting model projects annual rates of new intrastate conflicts to rise in our baseline scenario, but the same model also predicts a significant decrease in overall levels of intrastate conflict through 2040. After an initial increase to levels of intrastate conflict last seen near the end of the Cold War, these rates of intrastate conflict decline sharply, such that by 2020 the incidence of intrastate conflict declines from 35 ongoing conflicts to about 28 ongoing conflicts. These downward trends continue until 2040, when just over 20 intrastate armed conflicts are projected to be active, a level last seen around the year 2000.[5]

Our model projects this gradual decline in intrastate conflict incidence alongside the gradual increase in intrastate conflict onset shown in Figure 4.4 because many of the new intrastate conflicts projected by our model are predicted to be quite short, lasting only one to two years. Although some intrastate conflicts, such as ongoing insurgencies in India and the Middle East, have lasted for decades, this is not necessarily surprising, because most rebellions end within their first few years of activity. Because these new conflicts are projected to be short, overall levels of conflict decline, even as rates of new conflicts increase.

Figure 4.6 shows our regional forecasts of intrastate conflict. Not surprisingly, given our global projections, all regions experience either a stagnation or a decline in their levels of intrastate conflict incidence. The Americas and Europe see fairly stable low levels of intrastate conflict in our forecasting period, which is not surprising given those regions' relatively low levels of historical conflict. Our projections also suggest small declines in levels of intrastate conflict in Eurasia, West Africa, South Asia, and East/Southeast Asia, although intrastate conflict remains common in each of those regions. The projections suggest a more substantial reduction in intrastate conflict in the Middle East and East/Southern Africa, mirroring our global projections. Both regions are expected to benefit from increases in global wealth and from trends in democratization, which help fuel these declines, though intrastate conflict is still anticipated to be quite prevalent in these regions, which retain the highest incidence of intrastate conflict in the world.

Forecasts of U.S. Ground Interventions

These projections of interstate war and intrastate conflict form part of the demand signal for U.S. interventions abroad, which will be reflected in our forecasts of U.S. ground interventions. We provide four types of intervention forecasts—a set of overall forecasts that look at numbers of active interventions and associated ground troop levels at the global and regional levels, a set of forecasts for interventions into armed

[5] Many underlying projected trends are supportive of this overall decline in the incidence of intrastate conflict, including anticipated reductions in the number of youth bulges and relatively broad-based increases in levels of economic development.

Figure 4.6
Baseline Regional Forecasts of Intrastate Conflict Occurrence, 2017–2040

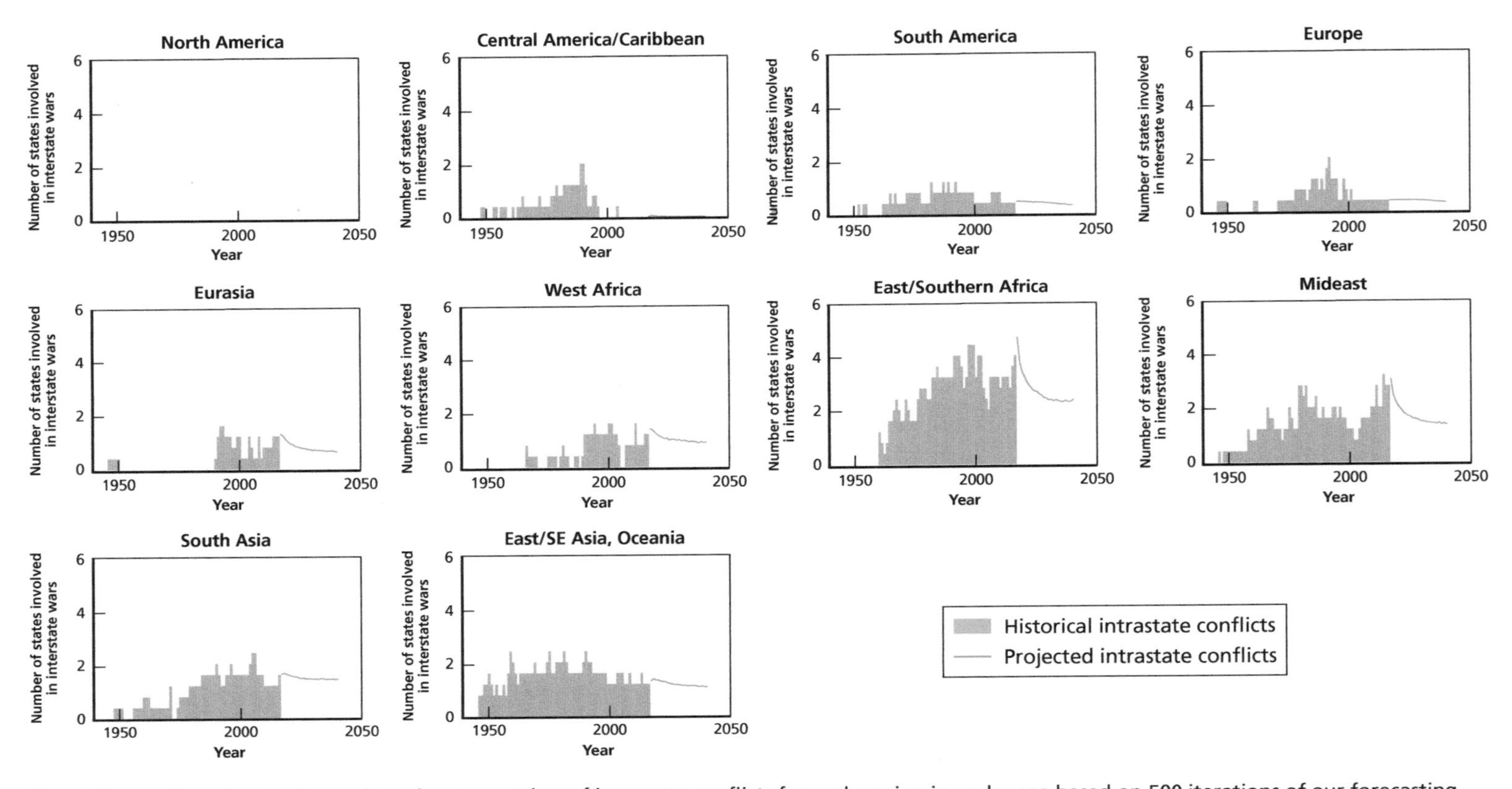

NOTE: The red lines denote the projected mean number of intrastate conflicts for each region in each year, based on 500 iterations of our forecasting model.

conflicts, a set of forecasts for stability operations, and a set of forecasts for deterrent missions.

As with our forecasts of armed conflict, we provide two summary measures of the results of our projection models. First, we provide the mean estimates of our projections from 500 iterations of our forecasting model, which provides our model's average expected trend in U.S. ground interventions and troops employed. Second, we provide the 10th and 90th percentile projections from those same 500 iterations of our forecasting model.

While our forecasting model projects an overall decline in the number of interventions as the most likely baseline trend, our projections also expect an increase in the number of U.S. ground forces required. Importantly, this increase is partly driven by two assumptions built into our model. First, we assume that ongoing U.S. ground interventions will continue at their 2016 force size, meaning that interventions ongoing as of 2016 do not increase or decrease in size during the course of our forecasts.[6] For example, while our models do project the likelihood that the U.S. intervention in Afghanistan will continue or end in each year, we do not assess the possibility that it may increase dramatically in size. Second, and similarly, we assume that new interventions require a static number of U.S. ground forces throughout their duration, using the force-sizing estimates detailed in Chapter Three. Like our estimates for U.S. forces required for ongoing interventions, this necessarily means that our model assesses a set number of forces required that does not vary over the course of an intervention. The projected increase in the number of ground troops employed in U.S. interventions is therefore driven, in part, by the anticipated ending of some interventions for which drawdowns have already taken place and their substitution (albeit likely in different countries) with new interventions whose average size is expected to be larger than those drawn-down troop levels.

Global Trends in U.S. Ground Interventions

Figure 4.7 shows our baseline projections of the total number of U.S. ground interventions in the 2017–2040 period. Under the baseline scenario, we expect the overall number of U.S. ground interventions to remain stable for about four years before beginning a slow decline, amounting roughly two fewer interventions in 2040 than in 2017, for a total of about seven ongoing interventions, on average.

Our percentile projections are particularly worth noting in this case. First, our 90th percentile projections emphasize that there is no guarantee that the total number of U.S. ground interventions will decline at all. Rather, the number of active U.S. ground interventions could increase by roughly one additional intervention for much of the period. Second, however, our average projections may understate the decline that

[6] The factors that may drive potential changes in size of interventions already ongoing are not currently well understood, and we judged separate models to project such changes to be beyond the scope of this effort.

Figure 4.7
Baseline Forecasts of Total U.S. Ground Interventions, 2017–2040

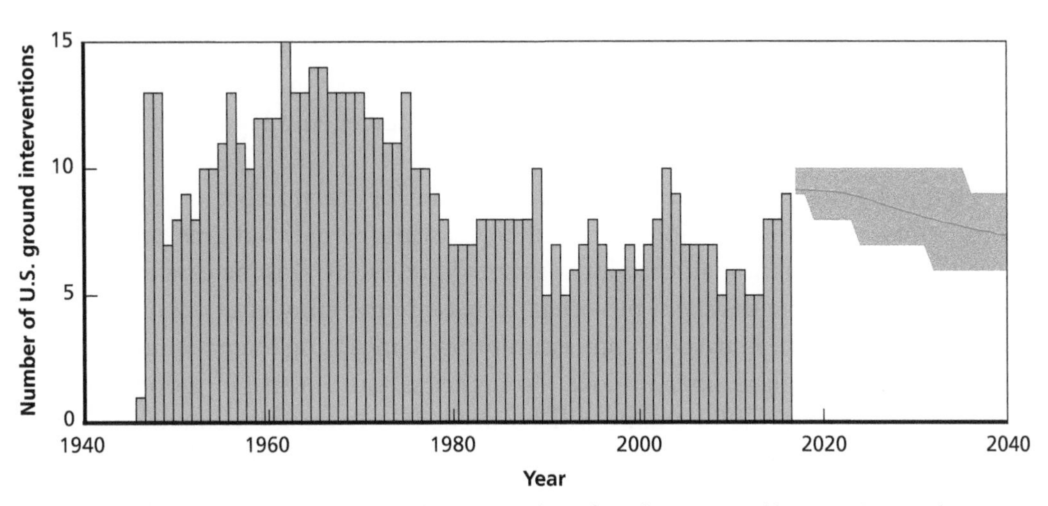

NOTES: The red line denotes the projected mean number of total U.S. ground interventions each year, based on 500 iterations of our forecasting model. The gray shaded area represents the range of forecasts bounded by the 10th and 90th percentiles of U.S. ground interventions each year, based on 500 iterations of our forecasting model.

Figure 4.8
Baseline Forecasts of Demands for U.S. Ground Intervention Forces, 2017–2040

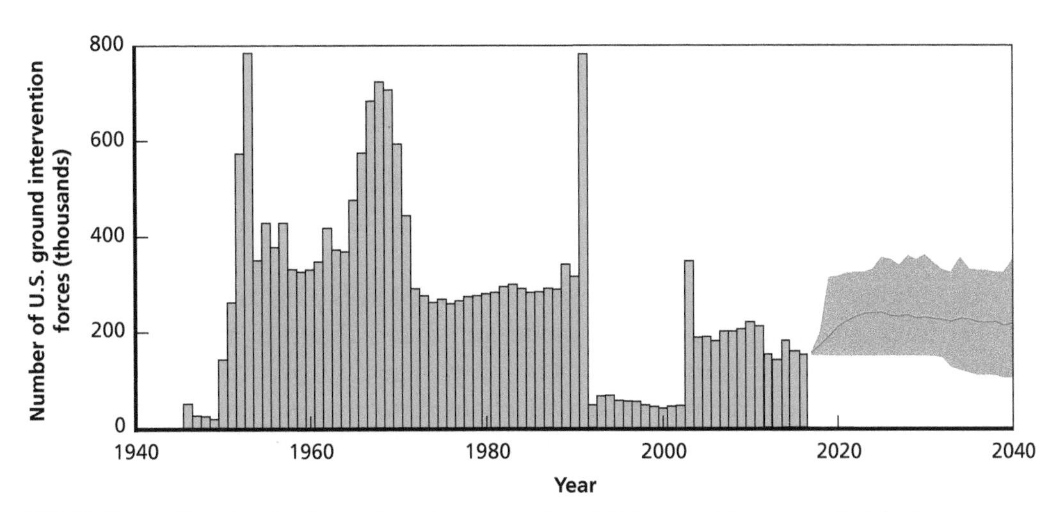

NOTES: The red line denotes the projected mean number of U.S. ground forces required for interventions each year, based on 500 iterations of our forecasting model. The gray shaded area represents the range of forecasts bounded by the 10th and 90th percentiles of U.S. ground forces required for interventions each year, based on 500 iterations of our forecasting model.

occurs, as shown by the 10th percentile projections that anticipate roughly six interventions by about 2030.

As a close read of Figure 4.7 illustrates, however, numbers of interventions have not historically provided a very robust or informative signal of the demand for U.S. troops, given how widely individual interventions may vary in size. Figure 4.8 therefore shows projected global demands for U.S. ground forces, and Figure 4.9 breaks this demand down to the regional level.

Figure 4.8 suggests a likely moderate increase in the number of U.S. ground troops employed in interventions over our forecasting period, but this projection is accompanied by substantial uncertainty. As noted previously, this increase in projected troops numbers despite the slight projected decrease in the number of ongoing ground interventions is driven, in part, but our assumptions concerning force requirements for ongoing and new ground interventions. Nonetheless, our mean projection line suggests that just over 200,000 troops will be needed to cover the demand for forces employed in U.S. ground interventions beginning around 2020. This Figure remains relatively stable between 2020 and 2040, declining only slightly. This represents an increase of about 50,000 troops over the number of U.S. ground forces deployed in interventions in 2016.

The percentile projections of this forecast, however, show that a wide range of future demands for U.S. ground forces in interventions is plausible, with an overall range of between about 100,000 and 450,000 U.S. ground troops required to meet intervention demands in different years through 2040. This degree of uncertainty is certainly large from a policymaking perspective. Planning to sustain more than 400,000 ground troops in interventions for over a decade would be very different from planning for 100,000. While we wish to emphasize the plausibility of either of these outcomes, we also emphasize that the average projection is more likely than either of the extremes. The degree to which policymakers plan for either the high or low end of these projections depends on the relative risk they prefer to accept, as will be discussed in greater detail in Chapter Five.

In addition to the approximate overall size of the forces projected, our results provide two additional insights. First, our model anticipates a short-term increase in U.S. ground troops employed in interventions in the average case, over the next five to ten years. Second, although the percentile projections show that our results have notable uncertainty throughout, the uncertainty in our projections becomes greater the further into the future they go. Whereas for the early 2020s, the difference between our 10th and 90th percentile projections is roughly 150,000 to 300,000, by the 2030s that range expands to cover roughly 100,000 to 400,000-plus troops. Military planners will perhaps be unsurprised that uncertainty increases the further into the future one goes, but the result is still worth emphasizing.

We can also provide some insight into how demand for U.S. ground forces may be expected to vary across regions, as detailed in Figure 4.9. Our baseline projections

Figure 4.9
Baseline Forecasts of Regional Demands for U.S. Ground Intervention Forces, 2017–2040

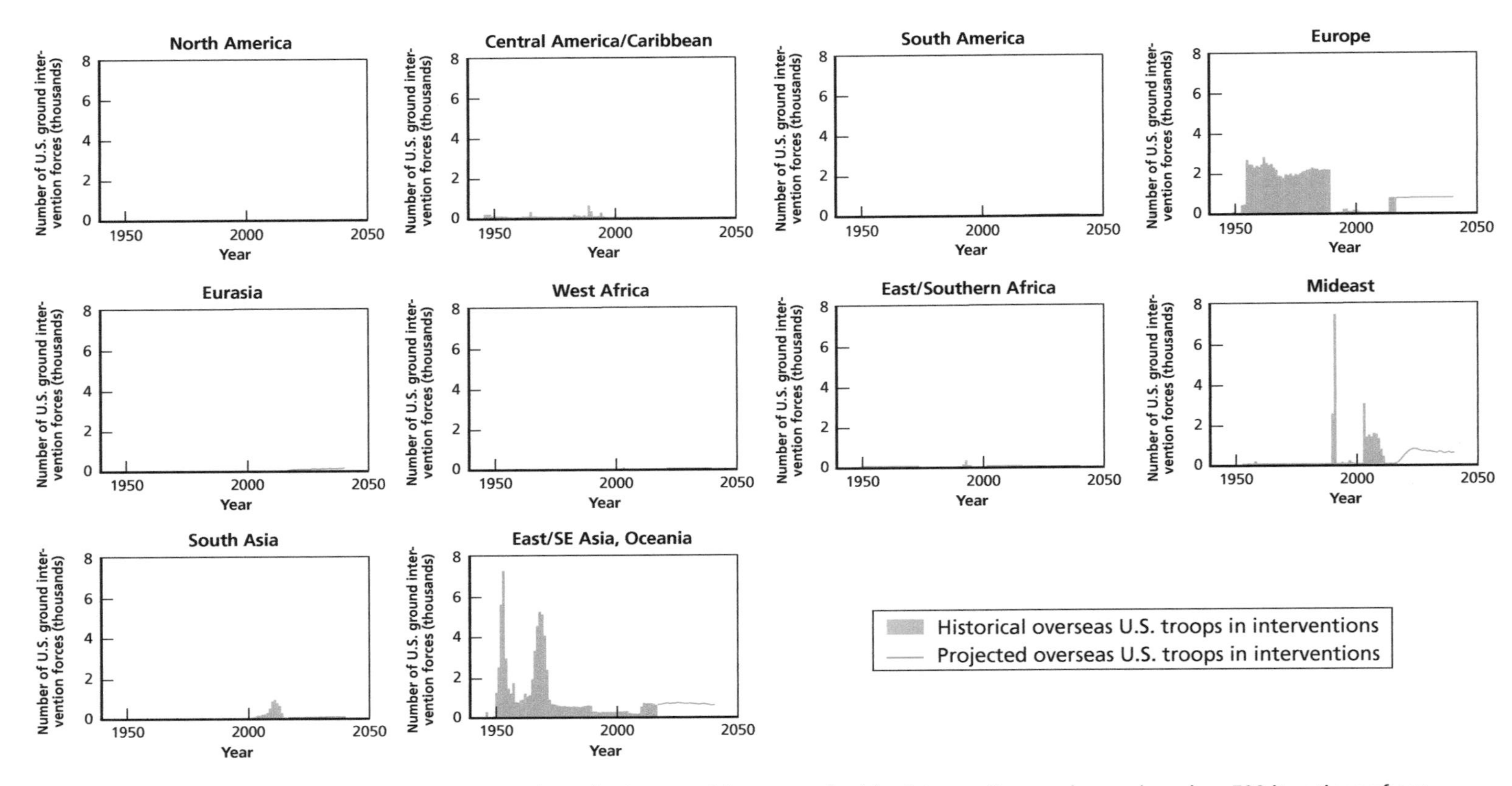

NOTE: The red line denotes the projected mean number of U.S. ground forces required for interventions each year, based on 500 iterations of our forecasting model.

are that the average employment of U.S. ground troops in interventions in the Americas and sub-Saharan Africa will remain quite rare throughout our forecasting period. The average number of ground troops employed in interventions in Europe, primarily deterrent missions, are expected to remain relatively stable. Forces expected to be employed in Eurasia in the average case remain low but do tick up slightly, a notable result given the historical lack of U.S. presence in this region. Demand for ground forces in South Asia is anticipated to remain low but steady. Forces in East/Southeast Asia, meanwhile, are anticipated to increase modestly in the 2020s in the average case, but overall to remain relatively stable, reflecting the continuation of deterrent missions in South Korea and Japan.

In contrast to the relative stability in the average number of U.S. forces employed in most regions, however, our projections do anticipate a sizable increase in demand for U.S. ground forces in the Middle East. In the average case, this would return the number of U.S. troops employed in this region to roughly half the peak of the Iraq War surge, though still notably below the large surges in forces that accompanied major combat operations such as the 1991 Gulf War. These projected increased forces are most likely to be employed in stability operations in the Middle East, with a post-conflict stabilization operation in Syria the most commonly predicted such mission.

U.S. Ground Interventions into Armed Conflict

As shown in Figure 4.10, our average forecasts reflect a consistent decrease in the number of U.S. ground combat missions in the 2017–2040 time frame. U.S. ground interventions into ongoing armed conflicts are rare overall, and the absolute size of this anticipated decrease in real terms is therefore relatively small. Overall, then, these results can be understood to reflect a moderately declining risk of U.S. involvement in armed conflicts.[7] The United States was involved in three combat missions in 2016—in Syria, Iraq, and Afghanistan—and our model projects that number to decrease by about one in the average case by 2040. The percentile projections emphasize, however, that both a sharper decline down to only one or a continuation at three interventions are also plausible. Either result may occur from the continuation of existing conflict interventions or from their cessation and U.S. ground intervention in a new conflict.

Although we expect that the number of U.S. ground interventions into armed conflicts will most likely decline, our projections also suggest that the number of U.S. forces involved in those conflicts will most likely increase. Figure 4.11 shows an

[7] This trend is, of course, informed by our projections of the incidence of armed conflict discussed above, which provide the opportunities for U.S. intervention in our models. All else equal, one would expect the overall reduction in the incidence of armed conflicts to translate into a smaller number of U.S. armed conflict interventions, and this is what we project. Trends in the variables that may make a U.S. intervention into any given armed conflict more likely (such as GDP per capita and regime type) are not necessarily in a direction that would suggest a reduced likelihood of intervention, though this may vary depending on where specifically the conflicts are forecasted to occur, and so the projected decrease in the number of armed conflict interventions is most likely driven by the reduced projected number of armed conflicts.

Figure 4.10
Baseline Forecasts of U.S. Ground Interventions into Armed Conflicts, 2017–2040

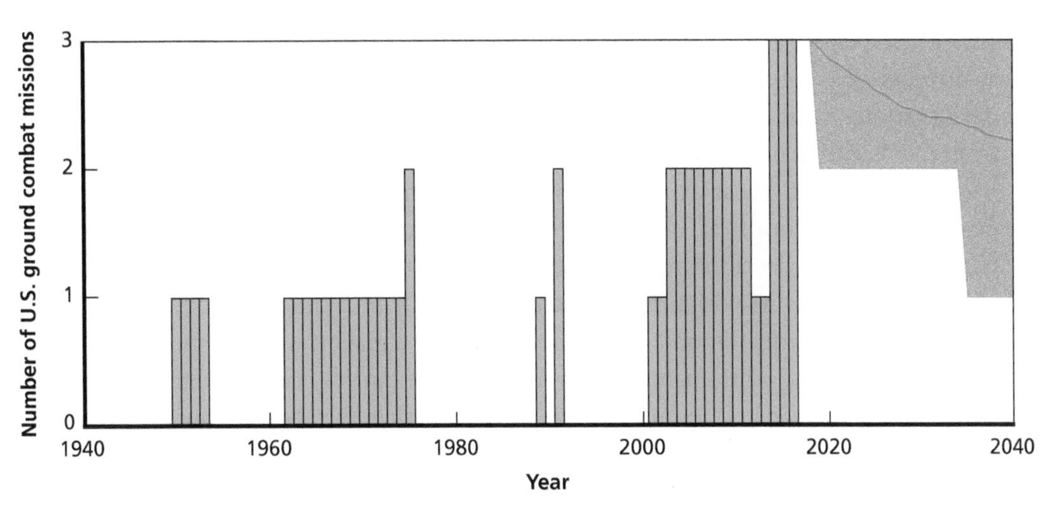

NOTES: The red line denotes the projected mean number of total U.S. ground force armed conflict interventions each year, based on 500 iterations of our forecasting model. The gray shaded area represents the range of forecasts bounded by the 10th and 90th percentiles of U.S. ground force armed conflict interventions each year, based on 500 iterations of our forecasting model.

Figure 4.11
Baseline Forecasts of Demands for U.S. Ground Combat Mission Forces, 2017–2040

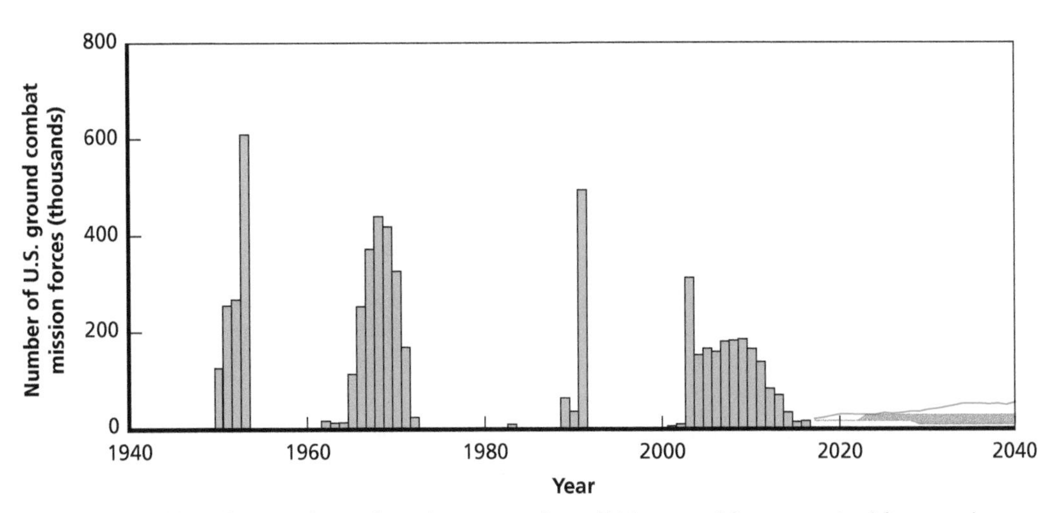

NOTES: The red line denotes the projected mean number of U.S. ground forces required for armed conflict interventions each year, based on 500 iterations of our forecasting model. The gray shaded area represents the range of forecasts bounded by the 10th and 90th percentiles of U.S. ground forces required for armed conflict interventions each year, based on 500 iterations of our forecasting model.

expected increase in the average number of troops involved in combat operations from about 15,000 in 2017 to closer to 75,000 by the 2030s.

Our percentile projections are particularly important for understanding these results. These projections, in contrast with the average case, suggest that a roughly constant number of troops in combat operations is the result in the large majority of our model iterations until 2040. The increase in the average projection is therefore driven by a relative handful of outlier cases with dramatic troop demands.

Taken together, these results emphasize the high uncertainty, and high consequences, of projections of U.S. involvement in combat operations. Although our baseline scenario projections to not suggest any anticipated increase in the number of combat operations in which the United States will become involved, the dramatically varying troop requirements of different specific conflicts are reflected in the large uncertainty regarding anticipated troop levels. For example, our model projects that a future conflict against Grenada would require very few troops—but in a conflict against North Vietnam, the projected demands are quite different. Our model suggests that larger-scale combat operations appear to be relatively unlikely for the next decade or so, but after that their plausibility appears to increase, and with it the uncertainty regarding the numbers of forces the United States is likely to need to devote to this category of intervention.

U.S. Ground Stability Operations

Our baseline forecasts suggest an increase in the number of U.S. ground stability operations in the average case into the mid-2020s, as shown in Figure 4.12. Although this increase is significant compared with the dearth of U.S. ground stability operations in recent years, if realized, this level would still be relatively small in comparison with the peak of U.S. involvement in stability operations in the 1990s and early 2000s.[8] The mean trend line then declines in the 2030s, albeit modestly.[9] The percentile projections of our forecasts emphasize that a range between zero and one stability operation during

---

[8] It is important to note that our data classify both post-2003 Iraq and post-2001 Afghanistan as combat operations for the U.S. military because they took place in the context of ongoing civil wars in those countries, rather than in their aftermath once hostilities had ceased, although of course both had elements of stability operation activities.

[9] This projected trend in the likelihood of a stability operation is likely driven by three key factors. First, because our model only projects an opportunity for a new stability operation when an armed conflict has ended, and no U.S. stability operations are currently ongoing, the opportunities for a potential intervention are expected to increase gradually over time, as more armed conflicts end. Second, our model highlights that the current level of zero U.S. stability operations is low based on historical trends and that the factors that make U.S. intervention in a postconflict environment more likely, such as low levels of GDP per capita, are still prevalent. So, a return to a more historically average level of stability operations would be expected. Third, however, some of the risk factors for a U.S. stability operation, such as low GDP per capita or low state population size, are expected to lessen over time, perhaps explaining the overall downward trend after roughly 2025. These projections, however, are dependent on the specific countries in which armed conflicts are forecast to occur, so isolating the specific factors most responsible for a complex trend such as the one observed in Figure 4.12 is difficult.

Figure 4.12
Baseline Forecasts of U.S. Ground Stability Operations, 2017–2040

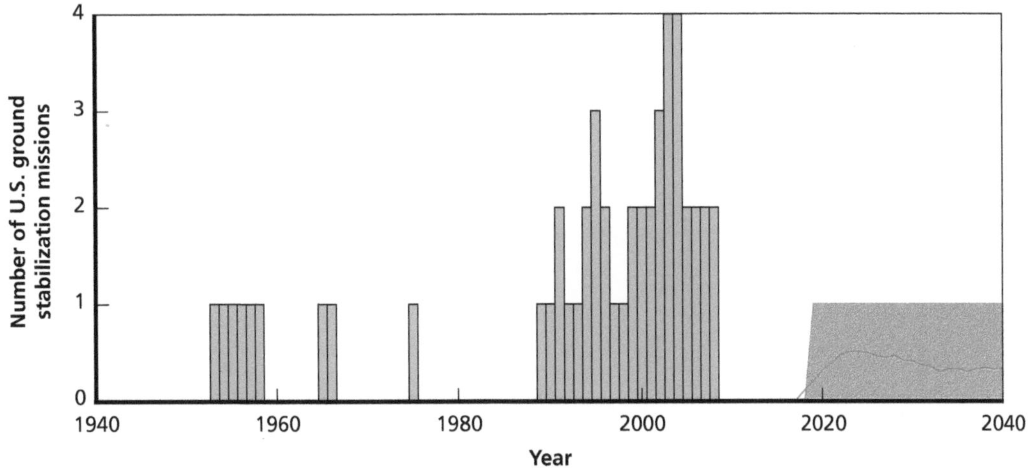

NOTES: The red line denotes the projected mean number of total U.S. ground stability operations each year, based on 500 iterations of our forecasting model. The gray shaded area represents the range of forecasts bounded by the 10th and 90th percentiles of U.S. ground stability operations each year, based on 500 iterations of our forecasting model.

the entire forecasting period is plausible. This suggests that the anticipated increase in the likelihood of U.S. involvement in a stability operation is likely to be limited, and dramatic increases in U.S. involvement in this activity are not anticipated in the baseline scenario, with no U.S. involvement in this activity remaining quite possible as well.

As shown in Figure 4.13, we expect the average number of U.S. forces involved in stability operations to follow a similar trajectory in the baseline scenario. The projections suggest an average increase beginning after 2017 to the mid 2020s to a level of about 75,000 before a gradual decline to under 50,000 by the mid-2030s. Once again, however, the percentile projections emphasize the wide range of plausible projections. According to our models, the number of U.S. ground troops involved in stability operations is likely to fall between 0 and 175,000 during the entire 2017–2040 period. As the average projection shows, we do expect the number of troops involved in stability operations to increase, in keeping with our expectation of an increase in the number of stability operations. But while the most likely outcome is a moderate increase, we cannot rule out a significantly larger increase, and there is still some probability of no increase at all. It is also worth noting that while we do not provide full regional projections for each intervention activity type for the sake of brevity, the most likely region for a new U.S. stability operation is the Middle East, as reflected in the expected overall increase in U.S. forces in that region shown in Figure 4.9.

Figure 4.13
Baseline Forecasts of Demands for U.S. Ground Stability Operations Forces, 2017–2040

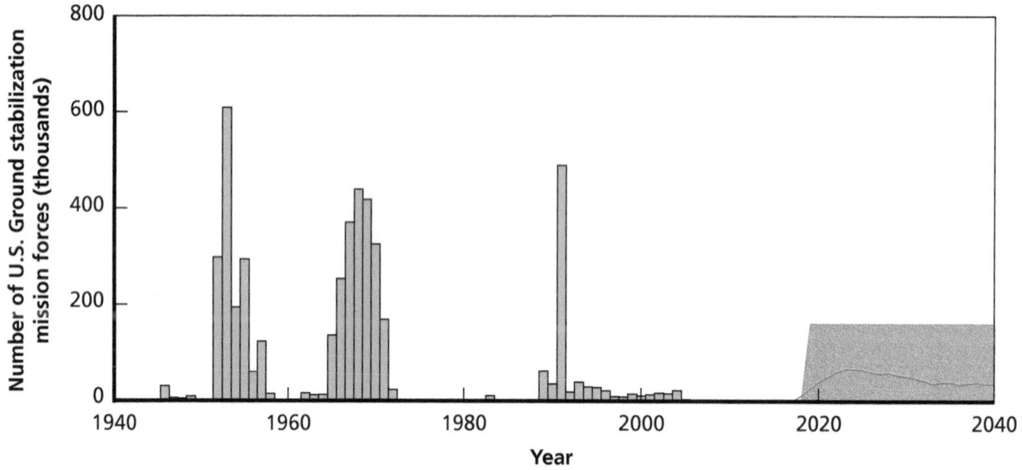

NOTES: The red line denotes the projected mean number of U.S. ground forces required for stability operations each year, based on 500 iterations of our forecasting model. The gray shaded area represents the range of forecasts bounded by the 10th and 90th percentiles of U.S. ground forces required for stability operations each year, based on 500 iterations of our forecasting model.

U.S. Ground Deterrence Missions

Figure 4.14 shows our forecasts of the number of U.S. ground deterrent interventions in the 2017–2040 period. Like our forecasts of armed conflict interventions, we expect the number of U.S. ground deterrent interventions to decrease over our forecasting period in the baseline scenario by a total of about one deterrent mission.[10] In substantive terms, this means that one of the United States' ongoing deterrent missions is projected to end, with the most likely candidate in our projections being the U.S. deterrent mission in the Sinai. This change in our mean projection is gradual, however, so this reduction appears more likely in 2040 than it does in the near future.

Our percentile projections provide the upper and lower bounds for this forecast. Interestingly, the upper bound of our forecast initially increases relative to current levels, suggesting a plausible increase of about one new deterrent intervention before eventually returning to current levels. In contrast, the lower bound of our projection suggests the potential for a sharper decline in active U.S. ground deterrent missions in the near future, suggesting that two active deterrent missions could plausibly end in the near future in some of our forecasts.

---

[10] This gradual decline would continue an overall trend observed from the height of the Cold War, notwithstanding a recent increase in deterrence missions over the past five years. Our models suggest a low likelihood of new deterrence missions, likely informed by an anticipated decline in relative U.S. military capabilities and a lack of new alliance commitments. But this decline is anticipated to be quite gradual, with most U.S. deterrence missions highly likely to persist to 2040 in our models.

Figure 4.14
Baseline Forecasts of U.S. Ground Deterrent Interventions, 2017–2040

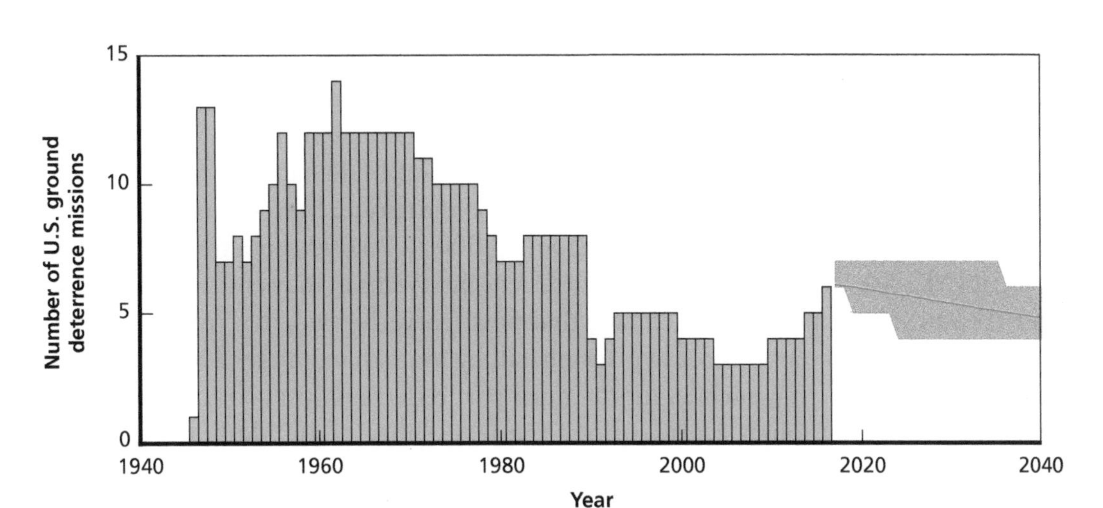

NOTES: The red line denotes the projected mean number of total U.S. ground deterrence missions each year, based on 500 iterations of our forecasting model. The gray shaded area represents the range of forecasts bounded by the 10th and 90th percentiles of U.S. ground deterrence missions each year, based on 500 iterations of our forecasting model.

Figure 4.15 shows the projected number of U.S. ground troops needed to support these demands. It is important to reiterate that, as discussed earlier, our model assumes

Figure 4.15
Baseline Forecasts of Demands for U.S. Ground Deterrent Intervention Forces, 2017–2040

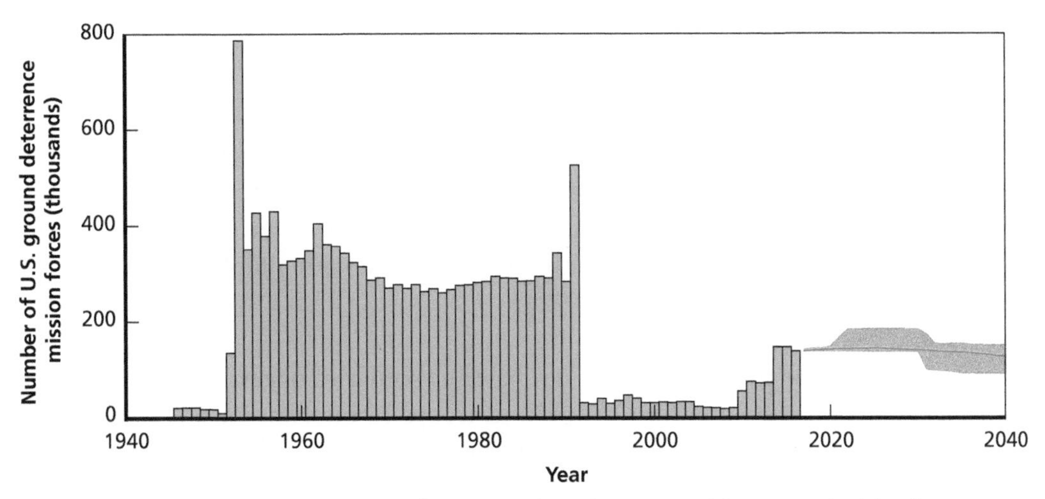

NOTES: The red line denotes the projected mean number of U.S. ground forces required for deterrence missions each year, based on 500 iterations of our forecasting model. The gray shaded area represents the range of forecasts bounded by the 10th and 90th percentiles of U.S. ground forces required for deterrence missions each year, based on 500 iterations of our forecasting model.

that ongoing interventions do not fluctuate in size from their 2016 levels, for as long as they are projected to continue. The limitation is particularly important for understanding the results of our deterrent intervention models, because deterrent interventions tend to be the longest-lasting. So, while our analysis suggests that some decrease in average demand for troops for deterrent interventions is likely, this is being driven by the expected likelihood that the ongoing deterrent mission in the Sinai will end sometime in the next decade or so, not by any reduction in demand for forces in other deterrent interventions that continue. Further into the future, however, our mean projections do suggest a further decline of about 15,000 to 20,000 more troops by 2040, reflecting the risk that another ongoing deterrent intervention may end over that time. Our 10th percentile projections reiterate that possibility, showing that troops committed to deterrent interventions could plausibly fall by more than half by 2040. The 90th percentile projection, meanwhile, for the most part reflects only a continuation of the status quo today, although it also reflects the plausibility of an additional deterrent intervention of moderate size in the mid-2020s.

U.S. Ground Heavy Forces

Our models made one additional projection for our baseline scenario: the number of heavy forces employed in U.S. ground interventions. As discussed in detail in Chapter Three and Appendix A, the share of heavy forces in an intervention may vary widely depending on the activities involved and the characteristics of the host state. Figure 4.16 provides a summary of our projected demand for heavy forces in U.S. ground interventions.

Comparing these projections with Figure 4.8 illustrates that demand for U.S. ground heavy forces in the average case is expected to be relatively proportional to demand for overall forces, at a ratio of roughly one to five. This result is perhaps expected given the assumptions made about this ratio for different types of interventions in Table 3.9, though it highlights that neither major combat operations (involving a higher ratio of heavy forces) or smaller-scale operations (and particularly stability operations) are expected to predominate in the future and drag this ratio more sharply in one direction or the other. Instead, a mix of contingencies is the most likely future, although, as the percentile projections show, futures with more interventions with heavier forces or fewer interventions with lighter forces do remain plausible.

Summary

Taken together, these projections from our baseline scenario suggest substantial uncertainty regarding the demand for U.S. ground troops in interventions, but with some clearer trends worth emphasizing. Our projections suggest a likely decline in the total number of intrastate conflicts at the global level but also a modestly increasing risk of interstate war. In both cases, these trends vary across regions, with the Middle East, Eurasia, and East/Southern Africa showing a higher likelihood of conflict. These

Figure 4.16
Baseline Forecasts of Demands for Heavy U.S. Ground Intervention Forces, 2017–2040

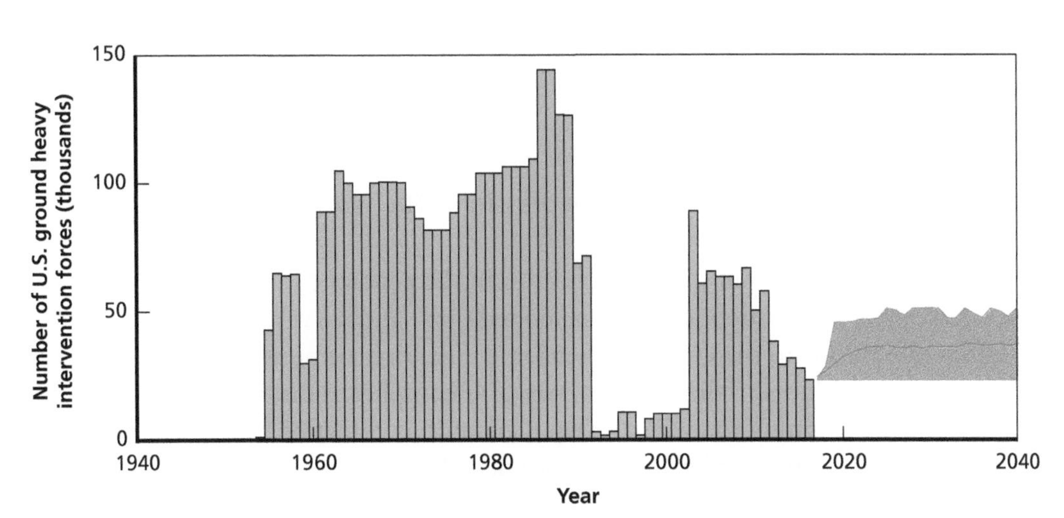

NOTES: The red line denotes the projected mean number of heavy U.S. ground forces required for interventions each year, based on 500 iterations of our forecasting model. The gray shaded area represents the range of forecasts bounded by the 10th and 90th percentiles of heavy U.S. ground forces required for interventions each year, based on 500 iterations of our forecasting model.

trends in conflict reflect only opportunities for U.S. ground interventions, however, and not the interventions themselves. On that front, we do expect a modest decline, in the average case, in the number of U.S. ground interventions, including deterrent interventions and interventions into armed conflicts. We also see potential, however, for an increase in stability operation interventions.

Despite this decreasing number of interventions in the average case, the average number of U.S. ground troops likely to be employed in interventions in the 2017–2040 period increases by 30 percent over 2016 levels, driven by an expected increase in the likelihood of a relatively large-scale stability operation and increasing demands from combat missions late in the period. Our projections of troop levels become substantially less certain in the 2030s, when the risk of involvement in large-scale combat and stability operations increases, although the lack of any such commitments remains plausible as well. Our estimates of regional demand highlight the Middle East, and to a lesser extent East/Southeast Asia and Eurasia, as the most likely sites in which these increased numbers of troops may be employed.

This set of estimates from our baseline scenario, however, reflects a future with few surprises in the strategic environment and does not reflect any shocks or changes in the international system that might alter the future more dramatically. The next four sections describe four alternative scenarios designed to substantially stress these projections by positing riskier futures and explore how demand for forces in these scenarios may diverge from the baseline case. In our discussion of each scenario, we present selected graphs illustrating the results of our projections that show the most relevant changes from the baseline scenario. A complete set of all results figures for each alternative scenario is included in Appendix C.

Forecasts of Armed Conflict and U.S. Ground Interventions in Alternative Future Scenarios

Alternative Future Scenario 1: Global Depression—The World Economy Fractures

Scenario Description

The financial crisis of 2008 shook confidence in the general consensus that fiscal and monetary policy could support a "great moderation" and avert an economic collapse of similar magnitude to the Great Depression.[11] The high amounts of public debt held by most advanced industrialized states may also complicate future efforts to provide a massive fiscal stimulus to avert a sharp economic contraction were such efforts to become necessary. Analysts have further argued that monetary policy remedies have been largely exhausted in the recovery from the 2008 financial crisis, which may limit the ability of policymakers to avoid periods of economic volatility in the future.[12] In addition, governmental indebtedness among major economies is likely to increase in the future as populations in North America, Europe, China, and Japan continue to age and retirement outlays increase.

In this alternative scenario, an economic crisis similar to the scale of the 2008 financial crisis occurs in the year 2025. Governments and central banks, however, no longer have the policy tools necessary to combat the sharp contraction in demand and plunging investor confidence. Faced with significant economic losses, economic nationalists come to power in many Organisation for Economic Co-operation and Development (OECD) countries and enact severe protectionist policies, which works to undermine much of the present globalized economy. Rival economic blocs form in North America, in Europe, around China, and around Japan.

This economic crisis also extends into the political realm and fosters a series of crises in the developing world. Many fledgling democratic governments are toppled, and these events pose legitimacy crises for many authoritarian regimes. China and Russia, in particular, are beset by turmoil caused by massive income inequalities, severe environmental degradation, and severe demographic imbalances.[13]

---

[11] Ben S. Bernanke, "The Great Moderation," remarks delivered at the meetings of the Eastern Economic Association, Washington, D.C., February 20, 2004; James H. Stock and Mark W. Watson, "Has the Business Cycle Changed and Why?" *NBER Macroeconomics Annual 2002*, Vol. 17, Cambridge, Mass.: MIT Press, 2003.

[12] National Intelligence Council, *Global Trends 2030: Alternative Worlds*, Washington, D.C.: Office of the Director of National Intelligence, December 2012; Lawrence H. Summers, "U.S. Economic Prospects: Secular Stagnation, Hysteresis, and the Zero Lower Bound," *Business Economics*, Vol. 49, No. 2, 2014; Coen Teulings and Richard Baldwin, eds., *Secular Stagnation: Facts, Causes and Cures*, London, UK: Centre for Economic Policy Research, August 15, 2014.

[13] William H. Overholt, *Asia, America, and the Transformation of Geopolitics*, New York, N.Y.: Cambridge University Press, 2008.

Changes to Key Factors

This scenario is designed to simulate effects similar to those that occurred in the Great Depression that began in 1929 and the ensuing economic and political turmoil of the 1930s. Consequently, many of the changes to our key factors parallel historical conditions of that period:[14]

- **Rate of economic growth:** Sharp declines in projected annual GDP growth rates, thereby reducing levels of GDP per capita.[15]
- **Extent of economic interdependence:** Rapid declines in levels of global trade, paralleling those experienced during the Great Depression.[16]
- **Turmoil in China:** Chinese growth rates stagnate and the country experiences two decades of lost growth. China's major trading partners also suffer as China is no longer able to act as an engine for the global economy.[17]
- **Exclusive economic trading blocs:** Large global economies leave the World Trade Organization system and create their own trading blocs, centered on the United States, China, the European Union, and Japan.[18]
- **Prevalence of consolidated democracies:** Many fledgling or weak democracies falter and revert to autocratic rule, similar to the experience of many European states in the 1920s and 1930s.[19]
- **U.S. Forward Presence:** The United States dramatically reduces its number of forward-deployed ground forces in response to growing fiscal constraints and growing isolationism.[20]

---

[14] These changes parallel an earlier scenario-development effort detailed in Watts et al., 2017a, pp. 87–90. The specific numerical changes reflect approximate values taken from an analysis of economic data from the 1930s.

[15] Specifically, a 4 percent reduction in projected annual GDP growth rates for each country for five years beginning in 2025.

[16] Specifically, a reduction in global trade flows of 55 percent over four years, or a reduction of 11 percent each year, beginning in 2025, followed by a gradual recovery of global trade back to originally projected levels in 2034.

[17] Beginning in 2025, Chinese GDP stops increasing through 2040. In addition, we model an additional 1 percent reduction in GDP growth rates from 2025 through 2040 for all of China's major trading partners, including the United States, Japan, Russia, South Korea, Taiwan, Australia, and most of the Association of Southeast Asian Nations (ASEAN).

[18] These new trading blocs begin in 2025. The U.S. trading bloc includes the United States, Canada, and Mexico. The China trading bloc includes China and North Korea. The EU trading bloc includes the existing EU member states. The Japanese trading bloc includes Japan, South Korea, and the existing ASEAN states.

[19] Beginning in 2025, we model a 4-point drop on the 20-point Polity scale for all states, excluding those that are established democracies. In 2030, this effect is removed, and these states revert to their original projections. (Monty G. Marshall, Ted Robert Gurr, and Keith Jaggers, *POLITY IV Project: Political Regime Characteristics and Transitions, 1800–2016—Dataset Users' Manual*, Vienna, Va.: Center for Systemic Peace, 2017.)

[20] The United States does not completely withdraw from any states where U.S. troops are already present, but the model reduced the number of U.S. forces deployed to each state by 50 percent after 2025.

Effects on Conflict and Intervention Forecasts

Compared with the baseline scenario, the Global Depression scenario shows a greatly elevated risk of interstate war but relatively little change in rates of intrastate conflict. Our full intrastate conflict projection results are provided in Appendix C and not discussed further here. The changes in interstate war, however, require a more detailed assessment. Figure 4.17 shows the total number of states involved in interstate wars in each year under this scenario.

Up until 2025, when we stipulate that the Depression occurs, the percentile projections range from 0 to about 4—larger than the current level but in line with levels in the pre-2003 period. After 2025, however, we see a notable increase. Both the average and percentile projections increase by roughly one additional state involved in interstate war by 2030. This higher projected level of interstate war is then maintained out to 2040. These projections represent a sizable increase in the risk of interstate war in the post-2025 period, although by 2040 the average and percentile projections have again largely converged with the baseline scenario.

At the regional level, many regions experience a sharp increase in interstate war under this scenario, as shown in Figure 4.18.

The most dramatic increases occur in the Middle East and Eurasia, where state involvement in interstate war spikes dramatically in the average case. However, even some ordinarily more-pacific regions, such as South America, see notable increases in state involvement in interstate war.

Figure 4.17
Global Depression: Forecasts of Interstate War Occurrence, 2017–2040

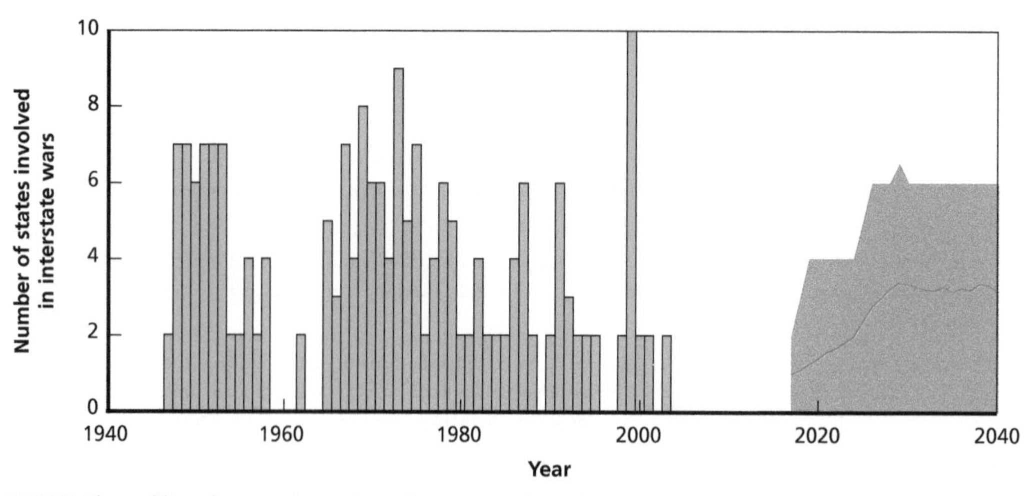

NOTES: The red line denotes the projected mean number of states involved in interstate war each year, based on 500 iterations of our forecasting model. The gray shaded area represents the range of forecasts bounded by the 10th and 90th percentiles of states involved in interstate war each year, based on 500 iterations of our forecasting model.

Figure 4.18
Global Depression: Regional Forecasts of Interstate War Occurrence, 2017–2040

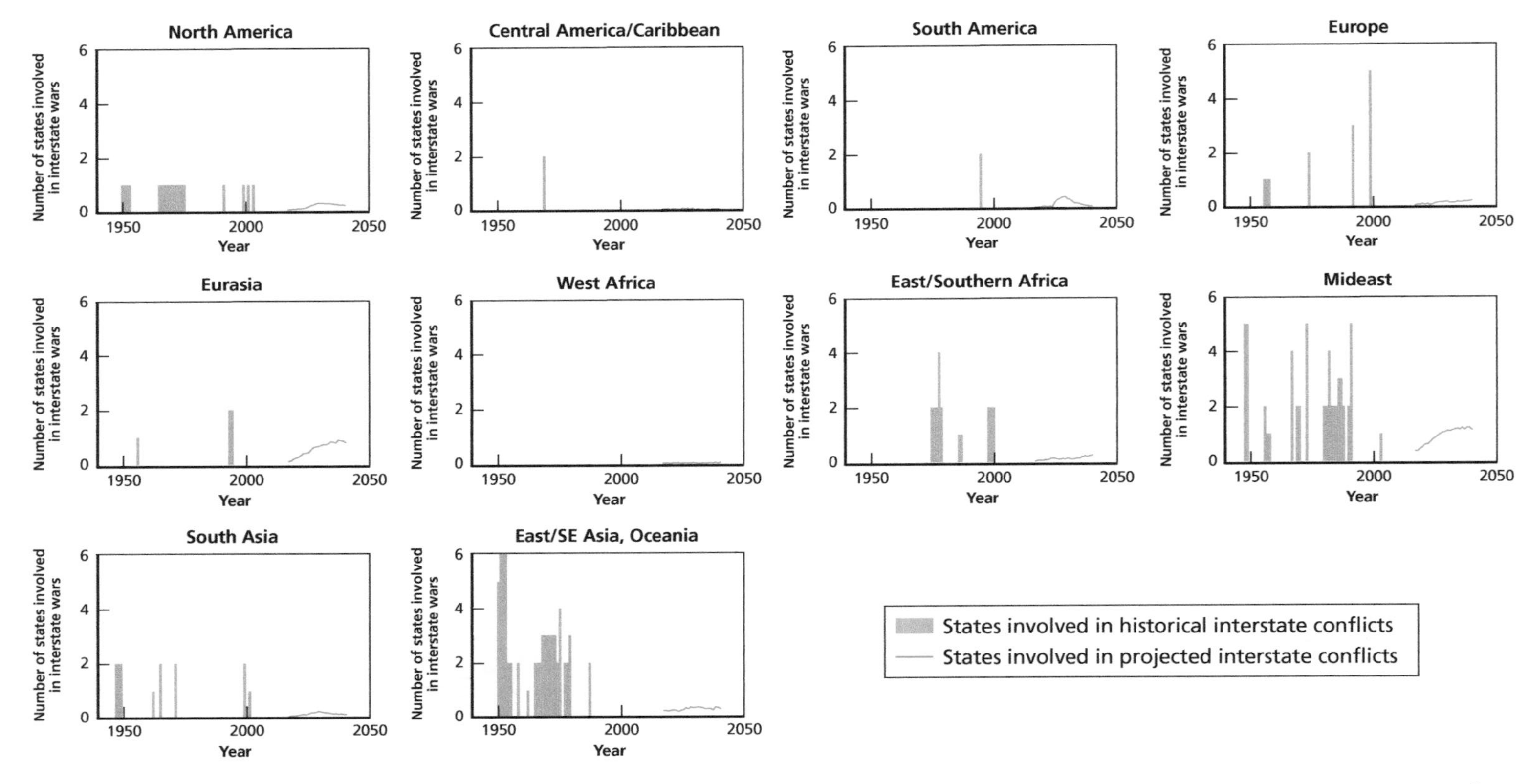

NOTES: The red lines denote the projected mean number of states involved in interstate wars in each region for each year, based on 500 iterations of our forecasting model.

For the global depression scenario, our models predict very limited change in the average case in the number of U.S. ground interventions compared with the present, as shown in Figure 4.19.

Importantly, however, this represents an increase relative to the baseline scenario, which expected a greater decline in the number of ground interventions over time in the average case. While the number of interventions would be expected to increase modestly in the Great Depression scenario, the demands for U.S. ground forces in these interventions would be expected to change much more dramatically.

Figure 4.20 shows an initial decline in U.S. troops, as deterrent forces from abroad are withdrawn as stipulated in the scenario design, followed by a sustained increase in forces to 2040. We also see a sharp increase by 2040 in the percentile projections that now range from 100,000 to 600,000. As will be discussed below, this trend reflects a sharp shift in the types of interventions to which the United States commits troops in the 2030s, away from deterrent interventions and towards combat interventions.

Our regional projections, shown in Figure 4.21, suggest that the most sizable increases in U.S. forces are likely in East/Southeast Asia and the Middle East. In these regions, U.S. deterrent forces are withdrawn in 2025, but then replaced by other demands. Notably, both are regions of high strategic importance for the United States and ones that appeared affected by the relatively greater potential for increased interstate war also observed in this scenario. By contrast, in Europe withdrawn U.S. deterrent forces lead to overall lower levels of U.S. troops in that region.

Figure 4.19
Global Depression: Forecasts of Total U.S. Ground Interventions, 2017–2040

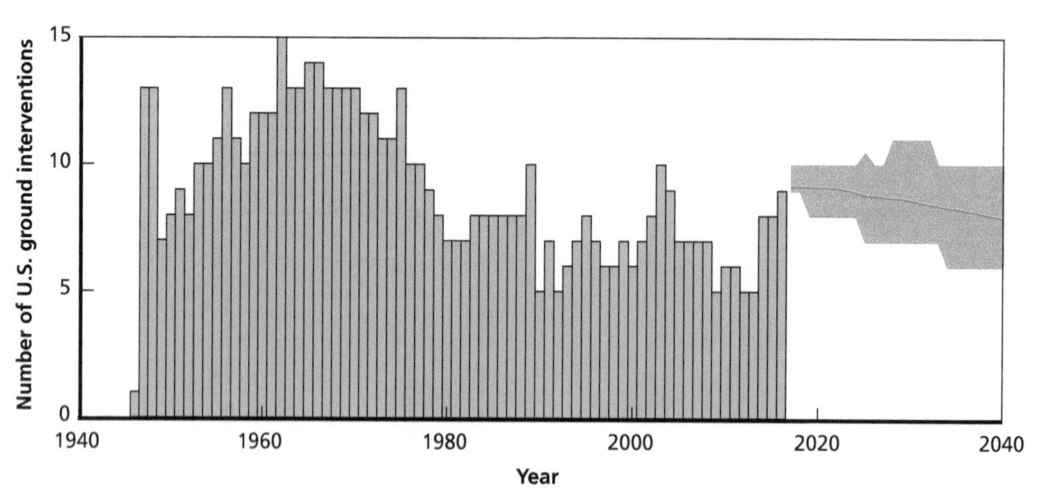

NOTES: The red line denotes the projected mean number of total U.S. ground interventions each year, based on 500 iterations of our forecasting model. The gray shaded area represents the range of forecasts bounded by the 10th and 90th percentiles of U.S. groundinterventions each year, based on 500 iterations of our forecasting model.

Figure 4.20
Global Depression: Forecasts of Demands for U.S. Ground Intervention Forces, 2017–2040

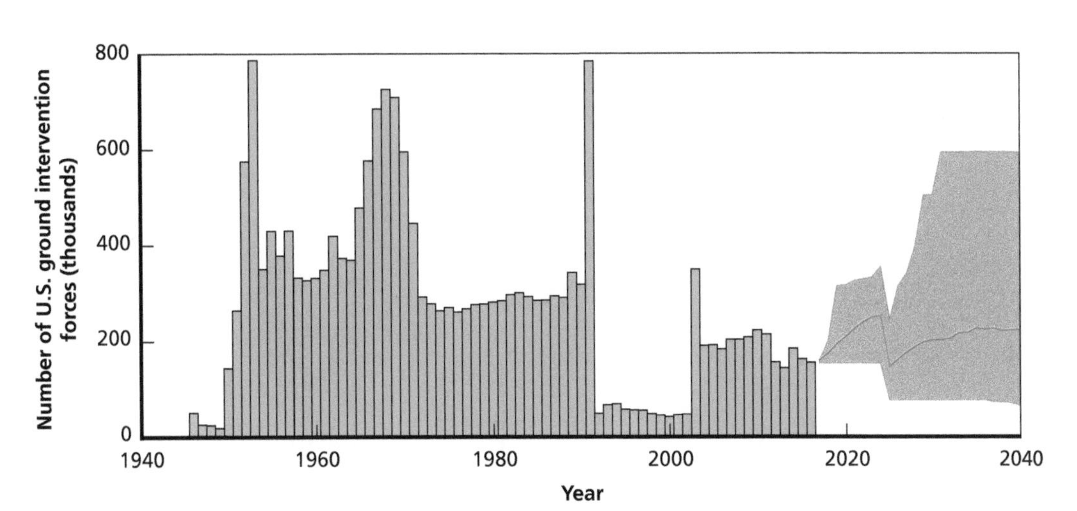

NOTES: The red line denotes the projected mean number of U.S. ground forces required for interventions each year, based on 500 iterations of our forecasting model. The gray shaded area represents the range of forecasts bounded by the 10th and 90th percentiles of U.S. ground forces required for interventions each year, based on 500 iterations of our forecasting model.

What is driving this change in interventions and demands for U.S. ground forces compared with the baseline? As Figures 4.22 and 4.23 illustrate, the increase is driven by a sharp rise in ground forces committed to armed conflict interventions and to some extent also an increase in commitments to stability operations. The number or type of deterrence missions, meanwhile, decline substantially, as stipulated in the scenario design.

This sharp increase in ground forces committed to armed conflict interventions in the average case would represent a return to a major combat operation nearly on the scale of Iraq. It is worth emphasizing the uncertainty in the percentile projections, however. At the 10th percentile, our projections would suggest very few if any ground forces committed to armed conflict interventions, while at the 90th percentile that would suggest a commitment considerably greater than the Vietnam War. While the exact location of such a conflict in our projections is highly uncertain, the most likely regions in which it would occur are East Asia and the Middle East.

U.S. ground forces committed to stability operations also increase somewhat over our baseline projections, although less dramatically than they do for armed conflict interventions. Commitments to stability operations increase again after 2025 in the average case, although even in this scenario the uncertainty remains substantial, ranging between zero and roughly 175,000 at the 10th and 90th percentiles, respectively.

Figure 4.21
Global Depression: Forecasts of Regional Demands for U.S. Ground Intervention Forces, 2017–2040

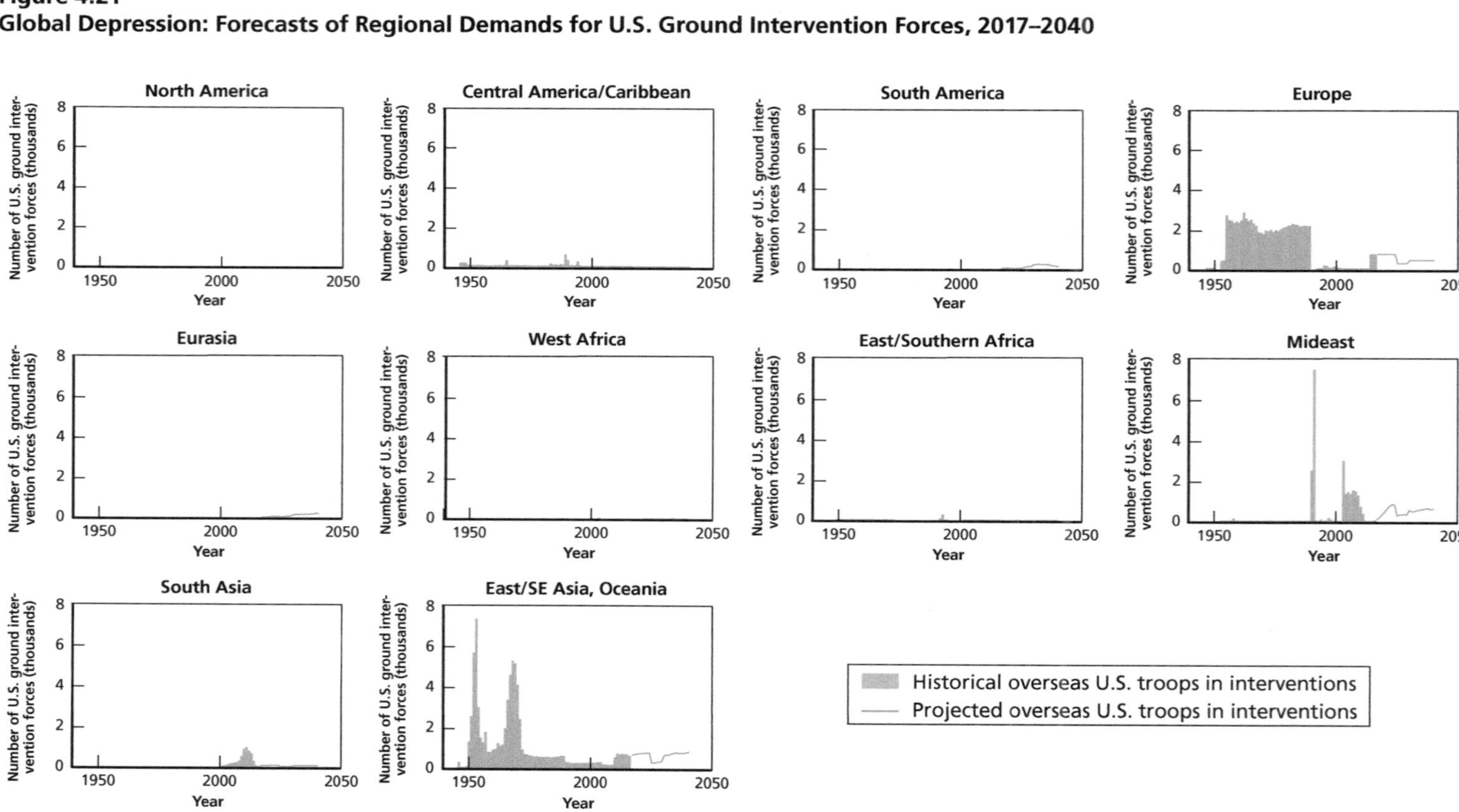

NOTE: The red line denotes the projected mean number of U.S. ground forces projected in each geographic region each year, based on 500 iterations of our forecasting model.

Figure 4.22
Global Depression: Forecasts of Demands for U.S. Ground Combat Mission Forces, 2017–2040

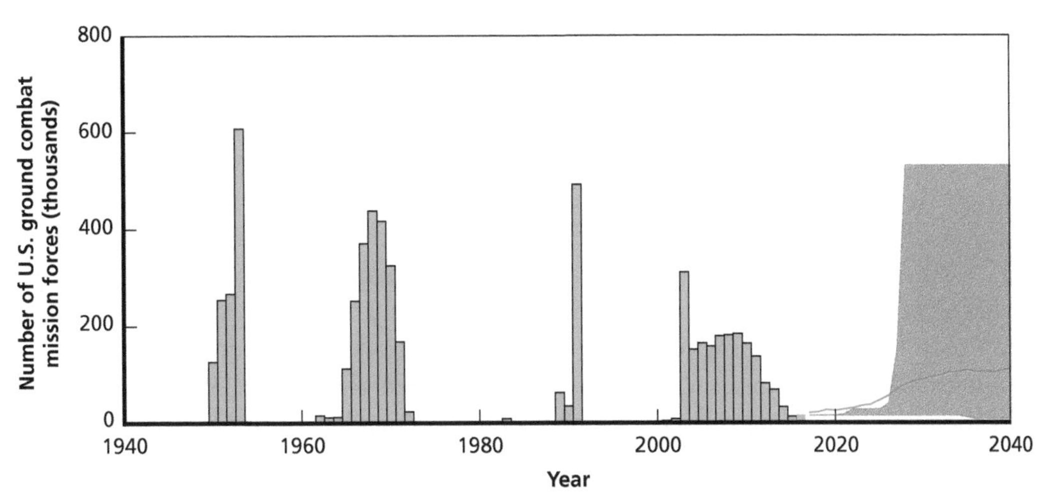

NOTES: The red line denotes the projected mean number of U.S. ground forces required for armed conflict interventions each year, based on 500 iterations of our forecasting model. The gray shaded area represents the range of forecasts bounded by the 10th and 90th percentiles of U.S. ground forces required for armed conflict interventions each year, based on 500 iterations of our forecasting model.

Figure 4.23
Global Depression: Forecasts of Demands for U.S. Ground Stability Operations Forces, 2017–2040

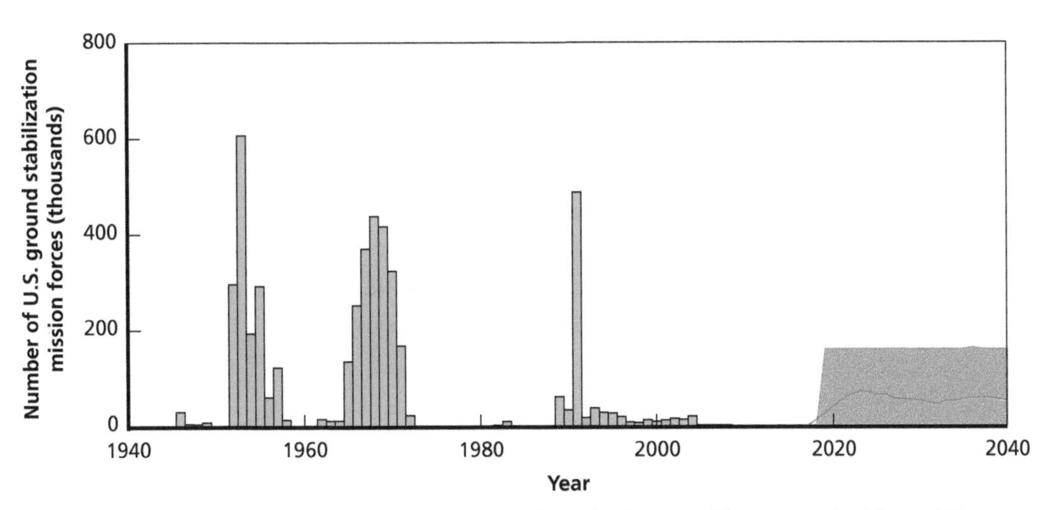

NOTES: The red line denotes the projected mean number of U.S. ground forces required for stability operations each year, based on 500 iterations of our forecasting model. The gray shaded area represents the range of forecasts bounded by the 10th and 90th percentiles of U.S. ground forces required for stability operations each year, based on 500 iterations of our forecasting model.

Alternative Future Scenario 2: Revisionist China

Scenario Description

This scenario is based on a dramatic expansion in Chinese efforts to revise the international system, underwritten by continued Chinese success in managing its internal challenges and prolonging its significant growth in economic and military power. As China continues to grow much more rapidly than the United States, Japan, and other major powers, its influence in regional and global politics similarly expands. The persistently strong growth in the Chinese economy permits the Chinese government to continue its military build-up in the Asia-Pacific region, which in turn leads to additional or more aggressively contested territorial claims throughout the region.[21]

China's continued rise also fuels significant changes across the global arena. China is hailed as a model for managed growth by many developing countries, leading to a rise in Beijing's diplomatic weight, which it uses to demand numerous changes in international institutions. Increasing tensions between the United States and China lead the United States to reduce its economic interdependence with China. Similarly, China's concerns about its remaining economic vulnerabilities, including dependencies on foreign energy sources, lead China to reduce its economic interdependence with the United States and rely more heavily on domestic sources of energy.

These moves ultimately lead to the formation of rival trading and military blocs, similar to those developed during the Cold War. This prolonged stand-off and significant shift in the global order causes the United Nations and similar intergovernmental organizations to lose diplomatic power in the global arena, leading to an erosion of international norms of peaceful conflict resolution.

Changes to Key Factors

This scenario roughly parallels the early period of the Cold War between the United States and the Soviet Union. This scenario again incorporates changes intended to be analogous to this period in order to project changes in our key factors:[22]

- **Exclusive economic trading blocs:** Two rival trading blocs develop: one centered on the United States and comprising most of the existing World Trade Organization, and one centered on China, which includes many of its close trading partners and states dissatisfied with the U.S. system.[23]

---

[21] Fareed Zakaria, *From Wealth to Power: The Unusual Origins of America's World Role*, Princeton, N.J.: Princeton University Press, 1999.

[22] This scenario again parallels a previously developed scenario detailed in Watts et al., 2017a, pp. 92–95. The specific numerical changes in variables were determined by examining analogous changes in these or related variables in the late 1940s and early 1950s.

[23] Members in the Chinese trading bloc in this scenario gradually grow between 2025 and 2034 to include states in Southeast Asia, Central Asia, South Asia, the Middle East, and Eurasia, notably including Russia and Iran. At its height, the Chinese trading bloc contains 30 states, selected to be those with the closest economic and stra-

- **Extent of economic interdependence:** All states experience a moderate decline in trade flows as a result of the bifurcation of the global trading system into rival economic blocs.[24]
- **Prevalence of consolidated democracies:** The process of democratization halts for many weak or fledgling democracies, as states that have not yet fully democratized hail the Chinese growth model and no longer see significant gains in continued democratization.[25]
- **Strength of international norms:** Support for international norms of peace conflict resolution decline to levels last seen in the early Cold War era, reflecting declines in the strength of intergovernmental organizations.[26]
- **Degree of territorial contestation:** The likelihood of territorial claims between states significantly increases.[27]

Effects on Conflict and Intervention Forecasts

The projections of interstate war and intrastate conflict in this scenario are similar to, though more pronounced than, the Global Depression scenario: a notably elevated risk of interstate war and a notably more modest increase in the risk of intrastate conflict. As shown in Figure 4.24, the mean number of states involved in interstate war rises sharply throughout the late 2020s and 2030s, while the percentile projections show a sizable increase at the 90th percentile, and even at the 10th percentile reach two states (or roughly one interstate war) by the late 2030s.

Looking at the regional interstate war projections suggests that the states involved in much of this increase will be concentrated in Eurasia and the Middle East, with some increases also seen among states in many other regions, including notably Europe, East/Southeast Asia, and North America, as shown in Figure 4.25. States in Europe and North America that become involved in interstate wars may not necessarily do so in their home regions, given their broader strategic interests and power projection capabilities, which are taken into account in our model.

These regional trends highlight the destabilizing nature of a potential split of the world into rival trading blocs, with the sharpest increases in the risk of war occurring in regions near China where some states have joined the Chinese-led bloc and others have not.

---

tegic relationships with China. However, not all states in these regions join, creating fault lines with heightened potential for conflict.

[24] Specifically, we model a decrease in trade of 14 percent beginning in 2025.

[25] The Polity values for all states in the international system remain static between 2025 and 2040.

[26] We model this as a 12-point decline in our measure of the strength of international norms—the percentage of states in each region that have committed to multiple treaties mandating the pacific settlement of disputes—which mirrors levels last seen in the 1950s.

[27] Specifically, we model a 26.5 percent increase in territorial claims beginning in 2025, paralleling the difference in such claims between the early 2000s and the 1980s.

Figure 4.24
Revisionist China: Forecasts of Interstate War Occurrence, 2017–2040

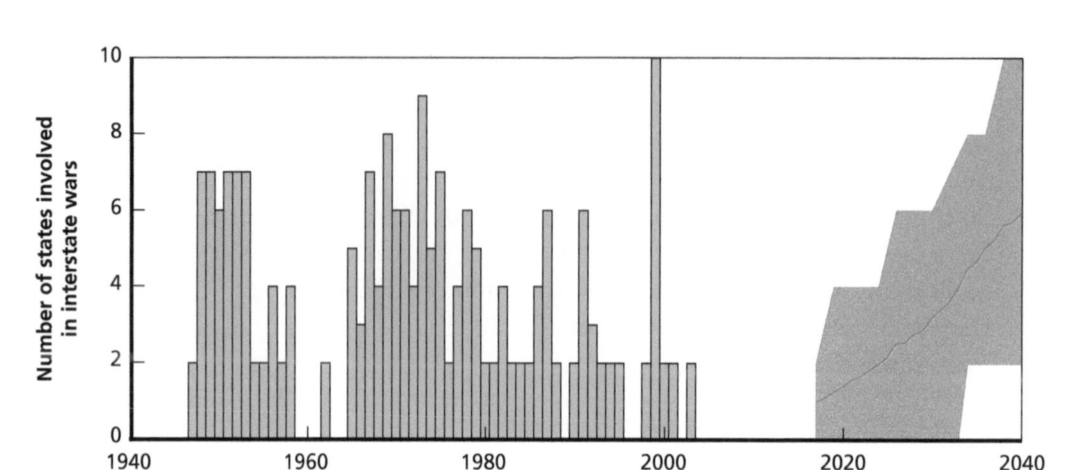

NOTES: The red line denotes the projected mean number of states involved in interstate war each year, based on 500 iterations of our forecasting model. The gray shaded area represents the range of forecasts bounded by the 10th and 90th percentiles of states involved in interstate war each year, based on 500 iterations of our forecasting model.

Turning to intervention projections, at the global level and considering all types of interventions, we find no significant change in the mean projected number of U.S. ground interventions compared with the baseline scenario. However, we do see a notable increase in the projected requirements for U.S. ground forces to support these interventions, suggesting that the mix of interventions the United States is likely to undertake is shifting in this scenario. As shown in Figure 4.26, the average number of U.S. ground forces employed in interventions in this scenario increases steadily, more dramatically than in the Global Depression scenario.

In the average case, these projections suggest that roughly 250,000 U.S. ground troops would be employed in military interventions by 2040, in comparison with roughly 200,000 in the baseline scenario. The 90th percentile projection, meanwhile, expands dramatically, indicating that while increases in U.S. forces employed in the average case are more moderate, this scenario is accompanied by substantially greater downside risk than the baseline scenario. The 10th percentile projections, meanwhile, remain roughly unchanged across the two scenarios.

At the regional level, as shown in Figure 4.27, the increase in U.S. forces deployed, to the extent it occurs, appears concentrated in the Middle East, Eurasia, and, to a small extent, East/Southeast Asia. The anticipated rises in forces employed in the average case in the Middle East and Eurasia correlate with the expected rise in the risk of interstate war in those regions noted above.

With regard to the types of interventions likely to be conducted in this scenario, compared with the baseline, we do not see major changes in projections of the num-

Figure 4.25
Revisionist China: Regional Forecasts of Interstate War Occurrence, 2017–2040

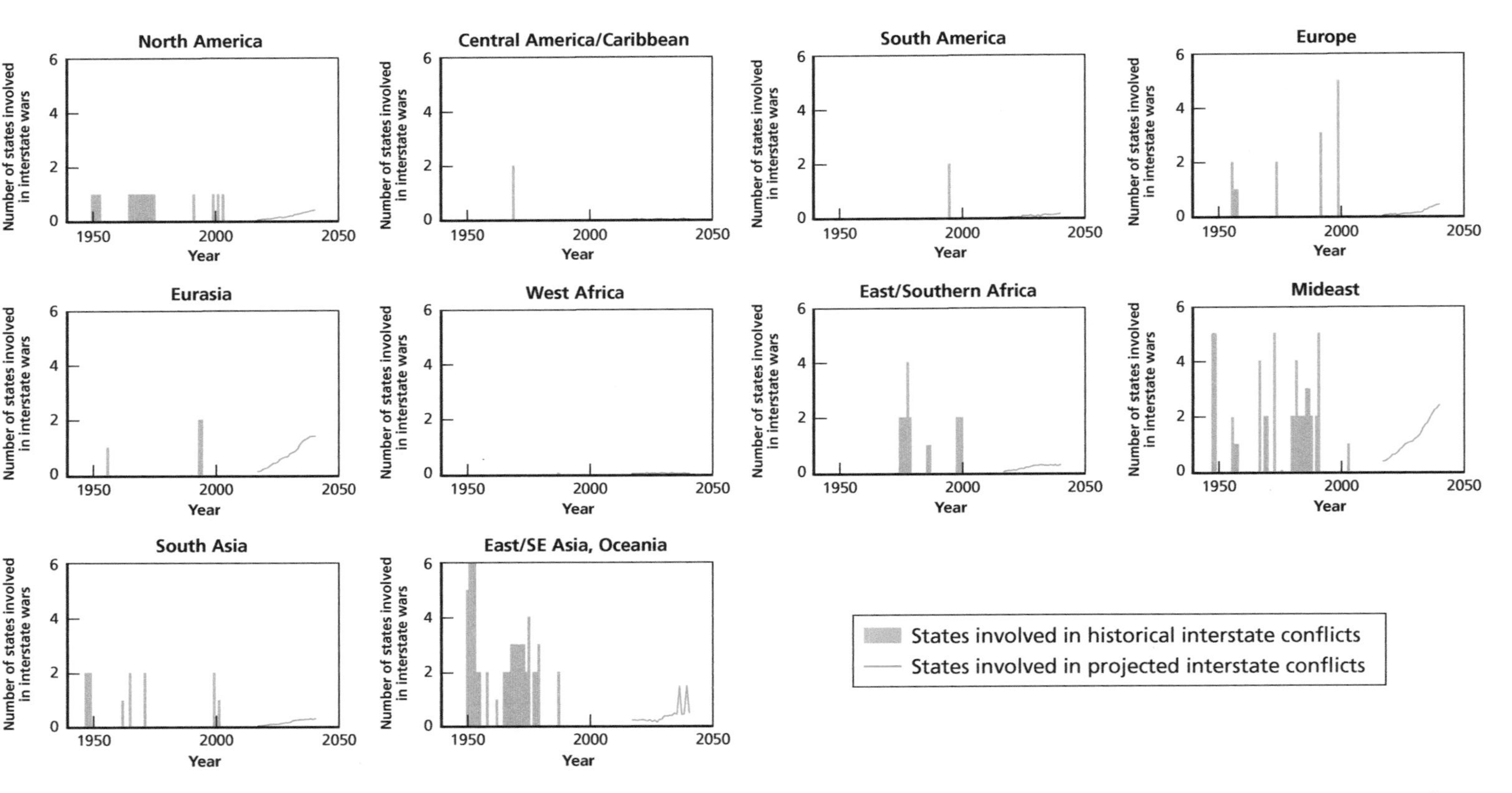

NOTE: The red lines denote the projected mean number of states involved in interstate wars in each region for each year, based on 500 iterations of our forecasting model.

Figure 4.26
Revisionist China: Forecasts of Demands for U.S. Ground Intervention Forces, 2017–2040

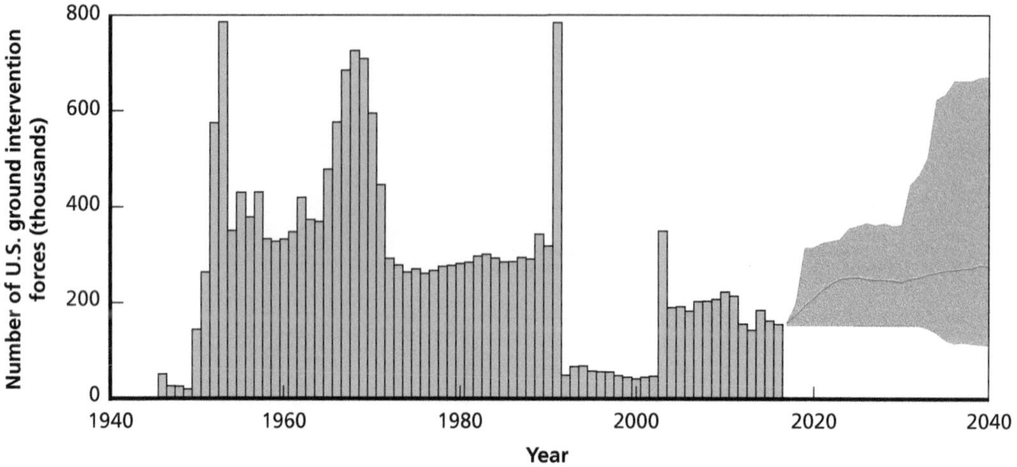

NOTES: The red line denotes the projected mean number of U.S. ground forces required for interventions each year, based on 500 iterations of our forecasting model. The gray shaded area represents the range of forecasts bounded by the 10th and 90th percentiles of U.S. ground forces required for interventions each year, based on 500 iterations of our forecasting model.

bers of armed conflict interventions, stability operation, or deterrent interventions. We do, however, see clearer increases in the number of ground troops committed to armed conflict interventions, as shown in Figure 4.28.

In this scenario we see a gradual increase in the average case in troops employed in armed conflict interventions from only about 25,000 in 2017 to more than 100,000 in 2040. The percentile projections expand more dramatically, from roughly 10,000 at the 10th percentile to over 500,000 at the 90th percentile. Given that the projections of forces committed to deterrence and stability operations do not increase notably in this scenario over the baseline, this suggests that the greatest increased risk for U.S. ground interventions in this scenario is armed conflict interventions into interstate wars, most likely in Eurasia, the Middle East, or East/Southeast Asia, although of course substantial uncertainty still accompanies these findings.

Alternative Future Scenario 3: Global Pandemic

Scenario Description

In 2015, the Global Challenges Foundation listed global pandemic as one of the top 12 threats of the future, noting that in the new highly mobile and interconnected era, diseases such as Ebola and dangerous strains of flu can travel quickly around the planet, infecting millions of people.[28] Recent events, including the 2014 Ebola crisis in Africa,

---

[28] Dennis Pamlin and Stuart Armstrong, *Global Challenges: 12 Risks that Threaten Human Civilization*, Global Challenges Foundation, 2015.

Figure 4.27
Revisionist China: Forecasts of Regional Demands for U.S. Ground Intervention Forces, 2017–2040

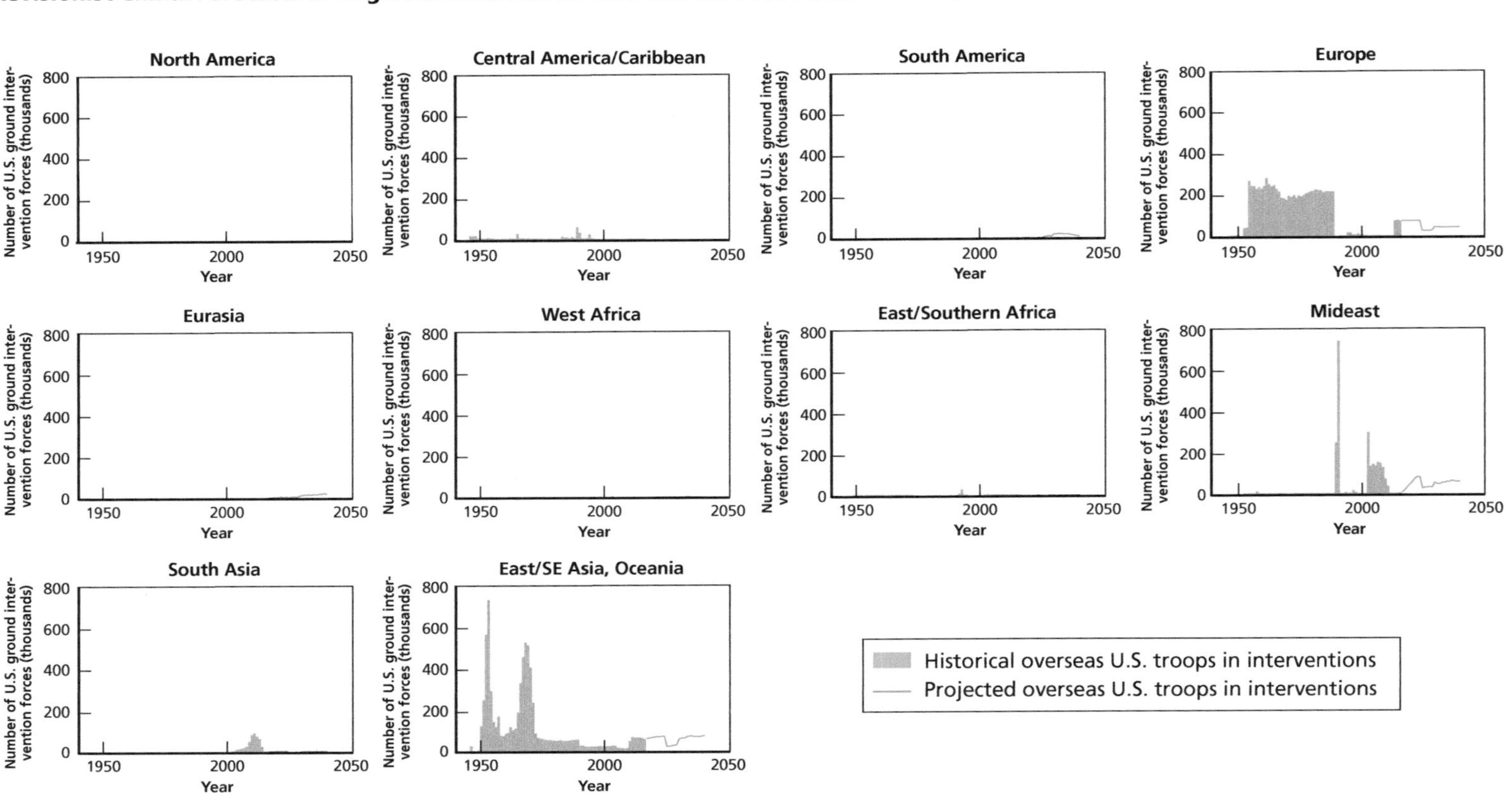

NOTE: The red line denotes the projected mean number of U.S. ground forces required for interventions each year, based on 500 iterations of our forecasting model.

Figure 4.28
Revisionist China: Forecasts of Demands for U.S. Ground Combat Mission Forces, 2017–2040

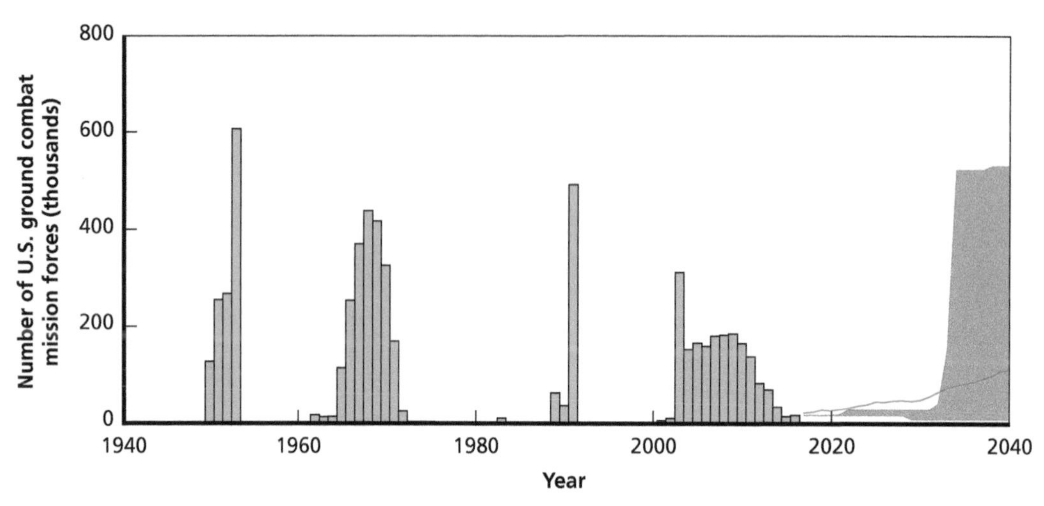

NOTES: The red line denotes the projected mean number of U.S. ground forces required for armed conflict interventions each year, based on 500 iterations of our forecasting model. The gray shaded area represents the range of forecasts bounded by the 10th and 90th percentiles of U.S. ground forces required for armed conflict interventions each year, based on 500 iterations of our forecasting model.

a serious avian flu outbreak in 2009, and diseases like the SARS epidemic in 2003, have demonstrated not only the speed with which epidemic-style outbreaks spread but the damage—economic and social—they can cause.

The most severe pandemic in recent history, the Spanish Flu in 1918–1920, led to reductions in GDP, trade flows, temporary shocks to the size of the workforce, and long-term health effects suffered by those born during the epidemic. The recovery from this epidemic was reasonably rapid, largely because the pandemic itself was fairly short (although there were two waves and the death tolls were high), and within five years GDP growth rates had largely returned to their original trajectory, at least for developed countries, such as the United States. One unusual characteristic of this epidemic was that it hit people ages 20–40 particularly hard, due to the way in which the disease attacked patient immune systems, in contrast with other epidemics that typically are most severe among young children and the elderly. As a result, a large percentage of those killed in the Spanish Flu were of working age, amplifying the economic effects of the disease.[29]

---

[29] Elizabeth Brainerd and Mark Siegler, *The Economic Effects of the 1918 Influenza Epidemic*, Paris, France: Centre for Economic and Policy Research, Discussion Paper No. 3791, 2003; Centers for Disease Control and Prevention, "History of the 1918 Flu Pandemic," 2018; Thomas Garrett, *Economic Effects of the 1918 Influenza Pandemic: Implications for a Modern-Day Pandemic*, St. Louis, Mo.: Federal Reserve Bank of St. Louis, November 2007.

In this alternative scenario, we assess the effects of a hypothetical severe pandemic outbreak that occurs in 2025, with characteristics similar to the Spanish Flu. The pandemic in our scenario is severe but is relatively short, lasting only one year. However, in our scenario it causes severe economic damage, a decline in global GDP, trade, and GDP per capita in nations across the globe.

The effects of such a global pandemic would be many. Fear of the disease would lead to absenteeism at work, causing reductions in economic output. National and regional quarantines would slow or stop trade flows, causing additional economic damage. Government resources would need to be spent on fighting the outbreak, and a large loss of life would also affect the supply of labor. Together, these changes would cause a significant shock to the international economy and to domestic economies around world, leading to potentially significant drops in GDP. Importantly, the economic and mortality effects of the epidemic are unlikely to be evenly distributed across the globe. Instead, less developed countries with lower state capacity, weaker economic institutions, and less well-developed health care systems are likely to suffer relatively greater consequences.[30] We also assume that, like the Spanish Flu, the outbreak that occurs affects people ages 20–40 much more severely than a typical disease.

Expectations about the recovery from such an epidemic vary and depend largely on the nature of the disease itself. A relatively short outbreak with a more limited death toll would be expected to do less economic damage and might allow for a more rapid recovery. On the other hand, a longer pandemic, with a greater loss of life would have longer-lasting effects—even when economic growth resumes, the recovery would be starting from a much lower point. Furthermore, there may be other, longer-term effects from a pandemic. For instance, it is possible that those born during the epidemic suffer long-term health consequences, which in turn could have longer-lasting serious economic consequences.[31] For the purpose of this scenario, we assume that, post-epidemic, it takes approximately five years for GDP per capita and GDP growth rates to return to their pre-epidemic levels. We do not include possible permanent effects, such as long-term damage to individual health.

Changes to Key Factors

Unlike our Global Depression and Revisionist China scenarios, which are characterized by long-term changes in the international system, the changes in our key factors

---

[30] For a discussion of the likely impacts from a pandemic, see, for example, Brainerd and Siegler, 2003; Andrew Burns, Dominque van der Mensbrugghe, and Hans Timmer, "Evaluating the Economic Consequences of Avian Influenza," World Bank, 2006; Garrett, 2007; Harvey Rubin, *Future Global Shocks: Pandemics*, Organisation for Economic Development/International Futures Programme on Future Global Shocks, January 2011; Alexandra Sidorenko and Warwick McKibbin, *What a Flu Pandemic Could Cost the World*, Washington, D.C.: Brookings Institution, April 28, 2009.

[31] Ryan Brown and Duncan Thomas, "On the Long-Term Effects of the 1918 U.S. Influenza Pandemic," unpublished manuscript, 2011.

for this scenario all represent severe but relatively short-lived effects on the international system. As noted above, we developed these changes in our key factors based on another historically disastrous pandemic, the Spanish Flu of 1918–1920:

- **Global population losses:** The pandemic causes a loss of 300 million people globally, representing 4 percent of the global population, with the effects spread unevenly across developed and developing states, with developing states suffering a higher share of losses than developed countries.[32]
- **Decreased international trade:** Global trade drops precipitously in the year of the pandemic as states seek to protect their borders from further contagion via shipping.[33]
- **Global GDP losses:** Severe population losses lead to significant declines in economic productivity for the duration of the pandemic, leading to losses in both overall GDP and GDP per capita from lost productivity and reduced trade.[34]
- **Declines in oil production:** Declines in international trade and economic outputs lead to a significant decline in global oil production, as demand for crude oil and oil-based products drops significantly.[35]

Effects on Conflict and Intervention Forecasts

The pandemic scenario does not lead to significant changes in conflict projections compared with the baseline. We see essentially no change in the number of ongoing interstate wars or intrastate conflicts, at either the global or regional levels. This applies to both the average and the percentile projections. A pandemic such as this would be a short-term economic and political catastrophe, but our models do not anticipate that these effects would to lead to notable changes in the risk of conflict, at least in the near term.

We do, however, observe differences between the pandemic scenario and the baseline in terms of numbers of U.S. ground interventions. Figure 4.29 shows the projected change in global interventions across activity types.

---

[32] We model this uneven loss as a 2 percent population loss beginning in 2025 for all countries with a GDP per capita above $14,000, and a 6 percent population loss for all countries with a GDP per capita below that threshold. We assume these losses are spread evenly across demographic groups, leading to no change in projected youth bulges.

[33] We model a 55 percent decline in all trade in 2025, which gradually increases until levels of global trade return to normal projected levels in 2030.

[34] States below a GDP per capita level of $14,000 experience a 13 percent loss in GDP per capita in 2025, while states above that level of GDP per capita experience a 9 percent decrease in GDP per capita in 2025. In both cases, levels of GDP per capita gradually increase until global GDP per capita returns to normal projected levels in 2030.

[35] Oil production levels in all states drops by 15 percent in all states in 2025, which gradually increases until global oil production returns to normal projected levels in 2030.

Figure 4.29
Global Pandemic: Forecasts of Total U.S. Ground Interventions, 2017–2040

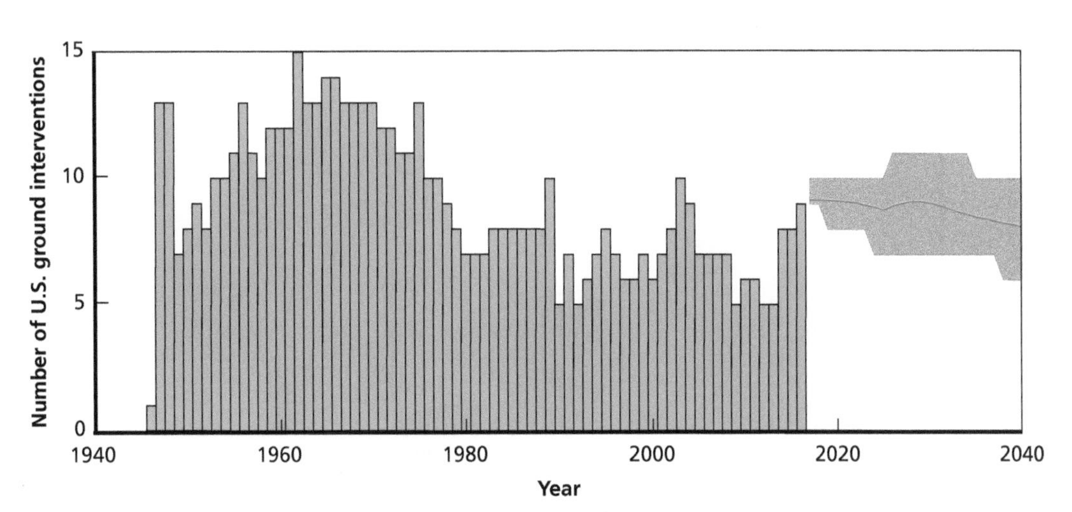

NOTES: The red line denotes the projected mean number of total U.S. ground interventions each year, based on 500 iterations of our forecasting model. The gray shaded area represents the range of forecasts bounded by the 10th and 90th percentiles of U.S. ground interventions each year, based on 500 iterations of our forecasting model.

Although the mean projection line begins on the same downward trajectory observed in the baseline scenario, we see a slight increase in the number of ground interventions starting in 2025, after the epidemic begins. This increase lasts for several years, until the gradual decline begins again. The mean projection line has a slightly higher endpoint, but the more important difference compared with the baseline is an increase in the 90th percentile projection, which rises by one intervention compared with the baseline for all years after 2025. Under the pandemic scenario, then, our models anticipate that the potential for an increase in total U.S. ground interventions is higher and the likelihood of a decline more limited.

The average number of ground forces required for U.S. interventions similarly rises modestly in this scenario compared with the baseline projection, as shown in Figure 4.30.

The 90th percentile projection increases substantially during the pandemic and its immediate aftermath, although by 2040 this projection is actually slightly lower than the baseline. The 10th percentile projection is relatively similar across the two scenarios, increasing only modestly in the aftermath of the pandemic. The regional projections of troops employed are also only modestly changed, with minor increases in the average projection in East/Southeast Asia and the Middle East.

The modest increases in the projected number of U.S. ground interventions are concentrated in deterrent interventions and, to a lesser extent, stability operations. Figure 4.31 shows a notable uptick in the average projected number of deterrent interventions following the pandemic, by roughly one additional intervention, persisting at

Figure 4.30
Global Pandemic: Forecasts of Demands for U.S. Ground Intervention Forces, 2017–2040

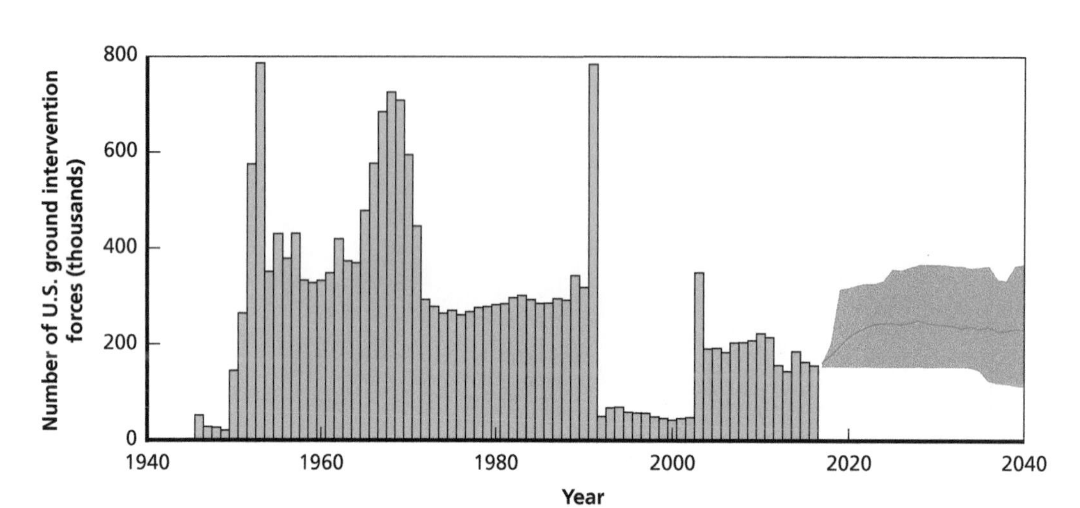

NOTES: The red line denotes the projected mean number of U.S. ground forces required for interventions each year, based on 500 iterations of our forecasting model. The gray shaded area represents the range of forecasts bounded by the 10th and 90th percentiles of U.S. ground forces required for interventions each year, based on 500 iterations of our forecasting model.

an elevated level to 2040. The 90th percentile projection also increases by an additional intervention for roughly a decade following the pandemic, while the 10th percentile increase is somewhat more short-lived, before returning to where it was in the baseline.

The projected number of ground troops committed to deterrence interventions in the average case also rises over the same period, though only modestly, suggesting that the new deterrent intervention or interventions undertaken in this scenario are relatively modestly resourced, as shown in Figure 4.32.

We see a more modest increase in the projected number of stability operation interventions in the aftermath of the pandemic, as shown in Figure 4.33, with the average number of ground interventions stabilizing after 2025 rather than declining further as in the baseline scenario. The percentile projections, however, remain unchanged.

The number of U.S. troops projected to be involved in stability operations after the pandemic also increases in the average case, but only modestly, as shown in Figure 4.34. The 10th and 90th percentile projections again remain unchanged.

Overall, this alternative scenario has relatively modest implications for U.S. ground interventions and demand for U.S. ground forces. The massive human toll generated by the pandemic is not projected to materially increase the numbers of armed conflicts in the world, or substantially affect U.S. ground interventions in these conflicts. Instead, the primary U.S. response to the scenario anticipated by our models would be a modest increase in deterrence missions and, to a lesser extent, stability operations. In our models, rather than withdrawing from the world in the face of this catastrophe, the United States would modestly increase its presence overseas to stabilize

Figure 4.31
Global Pandemic: Forecasts of U.S. Ground Deterrent Interventions, 2017–2040

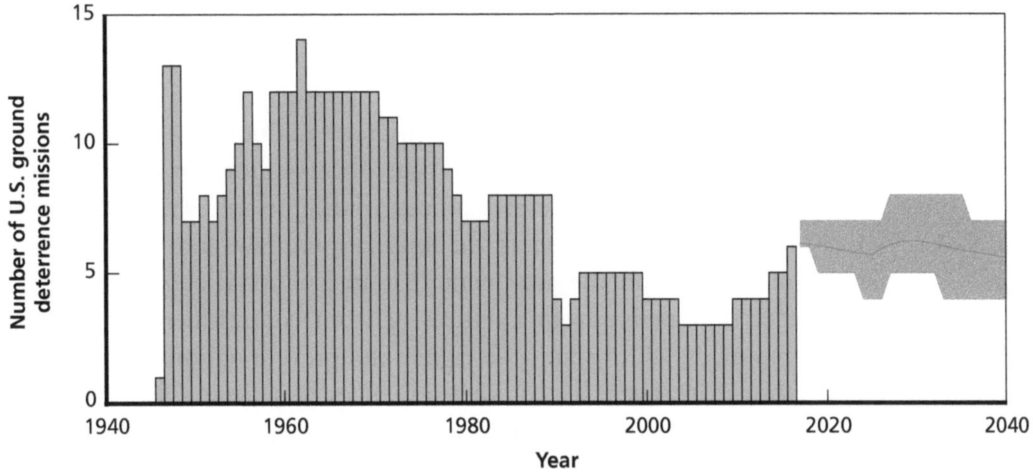

NOTES: The red line denotes the projected mean number of total U.S. ground deterrence missions each year, based on 500 iterations of our forecasting model. The gray shaded area represents the range of forecasts bounded by the 10th and 90th percentiles of U.S. ground deterrence missions each year, based on 500 iterations of our forecasting model.

key interstate and intrastate tensions that may arise in the aftermath of the pandemic. An alternative scenario that considers the opposite U.S. instinct, to retrench from the world, is discussed next.

Alternative Future Scenario 4: U.S. Isolationism—The United States Pulls Back from the World Stage

Scenario Description

Since the end of World War II, the United States has played a leading role in global affairs, through its involvement in international institutions such as the United Nations and the World Bank, its sizable deterrent force posture, and its frequent military interventions. U.S. involvement on this scale, however, is an active policy choice, and other alternatives exist. Engagement in international affairs is often juxtaposed against isolationism, but, in reality, there are at least two main versions of potential U.S. isolationism, each with distinct characteristics. In the first, the United States remains engaged in international institutions and supports international norms, but refrains from getting involved in military engagements, including both overseas deployments and military alliances. The theoretical foundation for such an approach is the argument that military retrenchment has significant economic and other benefits for the United States and that having a large overseas military presence and a web of military alliances does not provide significant additional security (or economic) benefits to the intervener, or at least not enough additional benefits to justify the costs and the risks. More specifically, fewer military commitments enable a smaller defense budget, which allows more extensive investment in the domestic arena—on things such as transportation and support to innovation. In addition, a reduction in costly military alliances reduces the risk that the United States will be pulled into unnecessary conflicts that do not directly affect its interests.[36] However, this version of intervention remains supportive of U.S. involvement in other forms of multilateral engagement, including multilateral diplomacy.

In the second model of isolationism, the United States maintains at least some of its overseas military presence but withdraws from other forms of international engagement, including participation in multilateral alliances, trading relationships, and international organizations, and reduces support for international laws and norms. This model of isolationism is based on the argument that international engagement and commitments, whether they derive from alliances or from commitment to international law or trade agreements, limit state sovereignty and prevent nations from pursuing their primary national interests. In this view, alliances may limit a state's pursuit of its interests; international institutions are a costly burden, especially for stronger,

Figure 4.32
Global Pandemic: Forecasts of Demands for U.S. Ground Deterrent Intervention Forces, 2017–2040

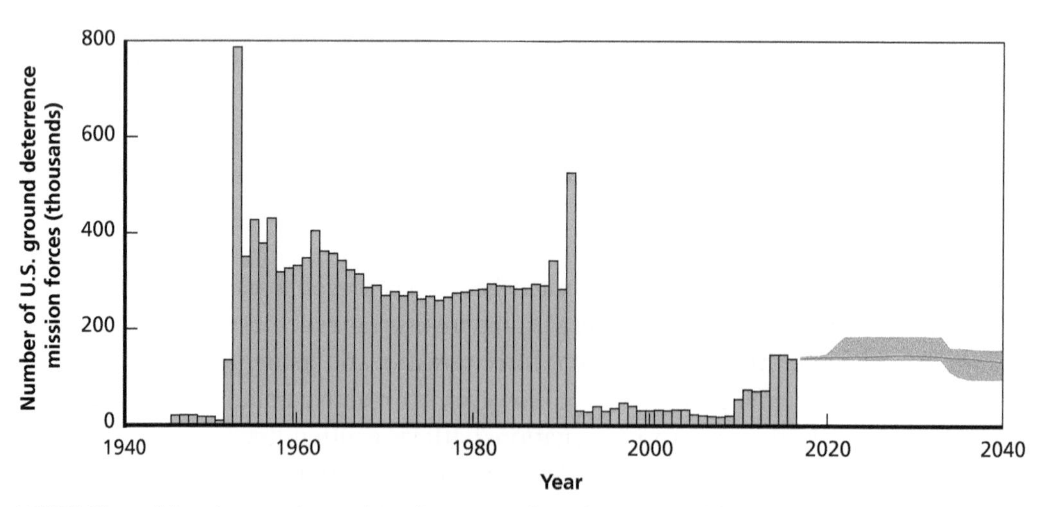

NOTES: The red line denotes the projected mean number of U.S. ground forces required for deterrence missions each year, based on 500 iterations of our forecasting model. The gray shaded area represents the range of forecasts bounded by the 10th and 90th percentiles of U.S. ground forces required for deterrence missions each year, based on 500 iterations of our forecasting model.

[36] Barry R. Posen, "Pull Back: The Case for a Less Activist Foreign Policy," *Foreign Affairs*, Vol. 92, 2013, p. 116.; Daniel W. Drezner, "Military Primacy Doesn't Pay (Nearly as Much as You Think)," *International Security*, Vol. 38, No. 1, 2013; Paul K. MacDonald and Joseph M. Parent, "Graceful Decline? The Surprising Success of Great Power Retrenchment," *International Security*, Vol. 35, No. 4, 2011; Ted Galen Carpenter, *A Search for Enemies: America's Alliances After the Cold War*, Washington, D.C.: Cato Institute, 1992.

Figure 4.33
Global Pandemic: Forecasts of U.S. Ground Stability Operations, 2017–2040

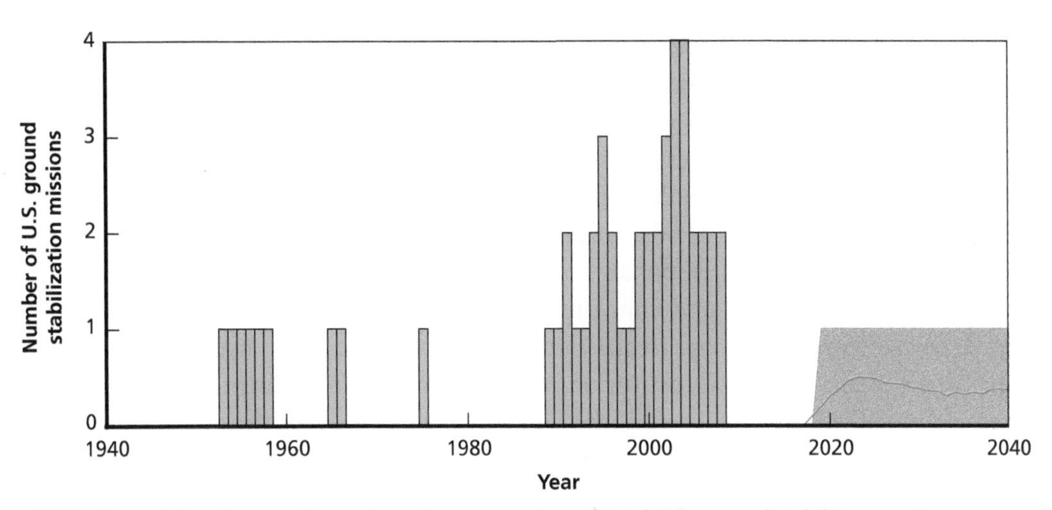

NOTES: The red line denotes the projected mean number of total U.S. ground stability operation interventions each year, based on 500 iterations of our forecasting model. The gray shaded area represents the range of forecasts bounded by the 10th and 90th percentiles of U.S. ground stability operation interventions each year, based on 500 iterations of our forecasting model.

Figure 4.34
Global Pandemic: Forecasts of Demands for U.S. Ground Stability Operations Forces, 2017–2040

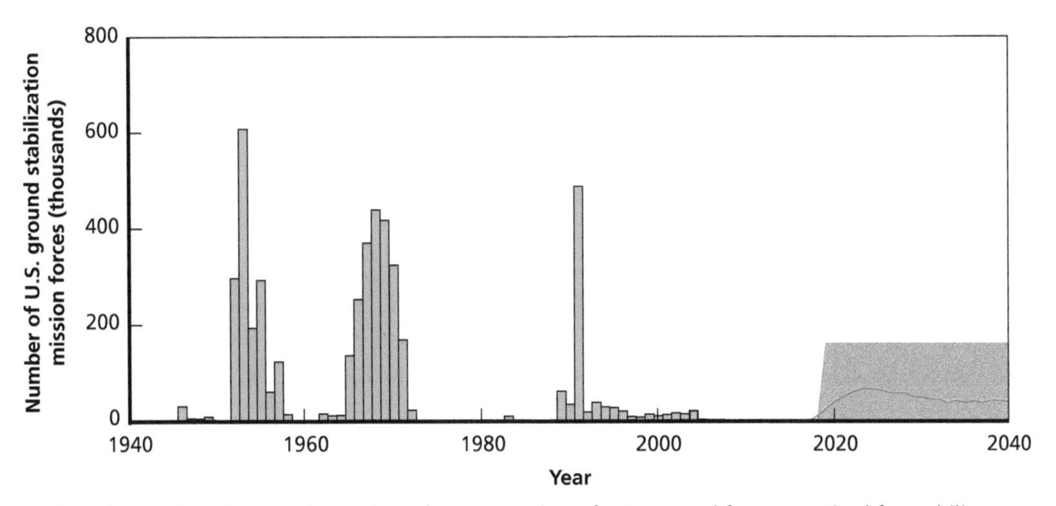

NOTES: The red line denotes the projected mean number of U.S. ground forces required for stability operation interventions each year, based on 500 iterations of our forecasting model. The gray shaded area represents the range of forecasts bounded by the 10th and 90th percentiles of U.S. ground forces required for stability operation interventions each year, based on 500 iterations of our forecasting model.

wealthier states such as the United States, and international law is an unnecessary constraint that limits freedom of action in ways that can compromise national security.[37] This version of isolationism emphasizes the primacy of U.S. sovereignty. As a result, this version of isolationism does not necessarily involve an immediate end to overseas military presence, which is viewed as an assertion of U.S. primacy, though it does likely reduce the scope of issues over which the U.S. military may be employed.[38] We model this version of isolationism to explore the implications of this type of approach toward international affairs and its implications for conflict and U.S. interventions.[39]

In this scenario, the United States enters a renewed period of isolationism starting in 2025, potentially in response to continued backlash against globalization, its unequal costs and benefits, and concerns over the growing costs of U.S. forward presence. Tensions with key European allies and with Korea and Japan over differing views of the future of multinational institutions and agreements (including trade agreements) create additional strains. As a result, the United States takes significant steps to reduce its involvement in these institutions. Most significantly, while current interventions continue, the United States abandons participation in key multilateral alliances, such as NATO; bilateral alliances, such as with Korea and Japan; and multilateral trade agreements and organizations, such as the World Trade Organization, effectively eliminating all entanglements with the exception of existing bilateral trade agreements. U.S. policymakers also limit their involvement in and security concerns over areas of the world to those most core to U.S. interests: Europe, East Asia, and the Americas.

This sharp reduction in the U.S. engagement in the international arena has serious implications for international trade, norms, and regional balance of power across the globe.

Changes to Key Factors

- **Exclusive U.S. trading bloc:** The United States pulls out of the World Trade Organization, opting instead to pursue a series of bilateral trade agreements across

---

[37] Carpenter, 1992; Duncan B. Hollis, "Private Actors in Public International Law: Amicus Curiae and the Case for the Retention of State Sovereignty," *Boston College International and Comparative Law Review*, Vol. 25, 2002; Stephen M. Walt, *Taming American Power: The Global Response to U.S. Primacy*, New York, N.Y.: Norton, 2006.

[38] For arguments about U.S. primacy, see Stephen G. Brooks, G. John Ikenberry, and William C. Wohlforth, "Don't Come Home, America: The Case Against Retrenchment," *International Security*, Vol. 37, No. 3, 2013; Samuel P. Huntington, "Why International Primacy Matters," *International Security*, Vol. 17, No. 4, 1993; William C. Wohlforth and Stephen G. Brooks, "American Primacy in Perspective," in David Skidmore, ed., *Paradoxes of Power*, New York, N.Y.: Routledge, 2015.

[39] Modeling of the alternative type of isolationism, which retains strong multilateral engagement, would be a useful additional scenario to consider but was beyond the scope of the current effort. Such an effort could, for example, stipulate that U.S. forward presence be eliminated altogether, and then the model could assess the potential effects on armed conflict in different regions. However, because the withdrawal of U.S. forward presence would be stipulated as a condition of the scenario, it would become a less interesting modeling exercise for the purposes of the present report, which is why we chose to focus on the alternative type of isolationism.

the international system, but leaving other states in the World Trade Organization system.[40]

- **U.S. economic activity:** The withdrawal of the United States from the globalized economy and the existing global trade structure leads to a significant decline in U.S. trade. This reduction, in turns, leads to an overall decrease in U.S. GDP growth.[41]
- **Strength of international norms:** Without U.S. leadership, support for international norms of peace and conflict resolution decline to levels last seen in the early Cold War era, also reflecting declines in the strength of intergovernmental organizations.[42]
- **Ending of U.S. formal alliances:** The United States drops all its existing formal alliances, including its leadership of NATO.[43]
- **Global nuclear umbrella:** The ending of long-standing U.S. alliances also withdraws the U.S. nuclear umbrella from former allies.[44]
- **Limitation of U.S. interests:** The United States substantially reduces its concern for security dynamics in regions outside its core interests.[45]
- **Regional hegemony:** The withdrawal of the United States from the world arena, most notably in the breakdown of the U.S. alliance system, causes the balance of power in many regions to shift.
- **Altered interventionist tendencies:** The United States does not immediately end its ongoing interventions, but it does pursue foreign policies that make the United States significantly less likely to undertake new interventions and significantly more likely to end ongoing interventions over time.[46]

---

[40] Beginning in 2025, the United States is placed in a different trading bloc from all other states in the international system.

[41] Beginning in 2025, levels of U.S. trade drop by 27 percent over a three-year period until levels of trade stabilize at new reduced levels. That reduced level of trade persists through 2040. Projections of U.S. GDP growth decrease by 2 percent beginning in 2025.

[42] We model this as a 12-point decline in our measure of the strength of international norms, the percentage of states in each region that have committed to multiple treaties mandating the pacific settlement of disputes, which mirrors levels last seen in the 1950s.

[43] Beginning in 2025, all existing U.S. alliances are removed, which is maintained through 2040.

[44] Former non-nuclear-possessing U.S. allies are no longer assumed to have the deterrent benefit of U.S. nuclear capabilities beginning in 2025.

[45] Specifically, we eliminate the dyads between the United States and non–Great Power states in sub-Saharan Africa, the Middle East, Eurasia, and South Asia in our interstate war model, while retaining these dyads in the Americas, Europe, and East/Southeast Asia. This effectively means that the United States cannot be projected to start new interstate wars in the omitted regions after 2025, though it can still intervene in them after the fact if the model projects them to after applying the other changes in this scenario.

[46] Beginning in 2025, our predictions of new U.S. ground interventions are deflated by 80 percent, whereas our predictions of the cessation of ongoing U.S. ground interventions are inflated by 80 percent, which mirrors

Effects on Conflict and Intervention Forecasts

The isolationist U.S. scenario has a moderate impact on the projected incidence of interstate war, while having no clear effect on the incidence of intrastate conflict. Figure 4.35 shows the projected upward trend in the average number of ongoing interstate wars in the years after 2025, when the change in U.S. policy occurs.

In comparison with the baseline scenario, the average number of states involved in interstate wars in 2040 increases from roughly three states to just over four states, or roughly an even chance of one additional ongoing interstate war per year. The percentile projections also increase moderately, including notably even the 10th percentile projection indicating two states involved in interstate wars by 2040.

Regionally, the increased incidence in state involvement in interstate war in these projections is widespread, as shown in Figure 4.36, with nearly all regions showing a moderate increase in the risk of states becoming involved in interstate war. The exception to this trend is in the Americas, with Central and South America relatively unaffected and a reduced risk of involvement in interstate war among states in North America.

These results highlight that, in our models and particularly in certain regions of the world, broader U.S. engagement likely has a pacifying effect on the risk of inter-

Figure 4.35
U.S. Isolationism: Forecasts of Interstate War Occurrence, 2017–2040

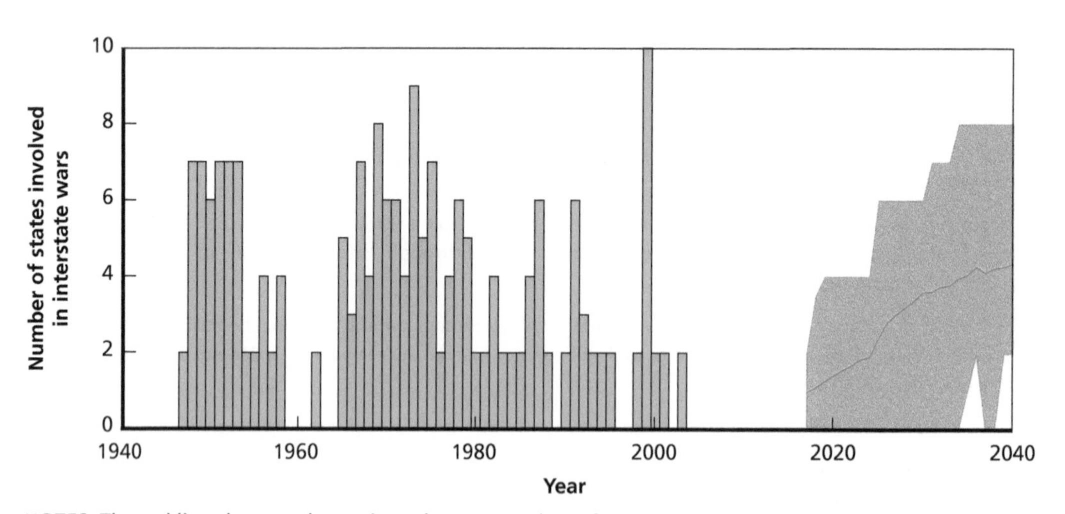

NOTES: The red line denotes the projected mean number of states involved in interstate war each year, based on 500 iterations of our forecasting model. The gray shaded area represents the range of forecasts bounded by the 10th and 90th percentiles of states involved in interstate war each year, based on 500 iterations of our forecasting model.

---

historical trends in U.S. interventions prior to World War II when the United States pursued significantly more isolationist policies.

Figure 4.36
U.S. Isolationism: Regional Forecasts of Interstate War Occurrence, 2017–2040

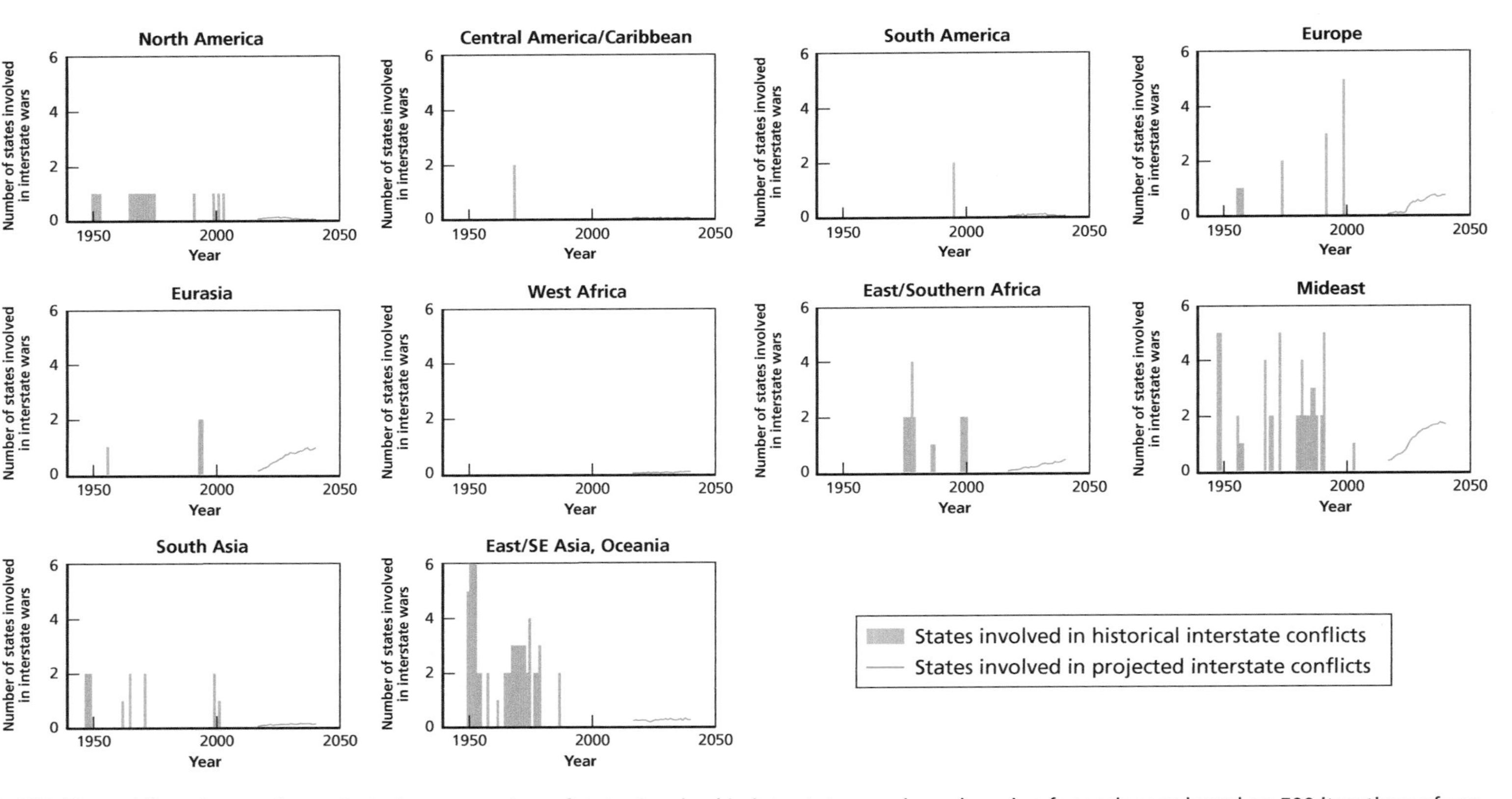

NOTE: The red lines denote the projected mean number of states involved in interstate wars in each region for each year, based on 500 iterations of our forecasting model.

state war, and a sudden pullback in such engagement would be expected to be destabilizing. However, the scale of these changes should also not be overstated. The risk of interstate war was projected to rise in the baseline scenario even with continued U.S. engagement, and in this scenario U.S. isolationism appears to accelerate this trend in several regions, such as the Middle East and Eurasia. More notable, perhaps, are the average projections of interstate war for Europe, which trend upward only in the U.S. Isolationism scenario.

In an isolationist scenario, we would expect the number of U.S. ground interventions to decrease. Figure 4.37 does show a decrease in the total number of U.S. ground interventions over the projection period, with a decrease in the average number of interventions of between one and two interventions each year by 2040 compared with the baseline scenario. The percentile projections also decline, by roughly one intervention per year by 2040.

That said, although the average projected number of ground interventions does decline to roughly the lowest numbers seen in the post-1945 period, the United States does remain quite militarily engaged in the world, relative both to other states and to its own behavior in the pre–World War II period. While it is possible that the scenario changes specified above are not sufficient to truly capture the changes in policy that would accompany a strong isolationist turn by the United States, it is also quite plausible that even in this scenario the United States would retain interests that would motivate continued military engagements when those interests are deemed to be at risk.

Figure 4.37
U.S. Isolationism: Forecasts of Total U.S. Ground Interventions, 2017–2040

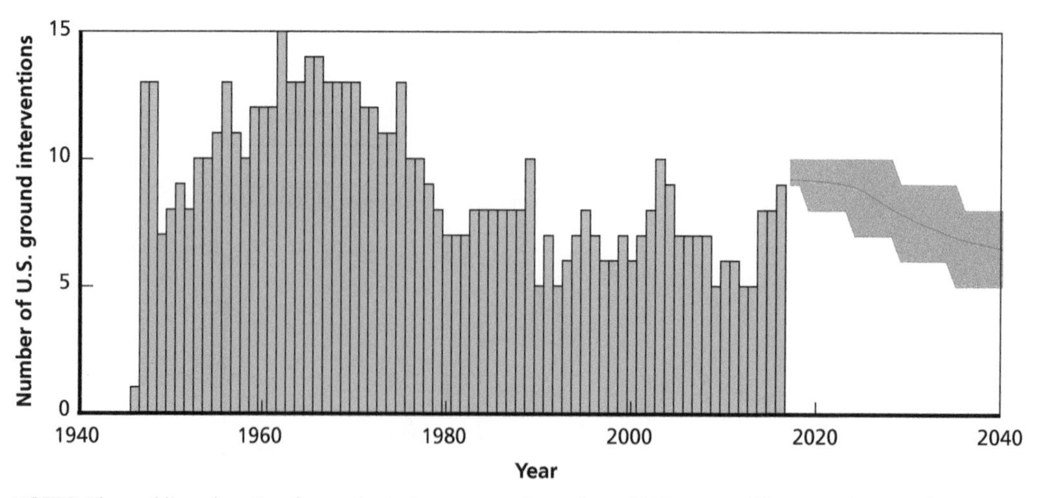

NOTES: The red line denotes the projected mean number of total U.S. ground interventions each year, based on 500 iterations of our forecasting model. The gray shaded area represents the range of forecasts bounded by the 10th and 90th percentiles of U.S. ground interventions each year, based on 500 iterations of our forecasting model.

The decline in U.S. ground interventions in this scenario is concentrated in a reduced number of projected deterrence interventions, while interventions into armed conflicts and stability operations remain relatively similar to the baseline scenario. Figure 4.38 shows the projected number of deterrence interventions, which declines by roughly one full intervention relative to the baseline, and roughly two interventions from the present day, in both the average and percentile projections. These projections suggest that in this scenario the United States would end at least one of its current deterrence missions in East Asia, in addition to the expected end of the Sinai deterrence mission also anticipated in the baseline scenario. The projected average number of troops committed to deterrence missions roughly parallels this decline.[47]

A look at the number of U.S. ground troops involved in military interventions in this scenario, shown in Figure 4.39, highlights that despite these changes in deterrence interventions, there is relative continuity of effort elsewhere. The average projected number of U.S. ground troops does decline relative to the baseline scenario, by roughly 30,000 troops. The 10th percentile projection remains relatively unchanged, but the 90th percentile projection decreases more substantially, by 100,000–150,000 troops in many years. This highlights that while the average case may be more modestly changed, there seems to be a clearly reduced risk of large-scale troop commitment spikes in this scenario.

Figure 4.38
U.S. Isolationism: Forecasts of U.S. Ground Deterrent Interventions, 2017–2040

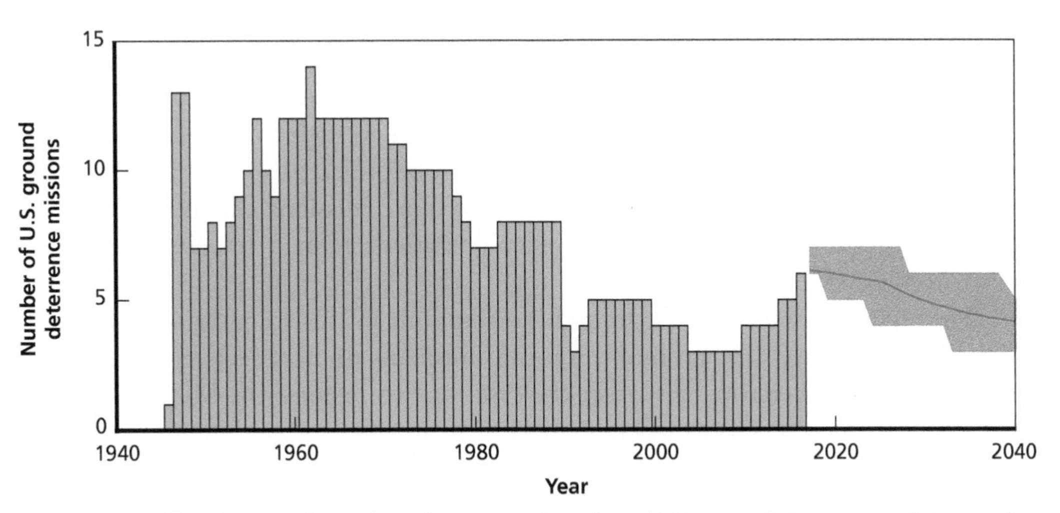

NOTES: The red line denotes the projected mean number of total U.S. ground deterrence missions each year, based on 500 iterations of our forecasting model. The gray shaded area represents the range of forecasts bounded by the 10th and 90th percentiles of U.S. ground deterrence missions each year, based on 500 iterations of our forecasting model.

---

47 The Figure showing this projection is omitted here for brevity but is shown in Appendix C.

Figure 4.39
U.S. Isolationism: Forecasts of Demands for U.S. Ground Intervention Forces, 2017–2040

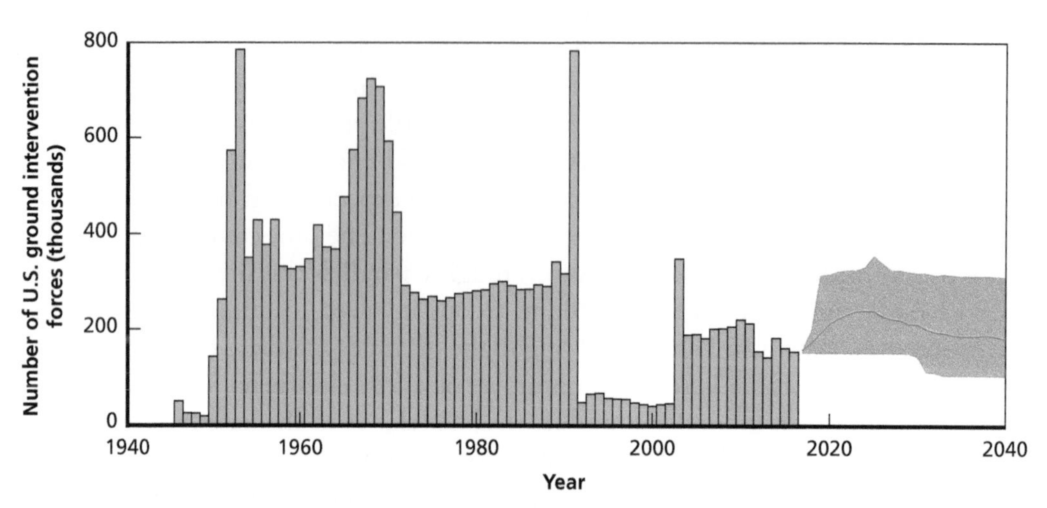

NOTES: The red line denotes the projected mean number of U.S. ground forces required for interventions each year, based on 500 iterations of our forecasting model. The gray shaded area represents the range of forecasts bounded by the 10th and 90th percentiles of U.S. ground forces required for interventions each year, based on 500 iterations of our forecasting model.

This reduction in "tail risk" can be seen most clearly in the projected number of U.S. ground troops employed in armed conflict interventions, shown in Figure 4.40.

While the average number of ground troops projected to be involved in armed conflict interventions declines modestly in this scenario, the more dramatic change is the sharp decline in the 90th percentile projection. Taken together, these changes suggest that some risk of involvement in large-scale combat missions persists, as reflected in the average projections, but the frequency with which such missions are projected is reduced, now occurring in fewer than 10 percent of projected simulations. As noted above, the number of troops committed to deterrence missions declines in this scenario, whereas those committed to stability operations remain relatively similar in comparison with the baseline.

The regional trends in U.S. ground interventions in this scenario are also worth noting. While average U.S. ground troop commitments to most regions are relatively unchanged in this scenario, this is not the case for two regions: the Middle East and East/Southeast Asia, as shown in Figure 4.41.

In the Middle East, there is a notable decrease in the projected number of U.S. forces committed to the region on average. In part this reflects some reduction in troops involved in deterrence, anticipating the ending of the Sinai mission, but the greater part of the reduction likely reflects a lower likelihood of involvement in armed conflict interventions in the region. In East/Southeast Asia, meanwhile, we see the opposite pattern. Whereas in the baseline scenario U.S. troops in East/Southeast Asia modestly decline over time, in this scenario this trend reverses, and it does so

Figure 4.40
U.S. Isolationism: Forecasts of Demands for U.S. Ground Combat Mission Forces, 2017–2040

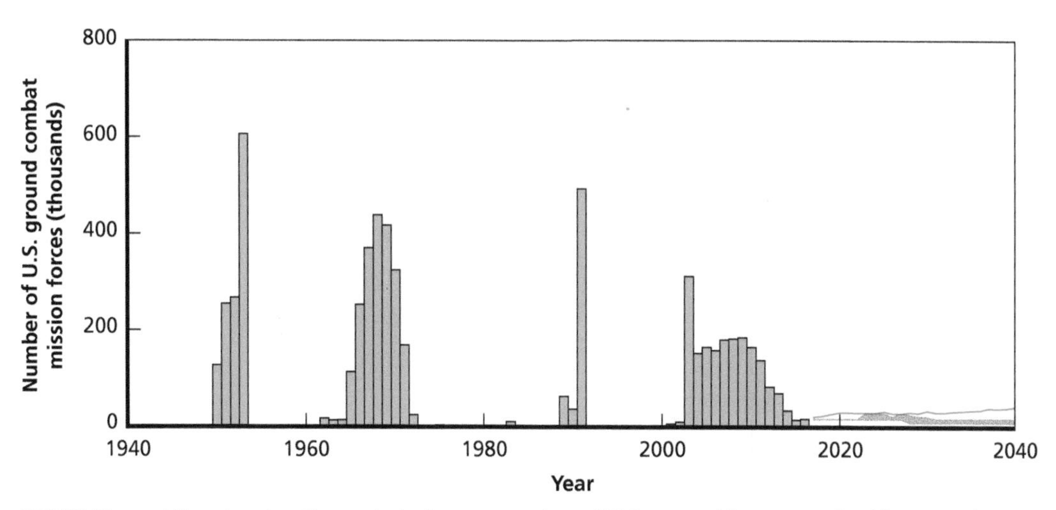

NOTES: The red line denotes the projected mean number of U.S. ground forces required for armed conflict interventions each year, based on 500 iterations of our forecasting model. The gray shaded area represents the range of forecasts bounded by the 10th and 90th percentiles of U.S. ground forces required for armed conflict interventions each year, based on 500 iterations of our forecasting model.

despite the anticipated decline in troops committed to deterrence missions. The average number of ground troops committed to deterrence missions in East Asia declines from roughly 60,000 in 2025 down to roughly 35,000 troops by 2040 in this scenario. At the same time, however, the average number of ground troops committed to armed conflict interventions in the region rises substantially, from roughly 3,500 in 2025 to more than 40,000 in 2040. As noted above in Figure 4.40, combat missions overall remain rare in this scenario. However, these results suggest that the risk of these large-scale missions increases substantially in East/Southeast Asia.[48]

The juxtaposition of a declining projected U.S. deterrent presence in East/Southeast Asia combined with a projected rising risk of a U.S. involvement in a major combat operation in the region seems to highlight the value of continued U.S. commitments to deterrence and stability in the region. Although these projections of course remain uncertain, the pattern they illustrate is worth noting, as is the fact that in this scenario a similar dynamic does not occur in Europe. In that region, despite a slight increase in the risk of interstate war in this scenario, the U.S. (formerly NATO) deterrent mission is projected to continue, and there is no notable increase in the number of ground troops anticipated to be committed to combat interventions.

[48] Although country-specific projections are particularly uncertain in our models, the states with the highest average projected number of U.S. troops committed to combat interventions after 2025 in this scenario are Myanmar, Thailand, Vietnam, and North Korea.

Figure 4.41
U.S. Isolationism: Regional Forecasts of Demands for U.S. Ground Intervention Forces, 2017–2040

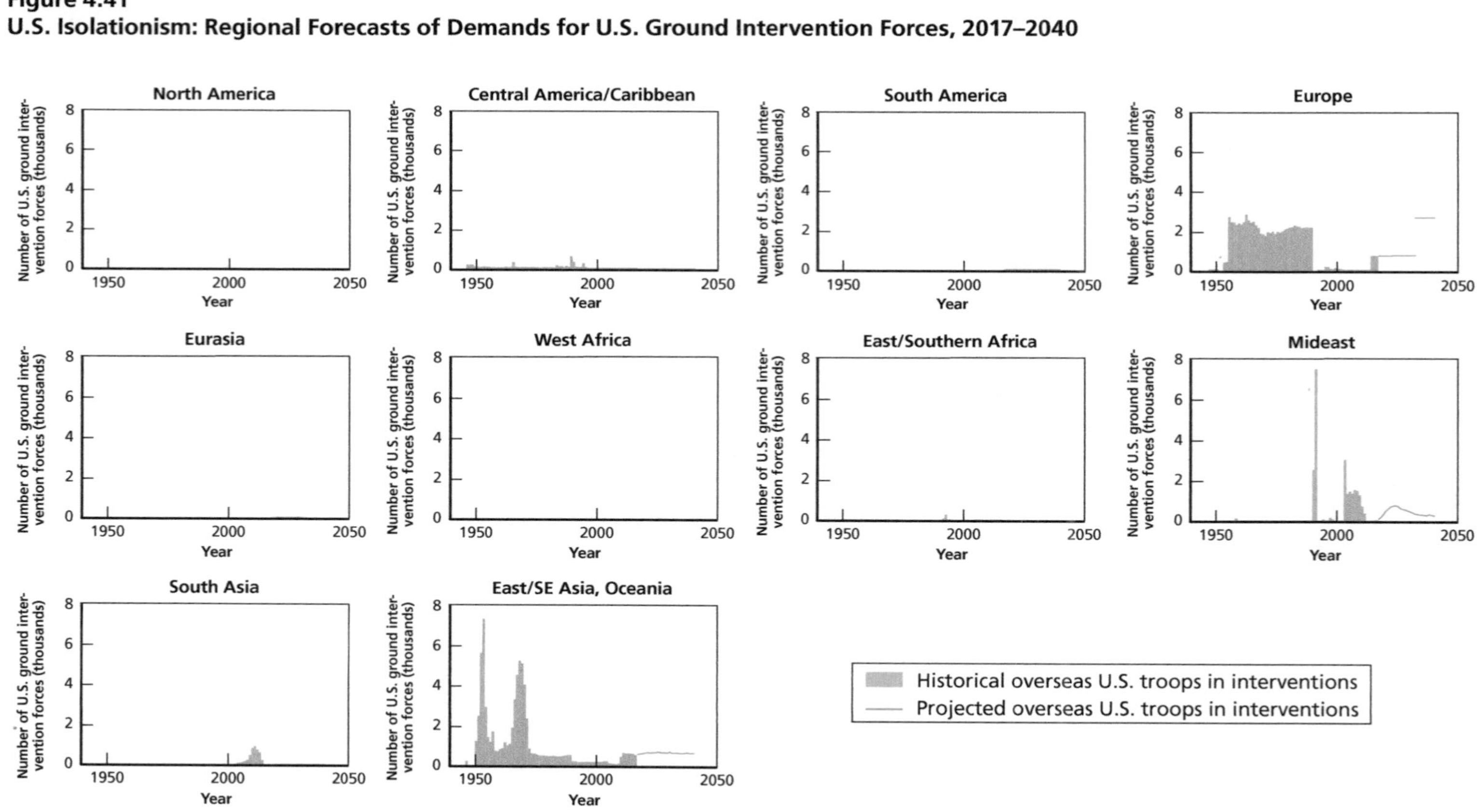

NOTE: The red line denotes the projected mean number of U.S. ground forces projected in each geographic region each year, based on 500 iterations of our forecasting model.

Summary

Overall, these alternative scenarios highlight the scale of potential changes that could occur in armed conflict trends and U.S. military intervention behavior as a result of different events or developments in the international system. While these were constructed to represent "worst-case" scenarios, they do remain plausible, based on historical analogs, and should therefore be understood to inform the degree of uncertainty attached to our baseline scenario projections. The concluding chapter summarizes our overall interpretation of our results and identifies resulting implications for policy.

CHAPTER FIVE

Conclusions and Implications for U.S. Army Force Planning

Summary of Projections

The conflict and intervention projections presented in the previous chapter vary across the five different scenarios presented, but we can make a number of cross-cutting insights. First, as noted throughout Chapter Four, many of the results of our models have relatively high levels of uncertainty associated with their specific projections. This uncertainty illustrates the limitations of the models, but it is also an accurate reflection of the challenges involved in such an effort. Indeed, claims to provide highly precise estimates of future wars, conflicts, and U.S. ground interventions 10, 15, or 20 years from now would simply not be credible. Notwithstanding this uncertainty, the trends in our projections can still provide insight into the possible shape of the future of use to U.S. policymakers and the U.S. Army in particular.

Turning first to trends in armed conflict, our analyses suggest that, although future levels of interstate war are likely to remain low by historical standards, the risk of interstate war appears to be likely to increase, in both our baseline scenario and in several of our alternative scenarios. In the baseline scenario, this increase in risk is concentrated in Eurasia and the Middle East and corresponds roughly to an increase of one additional ongoing interstate war each year. This would effectively return the frequency of interstate war back to what it was on average during the Cold War period, while still remaining well below levels of the pre-1945 era. In our alternative scenarios, this risk is largest in the Global Depression scenario, in which new conflicts emerge at the friction points between separate geopolitical and trading blocs, and in the Revisionist China scenario, in which conflict emerges between the group of states that become closely aligned with China and others that remain outside its orbit.

However, while our models consistently suggest some increase in the future likelihood of interstate war, they also consistently project a decrease in intrastate conflict from current levels, in the baseline scenario and across the alternatives. This decrease is generally spread across regions but is most pronounced in East/Southern Africa and the Middle East, the regions with the highest current levels of such conflict. Intrastate conflict remains prevalent, however, even with the anticipated declines, returning only to the level last seen in the late 1990s/early 2000s. Notably, this trend appears robust across scenarios, even the Great Depression and Global Pandemic scenarios, which we would expect to place substantial strain on many states.

Our projections of the number of future U.S. ground interventions and the troops anticipated to be employed in those interventions also vary across scenarios, but again some notable patterns emerge. In the baseline scenario, the total number of U.S. ground interventions is projected to decline slightly or remain the same, but this trend is accompanied by a projected increase in the forces required to meet the demands of these interventions. As noted elsewhere, this increase is partially a result of assumptions made in the model, and partially reflective of a replacement of current interventions that have drawn down notably from the peak troop levels, such as Afghanistan, with future interventions that are estimated to be notably larger. In the baseline scenario, for instance, our model suggests a demand that ranges from 100,000 to 425,000 ground troops between the 10th and 90th percentile projections, but with an average number of projected ground troops employed of more 200,000. While this is a broad range, it notably excludes both commitment levels at the Vietnam-era peak or the immediate post–Cold War valley. Further, the regional patterns of these anticipated interventions suggest that the deployed troops are most likely to be involved in interventions in the Middle East or Eurasia. The projections of the numbers of troops demanded should be viewed with particular caution, however, for reasons discussed above.

Notably, we do not forecast a large increase in the *numbers* of interventions in any scenario. Even in the Global Depression scenario, the average number of U.S. ground interventions is projected to remain relatively in line with the baseline scenario. We do, however, observe some sizable increases in U.S. ground forces committed to interventions, particularly compared with the present. As in the case of conflict projections, however, our estimates of demand for future U.S. ground intervention forces have a high degree of uncertainty, and the 10th and 90th percentile projections underscore the wide range of plausible projections. These projections are still valuable in a comparative sense, to understand how demand for U.S. ground intervention forces, and the trends in this demand over time, are likely to be affected by various assumptions in our five scenarios. We see the largest increases in demands for forces, for instance, with interventions into armed conflict and stability operations for the Global Depression scenario and interventions into armed conflict in the Revisionist China scenario. In terms of the regional patterns, despite the pronounced changes in other factors across scenarios, certain key regions persist as having the highest risk of future, larger U.S. ground interventions: the Middle East, Eurasia, and East/Southeast Asia. In the cases of the Middle East and East Asia, such interventions would continue long-standing historical patterns. The model's suggestion that substantial U.S. ground forces could be deployed to Eurasia, however, would be a notable strategic departure for the United States, and could involve heightened tensions with Russia, China, or both.

We can also compare across the different scenarios in a general sense to explore which types of external shocks are most likely to have significant implications for ground force demands. The scenario with the fewest implications for ground forces (or the fewest deviations from the baseline) is the Global Pandemic scenario. Although we do see a small increase in the projected number and size of deterrent interventions, force demand and conflict incidence do not vary as widely from the baseline as in other scenarios. Although there is much concern over the implications of a pandemic from economic and human welfare perspectives, our projections suggest that the implications from a conflict and intervention perspective may be more limited. Second, the effects of a global depression are among the most severe. We often think about economic collapse for its domestic effects on wages and employment, but our models clearly suggest that if severe economic downturn is accompanied by geopolitical dislocations and the establishment of rival trading blocs, it can also be associated with an increase in interstate war and a greater demand for U.S. ground combat forces. Should such a scenario occur in the future, it could pose a significant challenge for the United States. In a period of economic decline, there may be fewer resources to provide the military with the equipment, training, and personnel needed to be successful. This could complicate U.S. ground interventions in this context.

The Revisionist China scenario also suggests a more conflictual trajectory for the world and a higher demand for U.S. ground forces over the baseline. It is also especially relevant within the context of the National Defense Strategy and its focus on thinking about the potential for more direct great power competition. The practical implications for the U.S. ground of the Revisionist China scenario are broadly similar to the Global Depression scenario—although the increases in interstate war and U.S. interventions are more significant. Together, this similarity highlights that both trends in conflict and trends in intervention respond in similar ways to severe shocks that disrupt the current international system—that is, the current set of alliances, partnerships, and trade networks.

Finally, the results of our U.S. Isolationism scenario have interesting implications. There is a long-running debate about the extent to which U.S. deterrent missions, participation in multilateral institutions, and building strong alliances are effective ways to reduce conflict. Although contributing to that debate was not an objective of this report, our U.S. Isolationism scenario suggests that, at least in some regions, these factors do provide a pacifying effect and that risk of conflict may increase in their absence. In East/Southeast Asia, our results suggest that a drawdown in U.S. deterrence commitments may be more than compensated for by an increase in commitments to new combat missions in the region, such that U.S. ground forces in the region could actually increase despite the isolationism stipulated in the scenario.

Each of the scenarios in this report imagines a future that is more negative than the baseline. This is a limitation of our analysis. It is equally possible that the future will be more positive than the baseline scenario, with less conflict and fewer interventions. Future work should explore these scenarios as well to provide a more balanced assessment of possible futures.

Implications for the Army

Planning for the future, whether in the U.S. Army or elsewhere, is necessarily an exercise in risk management. Precise predictions about complex social phenomena such as armed conflict and military interventions decades into the future are impossible. The important implications for the Army of the analyses presented in this report are therefore less about specific force estimates (although those are worth noting) and more about identifying and understanding how different aspects of the future operating environment may function as drivers of demand for U.S. Army forces. The Army can use this understanding to take steps to manage future risk and ensure that it is sufficiently prepared to respond to the most likely or high-consequence contingencies. This section highlights several lessons drawn from our results that can inform Army planning in this manner.

Future Stability Operations

Although our different scenario projections diverge in many respects, our results are relatively consistent in finding a high likelihood that the United States will conduct a sizable stability operation at some point between now and 2040. In our baseline and alternative scenarios, our projections consistently highlight the significant likelihood of an increase of at least one new stability operation in the projection period. Our model functions by identifying the conditions under which the United States has historically conducted stability operations, and then by looking for future instances in which those conditions recur. Although the specifics of our projections, including the exact timing of such an intervention and where it will take place, remain highly uncertain, the recurrence at some point in the next two decades of conditions that have previously prompted the United States to undertake a stability operation seems quite likely.

Our expectation that a relatively sizable U.S. ground stability operation is likely to occur in the next 20 years is somewhat at odds with the current prevailing orientation of Army strategic thinking. In 2017, the Army published an updated version of Field Manual 3-0, *Operations*, which shifted focus away from the counterinsurgency and stability operations that had been the Army's focus since 2001 and increased the emphasis on large-scale ground combat with near-peer adversaries such as Russia and China.[1] This new focus reflects an assessment by Army and DoD leaders that the threat of large-scale ground combat with capable adversaries is increasingly likely, and that the potential for future large-scale stability operations like those in Iraq and Afghanistan is dwindling.[2] This shift in focus has already led to a plan to eliminate the

---

1 Field Manual 3-0, *Operations*, Washington, D.C.: Department of the Army, October 2017.

2 "Why Soldiers Can Now Pretty Much Say Goodbye to Counter-Insurgency Training," We Are the Mighty, October 2017.

Peacekeeping and Stability Operations Institute, which had been focused on studying and preparing the U.S. Army for irregular warfare.[3]

Our results suggest that it is unlikely that the era of large U.S. ground stability operations has passed for good. The U.S. Army has used the experience in Iraq and Afghanistan to build strong expertise in how to execute complex stability operations in urban and other environments. The Army has developed training simulations and exercises to prepare its forces to operate in these environments and has acquired the equipment needed to support these operations.[4] Letting this institutional capacity and knowledge atrophy would significantly undermine the Army's ability to successfully conduct these types of operations in the future and reduce the Army's flexibility and responsiveness should such a demand arise again. To ensure that it has sufficient capacity to support a large-scale stability operation, the Army would likely need to bring in or retain sufficient numbers of qualified personnel and might even develop a surge strategy to hedge against risk.

Great Power Competition and Deterrence

As noted elsewhere, our force projections for future deterrent interventions predict only relatively modest changes (some increases, some decreases, depending on the scenario) from current demands. A potential new deterrent intervention in the Philippines is, for example, a frequent projection in many of our model iterations, and the present deterrent mission in the Sinai is typically the most likely such mission to be projected to end, depending on the scenario. However, it is important to note that our models are, by design, silent on one of the largest possible drivers of an increase in deterrent force demands: an increase in the size of current deterrent deployments. As discussed above, our models assume that all ground interventions ongoing in 2016 retain their 2016 size throughout the remainder of the time when the model projects those current interventions to continue. This was a necessary modeling simplification, but it has significant implications, especially given the current focus on great power competition, and concern over the threat presented by revisionist powers in Russia, China, and possibly Iran, as expressed in the National Defense Strategy.[5] One possible response to the perceived threat presented to U.S. interests and allies by rising adversaries would be to increase the size of deterrent interventions intended to safeguard these interests and to place a check on adversary ambitions. For example, there are ongoing discussions about the appropriate size of a U.S. ground deterrent force in Eastern Europe, with some arguing that more forces are needed. The size of the current intervention has been

---

[3] Tammy Schultz, *Tool of Peace and War: Save the Peacekeeping and Stability Operations Institute*, Washington, D.C.: Council on Foreign Relations, July 31, 2018.

[4] See for example, Geoff Manaugh and Nicola Twilley, "It's Artificial Afghanistan: A Simulated Battlefield in the Mojave Desert," *The Atlantic*, May 18, 2013.

[5] U.S. Department of Defense, 2018.

increased multiple times, including in 2016 and in 2017 with new equipment.[6] Similar discussions surround deterrent deployments in the Asia-Pacific region.[7]

It seems quite plausible, then, that another major driver of future force demands will be decisions made about the size of ongoing deterrent deployments in places like Eastern Europe and the Asia-Pacific region. Our model does not capture this increasing demand, but we mention it here both as a limitation and to highlight it as another consideration for Army force planners as they work to estimate future force requirements, develop force plans, and make short- and long-term force posture decisions.

Combat Interventions and Force Demands

Just as our forecasts highlight the plausibility of a future large-scale U.S. ground stability operation, our projections also suggest that there is a good chance of a sizable U.S. ground combat intervention into an ongoing armed conflict over the next two decades, particularly later into that period. We see this risk clearly in the baseline scenario, and it is even more dramatic in some of our alternative scenarios. More importantly, however, the implications of such an intervention for force demands would be substantial. As noted elsewhere, even where we expect the numbers of U.S. ground interventions into armed conflict to decline, such as in the baseline scenario, we still see sharp increases in the number of troops that could plausibly be committed to such interventions. In 90th percentile projections in the baseline scenario, armed conflict interventions are expected to require more than 150,000 ground troops in the early 2030s, while the average demand across all model iterations is roughly 75,000 troops. The projected demand for ground combat forces is dramatically higher under some of the alternative scenarios. In the Global Depression scenario, for example, demand for ground combat intervention forces is predicted to exceed 500,000 in the 90th percentile projection by 2040, with the average projection exceeding 100,000. These results are in keeping with historical patterns, as well. It has been the large combat interventions that have most significantly increased force demands in the past. Combat interventions are often shorter than stabilization and deterrent missions, but they have tended to be consistently larger.[8]

The need to undertake such a massive combat intervention is by no means a certainty. The 10th percentile projection for ground forces devoted to armed combat interventions remains much lower, in the low tens of thousands, even in the Revisionist China scenario. For Army planners, however, the continued, sizable risk of a large-scale

---

[6] Jen Judson, "Deterring Russia: U.S. Army Hones Skills to Mass Equipment, Troops in Europe," *Defense News*, March 17, 2017; Andrew Tilgham, "More U.S. Troops Deploying to Europe in 2017," *Military Times*, February 2, 2016.

[7] Mike Yeo, "Incoming US Pacific Command Chief Wants to Increase Presence Near China," *Defense News*, April 23, 2018.

[8] Kavanagh et al., 2017.

combat intervention is important, for a few reasons. First, the possibility that Army forces will be called on to engage in major combat operations is consistent with the increasing focus on great power competition and the Army's growing focus on the risk of large-scale conventional warfare. Our projections reinforce the value of the Army's strategic focus on these areas.

Second, our projections—and the review of similar, historical operation sizes that inform them—provide some insight into the size of possible force demands for combat interventions and can inform the decisions of Army leaders about training, force structure, recruiting, and retention. Notably, the average demand for combat intervention forces under the baseline scenario is small enough that the Army likely has sufficient capacity already. On the other hand, sustained combat force demands of 100,000, or perhaps many more, in an alternative strategic environment such as the Global Depression, would place significantly more strain on the Army. Policymakers will need to assess the likelihood of such a dramatic worsening in the strategic environment, and the risk they are willing to accept, or not, in preparing for it.

Future Demand for Heavy and Light Forces

In addition to projecting overall force demands, our analysis considered the likely future demand for both heavy and light ground forces. Generally speaking, our results forecast a relatively consistent ratio of heavy to light ground forces, where heavy forces made up roughly one-fifth of the total number of ground forces required for future military interventions. This result is driven both by our assessment of the historical utilization of heavy forces in U.S. ground interventions of different activity types and characteristics as and by the frequency with which those different types of interventions are projected to occur in the future. In most of our results, including the baseline scenario, we projected that future interventions are likely to have higher demands for heavy ground forces than the current mix of deployed forces today. In the baseline scenario, the 90th percentile projection suggests a demand for heavy forces equal to the peak number of heavy ground forces deployed in Iraq and Afghanistan, with the average projection roughly two-thirds of that number. This average projection represents an increase of roughly 10,000 additional heavy forces employed in U.S. ground interventions compared with today. It should be emphasized that these increases would occur on top of any potential increase in the size of heavy forces in existing deterrent interventions, which, as discussed above, our model does not reflect. The 10th percentile projection, meanwhile, is roughly consistent with or slightly below the number of heavy forces already committed to interventions today. These projections of heavy and light forces, however, should be understood to be accompanied by substantial uncertainty.

The question of whether the Army could properly resource such an increased demand with the current mix of heavy and light forces in the continental United States is beyond the scope of this study. However, it is worth noting that during the peak of the wars in Iraq and Afghanistan, the United States struggled to meet the demand for heavy forces. Given the potential for an increase in demand for heavy troops to that level, it may make sense for the Army to explore increasing its heavy force capability at a minimum as a hedge against the risk that it might face that level of heavy force demand in the future. This could mean ensuring that sufficient numbers of personnel are trained in necessary occupations and might also suggest the need for careful and strategic investment to fill any equipment gaps or to support innovations and upgrades. Importantly, this is the direction that the Army is already moving. The fiscal year 2019 Army budget reveals a shift away from aviation and toward artillery and combat vehicles that are intended to rebuild and augment the Army's current heavy forces.[9] The development of Army Futures Command, which is focused on ensuring the Army has the technology, equipment, and materiel it needs to meet future force requirements, can support efforts to ensure a sufficiently capable heavy force structure.[10] The Army Futures Command is not solely focused on heavy forces, but many of its modernization-focused efforts should strengthen and enhance the capabilities of heavy forces in ways that may augment their actual numbers.

Risks of U.S. Isolation

While the results of each of our scenarios provides useful insights, our U.S. Isolationism scenario highlights a potentially important dynamic given contemporary policy debates. In the wake of recent wars in Iraq and Afghanistan, it is not surprising that many in the policy community have expressed an interest in pulling back and reducing U.S. military activity overseas.[11] There are also concerns among some policymakers and segments of the public about the constraints and entanglements of alliances, multinational institutions, and multilateral trade agreements, and there are those who suggest the United States would be better off without these commitments and relationships.[12] Our projections for this scenario, however, provide a note of caution regarding the potential effects of such changes. Our models suggest that a broad-based reduction in U.S. engagement internationally may indeed increase the risk of interstate war.

While some of these wars may occur in regions where U.S. policymakers would no longer feel compelled to intervene, such as the Middle East, this may not necessarily be the case everywhere. Our scenario generally projected a modest decline in U.S.

---

[9] Daniel Wasserby, "Pentagon Budget 2019: US Army Procurement Shifts from Aviation to Artillery, Combat Vehicles," *Jane's Defense Weekly*, February 13, 2018.

[10] For more on Army Futures Command, see Army Futures Command Task Force, "Army Futures Command," March 28, 2018.

[11] For an example of relevant arguments, see Joseph Parent and Paul MacDonald, "The Wisdom of Retrenchment: America Must Cut Back to Move Forward," *Foreign Affairs*, Vol. 90, No. 6, 2011.

[12] For an example of relevant arguments, see Bonnie Kristian, "Trump Is Right: It's Time to Rethink NATO," *Politico*, August 3, 2016; Ana Swanson, "Once the WTO's Biggest Supporter, U.S. Is Its Biggest Skeptic," *New York Times*, December 10, 2017.

ground troop commitments overseas. However, it also projected that in East/Southeast Asia a decline in U.S. ground troop commitments to deterrence missions was more than offset by an increase in the average projected number of U.S. ground troops committed to combat missions in that region, resulting in an overall increase in U.S. troop presence there. While it is difficult to isolate whether the increased risk of and commitment to combat missions in the region was driven by the reduction in deterrence commitments specifically or to other aspects of the scenario such as reductions in U.S. alliance commitments or support for international institutions and norms, this correlation is still worth bearing in mind, and it highlights that not all efforts to reduce U.S. commitments overseas may prove to be cost-effective in the long run.

Potential Model Improvements for More Accurate and Dynamic Forecasts

The analyses summarized in this report provide U.S. Army force planners with robust and dynamic forecasts of future trends in armed conflict and subsequent demands for U.S. ground forces. In addition, our analytical approach, which combines statistical modeling with qualitative assessments, also provides Army force planners with a transparent and rigorous process for developing, combining, and assessing new force-planning constructs that are grounded in global trends and historical precedents. While these analyses offer a significant improvement over past approaches to forecasting future trends in conflict and interventions, there is still room for further improvements to develop forecasts that are even more helpful to policymakers while taking full advantage of available historical and projected trends.

Importantly, future forecasting approaches could seek to incorporate more dynamic variables of conflict and interventions into the statistical models that underlie our forecasts. Because we are constrained to variables that can be projected into the future, our statistical models of conflict and interventions rely, with some exceptions, on more structural variables that change relatively slowly and do not utilize some variables that may shift quickly or dramatically in response to decisions by different actors, but that have been shown to be important and significant predictors, such as refugee flows, battle-related deaths, and external support for combatants.[13] Unfortunately, projecting such variables in the context of future conflicts or interventions is quite difficult. Our implementation of U.S. ground force characteristics into our forecasting model using descriptive statistics of historical trends, however, opens the door for potentially using similar approaches to inject rough approximations of those sorts of dynamic variables into our models.

---

[13] Kavanagh et al., 2017.

Similarly, our forecasting approach leaves open a significant opportunity to provide more dynamic forecasts of the sizes of U.S. ground forces required for future interventions. As detailed in Chapter Three, our current forecasting approach sets the number of U.S. ground forces required for new interventions at their historical average and the number of U.S. ground forces required for ongoing forecasted interventions at their 2016 levels. This approach resulted in two underlying trends in our forecasts. First, the number of U.S. ground forces required for interventions ongoing after 2016 are effectively fixed at their 2016 levels, meaning that those interventions could never result in force requirements greater than, or less than, their 2016 levels. Second, the number of ground forces required for new forecasted interventions remains static for the duration of the intervention, meaning that our forecasts cannot adjudicate surges or withdrawals of troops during the course of an intervention. While this approach was reasonable from a modeling standpoint, it does not fully reflect the varying demands for U.S. ground forces over the course of ongoing interventions. To make our forecasts more realistic in this regard, future approaches could incorporate more dynamic qualitative approaches or statistical models of the variation in U.S. ground troop demands throughout ongoing interventions.

Future forecasting efforts could also investigate other statistical models that may yield more accurate forecasts based on our projected key factors. This analysis utilizes theoretically driven statistical models of armed conflict and U.S. ground interventions to drive our forecasts.[14] As discussed in Chapter One, however, other statistical approaches, such as nonparametric machine-learning models, could potentially do a better job of predicting armed conflicts and interventions than these theoretically driven approaches.[15]

One key technical addition that could also be made is to conduct more dynamic out-of-sample tests on each of the component statistical models to assess each for their accuracy as forecasting tools. Specifically, this would involve constructing each model on a "training" set of historical data and then testing it on a different set of "out-of-sample" historical data in order to more rigorously assess the ability of each statistical model to accurately predict conflicts and interventions that we already know have occurred while also accounting for the historical rarity of both armed conflicts and interventions. We paid close attention to the overall ability of each model to predict outcomes in our model development process, but we were not able to determine a satisfactory way to apply this more rigorous testing procedure.[16] If an appropriate proce-

---

[14] Kavanagh et al., 2017; Watts et al., 2017a.

[15] Muchlinski et al., 2016.

[16] Our difficulty lay in determining how best to segment the historical data into appropriate training and test data, given the relative rarity of both interventions and, to a lesser extent, conflicts, and the potential for discontinuous trends across different historical periods. As an initial alternative, Appendix B provides in-sample model validation tests using receiver operating characteristic curves and associated area under the curve assessments.

dure could be identified, the results of this test would allow us to potentially revise the model specification to improve accuracy and would also help us to better quantify our degree of confidence in the forecasts themselves. Additionally, out-of-sample model validation tests could be paired with automated model building techniques to develop more dynamic statistical models that iteratively update based on changes in our key factors over time. This approach could be leveraged to create dynamic forecasts that potentially reduce uncertainty in our forecasts in far-term projections.

Finally, to make these analyses more applicable to broader DoD planning efforts, in addition to U.S. Army analyses, our forecasting approach could be expanded to include forecasts of future U.S. Air Force and U.S. Navy interventions. This expansion could be accomplished using many of the same processes developed for this research project; future work would primarily need to develop statistical models of U.S. Air Force and U.S. Navy intervention onset and cessation like those developed for U.S. ground interventions, as described in Chapter Three. Expanding this forecasting approach to include all U.S. military services also opens the opportunity for comparative analyses of alternative futures in which the United States relies more or less heavily on different services for different types of interventions and the resulting impact on future projections of armed conflict or interventions more broadly. Additionally, this type of comparative forecasting, in which forecasts are built around differing deployments of each military service, could provide new opportunities for assessing impacts on force readiness from alternative concepts of force employment.

These initial model validation tests suggest that our statistical models do provide substantially better than at-random classification.

APPENDIX A

Historical Force Requirements Coding Methodology and Notes

This appendix details the coding methodology, underlying data and assumptions, and other details of the historical force requirement analysis summarized in Chapter Three. It is divided into three sections. The first section describes the process by which we scoped the universe of historical cases meriting inclusion in the dataset. The second section details the variables that were applied to each historical case, as well as the coding rules and sources used to populate these variables, as summarized in Table 3.8. Finally, we present a more detailed version of the results of the historical force requirement data, broken down at the individual case level and organized by the operational environment typology that underlies the summary statistics described in Chapter Three.

Defining the Case Universe

To begin our development of rules of thumb for U.S. ground force requirements in forecasted interventions, we proceeded by first defining the universe of relevant historical cases. We drew on extensive data on U.S. ground interventions collected in previous RAND Arroyo Center studies, namely the RAND U.S. Ground Intervention Dataset (RUGID).[1] Covering the period 1898–present, this dataset defines a ground intervention to include any deployment of U.S. ground troops on the territory of another country that met or exceeded a manpower threshold of 100 "person years" (i.e., a minimum of 100 troops deployed for at least one year, or, equivalently, a larger number of boots on the ground for a shorter period of time, such as 200 troops deployed for at least six months, or 400 troops deployed for three months, etc.). RUGID excludes troops stationed overseas as part of the permanent U.S. global defense posture, unless they also served an explicit intervention function, namely a long-term strategic deterrence purpose.

[1] For a more detailed discussion of the data collection and coding methodology of RUGID, see Kavanagh, et al., 2017.

As discussed in more detail later in this appendix (see coding notes on "Intervention Size"), in many instances, long-term deployments of U.S. ground forces in a given country or theater of operations have been divided into contiguous—and sometimes overlapping—cases, because the overarching purpose or nature of the intervention fundamentally changed over time. For instance, the brief combat phases of the U.S. invasions of Afghanistan and Iraq in 2001 and 2003, respectively, are coded as distinct cases from the subsequent multiyear stability operations missions that followed. Similarly, in the aftermath of World War II, ongoing stabilization missions during the U.S. occupations of West Germany and Japan overlap with the onset of the decades-long Cold War deterrence operations in Europe and East Asia.[2] In total, these primary criteria yielded a list of 98 U.S. ground interventions since 1898.

Second, in the course of the present report, the initial RUGID case universe was subsequently circumscribed to exclude any interventions that did not at least partially fall within one of three typological bins; as described in more detail later in this appendix (see "Activity Type" variable coding notes), the U.S. ground force presence had to be involved in deterrence, stabilization, and/or conventional combat/counterinsurgency activities to merit inclusion in the current quantitative analysis. Approximately two dozen cases contained in RUGID were thus excluded because the ground forces were exclusively involved in other activity types, such as providing humanitarian assistance or natural disaster relief, conducting small training and advising missions, and/or performing limited security operations, such as protecting U.S. embassies and personnel during ongoing armed conflict.[3] Furthermore, consistent with RUGID, interventions involving only naval or air forces have been excluded from the present dataset.

Third, a handful of deterrence cases contained in RUGID were dropped from the case universe if they met one of two criteria: (1) the case terminated with the United States' entrance into—and eventual victory in—an interstate war, or (2) the case was an instance of immediate (versus extended) deterrence.[4] The former category of cases

---

[2] In the original version of RUGID, the U.S. Cold War deterrence cases in West Germany and Japan were coded as beginning coterminously with the conclusion of the preceding U.S. post–World War II military occupations in 1955 and 1952, respectively, and the U.S. Cold War/post–Cold War deterrence case in South Korea was coded as beginning coterminously with the conclusion of the preceding U.S. post–Korean War military occupation in 1957. In this iteration, the respective U.S. Cold War deterrence cases in West Germany/Europe and Japan/Asia-Pacific are coded as beginning in 1946, and the Cold War/post–Cold War deterrence posture in South Korea is coded as beginning in 1953.

[3] Historically, U.S. interventions outside of these activity types have generally been limited in size and therefore less likely to bear on the effort in this report to project overall demands for U.S. forces.

[4] The deterrence cases meeting the first exclusion criteria were World War II Deterrence in the Atlantic: Bases-for-Destroyers/Lend-Lease Bases (1940–1945); World War II Deterrence in the Atlantic: Protectorate of Iceland (1941–1945); WWII Deterrence in the Atlantic: Protectorate of Greenland; U.S. Show of Force in Panama/Operation Nimrod Dancer (1989); and U.S. Deterrence Posture in the Persian Gulf/Operation Desert Shield (1990–1991). The deterrence cases meeting the second exclusion criteria were: U.S.-Mexico Border Wars (I) (1911) and Guantanamo Bay Reinforcement During Cuban Missile Crisis (1962).

was dropped because they yielded problematic estimates of the duration of deterrent deployments (such deterrent deployments ended because the United States destroyed the potential aggressor being deterred, not because the deterrence mission itself succeeded or failed). The latter category of cases was dropped because of potential conceptual and empirical confusion that could result from including cases of immediate and extended deterrence together. (The social science literature on extended deterrence makes clear that it is qualitatively different from immediate deterrence. Moreover, it is very difficult empirically to distinguish the deterrent effect of U.S. forces mobilized within the borders of the United States for a specific crisis—such as border clashes with Mexico—from the deterrent effects of U.S. forces stationed within the United States more generally).

Finally, additional discussion is required regarding changes made to the RUGID case universe in the coding of U.S. interventions in the Balkans in the early 1990s. In the original iteration of RUGID, the activities of U.S. ground forces in the region following the breakup of Yugoslavia were coded as a single stability operation case covering the period 1992–2008 and including associated ground deployments to the geographical locations of Bosnia, Croatia, Macedonia, and Hungary. Subsequently, in the second iteration of RUGID, the case was recoded to include air and naval assets employed—namely those involved in air strikes and enforcing the Bosnia no-fly zone and arms embargo; air and naval activity types were thus also coded to reflect combat and deterrence missions. For this study, the Balkans case was again recoded to reflect a more granular perspective of the diversity of U.S. ground force activity types in different stages in different countries from 1992 through 2008. More specifically, the interrelated force deployments to the Balkans were broken down into five distinct cases:

1. *Bosnia, Phase 1 (July 1992–August 1995):* This case is considered strictly an air/naval intervention, including activities focused on enforcing the no-fly zone and sanctions regime/arms embargo, as well as massive airlifts of humanitarian aid (Operations Deny Flight, Sky Watch, Sharp Guard, and Provide Promise). It is therefore excluded from the forecasting demand/historical force requirement dataset because it did not involve U.S. boots on the ground in Bosnia (except for a small special operations forces footprint).
2. *Bosnia, Phase 2 (August 1995–December 1995)*: This case constitutes a separate, contiguous intervention beginning with the NATO air war in Bosnia through the conclusion of the Dayton Accords. It is also excluded from the forecasting demand/historical force requirement dataset because it did not include U.S. ground forces in Bosnia (again, except for a small special operations forces footprint).
3. *Bosnia, Phase 3 (December 1995–2008):* Beginning with deployment of U.S. ground forces to Bosnia in December 1995, this case encompasses the various post-Dayton, NATO/UN/EU peacekeeping and stability operations in which

U.S. forces participated (IFOR [Implementation Force], SFOR [Stabilisation Force in Bosnia and Herzegovina], UNMIBH [United Nations Mission in Bosnia and Herzegovina], EUFOR [European Union Military Operation in Bosnia and Herzegovina]/Althea). It also includes U.S. troops deployed to Hungary and Croatia, where IFOR/SFOR Headquarters and logistics/supply hubs were located. This stabilization case is included in the forecasting demand/historical force requirement dataset.

4. *Croatia (November 1992–December 1995):* Though U.S. ground forces did not deploy to Bosnia under the mandate of UNPROFOR [United Nations Protection Force], they did deploy to Croatia in November 1992. Although the broader UNPROFOR mission (later rehatted as UNCRO [United Nations Confidence Restoration Operation in Croatia] and then UNTAES [United Nations Transitional Administration for Eastern Slavonia, Baranja and Western Sirmium]) was involved in stability operations in Croatia, U.S. ground forces in-theater during this period were involved almost exclusively in humanitarian activities—namely, operating an Army hospital unit in Zagreb. This humanitarian assistance case is thus excluded from the forecasting demand/historical force requirement dataset by virtue of ground force activity type.
5. *Macedonia (July 1993–February 1999):* A limited number of U.S. troops also deployed to Macedonia under the mandate of UNPROFOR (later rehatted as UNPREDEP [United Nations Preventive Deployment Force]) to serve as an interpositional force to deter armed conflict from spilling over into Macedonia. This deterrence case is included in the forecasting demand/historical force requirement dataset.

Taken together, these coding parameters yielded a case universe of 65 historical ground interventions out of the 98 originally contained in RUGID relevant to the modeling of future U.S. ground force demands. Table A.1 contains a complete list of these cases included in the Historical Force Requirement Dataset.

Variables in the Historical Force Requirement Dataset

For each intervention in the historical force requirement dataset ($N = 65$), we coded several different dimensions. As described in greater detail below, some variables were coded for every case (average troop numbers, intervention duration, etc.), whereas other operating environment variables were coded only for certain intervention typologies (adversary strength, threat index, etc.). These variables are summarized in Table A.2. The coding rules, case anomalies, and data sources are then discussed in more detail in the remainder of this methodology section.

Table A.1
Historical Force Requirement Dataset: Case Universe

Case ID	Intervention Name	Start Year	End Year	Location	Activity Type(s)
1	Spanish-American War	1898	1898	Spanish colonial territories (Cuba, Puerto Rico, Philippines, Guam)	Combat
2	Boxer Rebellion	1900	1900	China	Combat, Stabilization
3	U.S. Occupation of Panama	1903	1915	Panama	Stabilization
4	Cuban Pacification Intervention	1906	1909	Cuba	Stabilization
5	U.S. Peacekeeping Force in Cuba	1912	1912	Cuba	Stabilization
6	Marines Landing During Nicaraguan Revolution	1912	1925	Nicaragua	Stabilization
7	U.S. Occupation of Veracruz	1914	1914	Mexico	Stabilization
8	U.S. Deterrent Posture in Panama Canal Zone	1915	1989	Panama	Deterrence
9	U.S. Occupation of Haiti	1915	1934	Haiti	Stabilization
10	U.S. Invasion of Mexico: Pershing's Expedition	1916	1917	Mexico, United States	Combat
11	U.S. Occupation of Dominican Republic	1916	1924	Dominican Republic	Stabilization
12	World War I	1917	1918	Europe (France, Germany, etc.)	Combat
13	U.S.-Mexican Border Wars (II)	1918	1919	Mexico, United States	Combat
14	American Expeditionary Forces in Vladivostock and Archangel	1918	1920	Russia/Soviet Union	Combat
15	Allied Occupation of the Rhineland Post–World War I	1918	1923	Germany	Stabilization
16	U.S. Occupation of Nicaragua	1926	1933	Nicaragua	Stabilization
17	"China Marines" Deployment	1927	1941	China	Stabilization

Table A.1—continued

Case ID	Intervention Name	Start Year	End Year	Location	Activity Type(s)
18	World War II (Asian/Pacific Theater)	1941	1945	Asia/Pacific (Philippines, Japan, Australia, New Zealand, Hong Kong, Wake Isle, Malaya, Singapore, Dutch East Indies, Gilberts, Marshalls, Marianas, Palaus, Borneo, Manchuria, India/Burma, Aleutians, PNG, Guadalcanal, Solomons, E. Mandates, Bismarks, Leyte, Luzon, Ryukus/Okinawa, Iwo Jima, Guam)	Combat
19	World War II (European/Mediterranean/North African Theater)	1941	1945	Europe, the Mediterranean, N. Africa (France, Germany, United Kingdom, Morocco, Algeria, Tunisia, Belgium, Netherlands, Denmark, Italy, Greece, Turkey, etc.)	Combat
20	U.S. Cold War Deterrent Posture in Saudi Arabia	1945	1989	Saudi Arabia	Deterrence
21	U.S. Occupation of W. Germany (Post–World War II)	1945	1955	Germany	Stabilization
22	Allied Occupation of Austria (Post–World War II)	1945	1955	Austria	Stabilization
23	U.S. Occupation of Japan (Post–World War II)	1945	1952	Japan	Stabilization
24	U.S. Occupation of South Korea (Post–World War II)	1945	1949	South Korea	Deterrence, Stabilization
25	Marines in Northern China (Post–World War II)	1945	1949	China	Stabilization
26	U.S. Cold War Deterrent Posture in Europe	1946	1989	West Germany, United Kingdom, France, Belgium, Netherlands, Greece, Turkey, Spain, Italy, Greenland, Iceland	Deterrence
27	U.S. Cold War Deterrent Posture in Libya	1948	1970	Libya	Deterrence
28	Berlin Airlift	1948	1949	Germany (Berlin)	Deterrence
29	U.S. Cold War Deterrent Posture in Ethiopia/Eritrea	1950	1973	Ethiopia/Eritrea	Deterrence
30	U.S. Cold War Deterrence in Taiwan	1950	1979	China (Taiwan)	Deterrence

Table A.1—continued

Case ID	Intervention Name	Start Year	End Year	Location	Activity Type(s)
31	Korean War	1950	1953	South Korea, North Korea	Combat
32	U.S. Cold War Deterrent Posture in Morocco	1951	1977	Morocco	Deterrence
33	U.S. Cold War Deterrent Posture in Asia/Pacific	1946	1989	Japan, Philippines	Deterrence
34	U.S. Cold War Deterrent Posture in Iran	1953	1978	Iran	Deterrence
35	U.S. Military Government in S. Korea	1953	1957	South Korea	Stabilization
36	U.S. Deterrent Posture in S. Korea	1953	Ongoing	South Korea	Deterrence
37	Lebanon Crisis of 1958 (Operation Blue Bat)	1958	1958	Lebanon	Stabilization
38	Vietnam War	1962	1975	Vietnam, Thailand, Cambodia	Combat, Stabilization
39	Laos Crisis/U.S. Show of Force	1962	1962	Laos	Deterrence
40	U.S. Occupation of Dominican Republic	1965	1966	Dominican Republic	Stabilization
41	Multinational Force and Observers (MFO) in Sinai	1982	Ongoing	Egypt	Deterrence
42	Lebanese Civil War	1982	1984	Lebanon	Stabilization
43	U.S. Demonstrations of Force/Cold War Deterrence in Honduras	1983	1992	Honduras	Deterrence
44	U.S. Invasion of Grenada	1983	1983	Grenada	Combat, Stabilization
45	U.S. Invasion of Panama	1989	1990	Panama	Combat, Stabilization
46	U.S. Peacekeeping Force in Panama	1990	1994	Panama	Stabilization
47	Persian Gulf War	1991	1991	Kuwait, Iraq, Saudi Arabia, United Arab Emirates	Combat
48	U.S. Peacekeeping Force in Kuwait	1991	1991	Kuwait	Deterrence, Stabilization

Table A.1—continued

Case ID	Intervention Name	Start Year	End Year	Location	Activity Type(s)
49	U.S. Deterrent Force in Turkey and Northern Iraq	1991	2003	Turkey, Iraq	Deterrence
50	Multinational Bosnian Peacekeeping Force	1995	2008	Bosnia	Stabilization
51	Multinational Peacekeeping Force in Somalia	1992	1995	Somalia	Stabilization
52	U.S. Deterrent Force in the Persian Gulf	1992	2003	Kuwait, Saudi Arabia, Bahrain, Iraq	Deterrence
53	Multinational Deterrent Force in Macedonia	1993	1999	Macedonia	Deterrence
54	U.S. Peacekeeping Operations in Haiti	1994	1996	Haiti	Stabilization
55	Multinational Peacekeeping Force in Kosovo	1999	Ongoing	Kosovo	Stabilization
56	U.S. Overthrow of the Taliban (combat phase)	2001	2001	Afghanistan	Combat
57	U.S. Occupation of Afghanistan	2001	Ongoing	Afghanistan, Pakistan, Uzbekistan, Kyrgyzstan	Combat, Stabilization
58	Joint Task Force Horn of Africa	2002	Ongoing	Djibouti	Stabilization
59	U.S. Invasion of Iraq (combat phase)	2003	2003	Iraq	Combat
60	U.S. Occupation of Iraq	2003	2011	Iraq, Kuwait, Turkey, Saudi Arabia	Combat, Stabilization
61	U.S. Peacekeeping Force in Liberia	2003	2003	Liberia	Stabilization
62	Multinational Peacekeeping Force in Haiti	2004	2004	Haiti	Stabilization
63	U.S. Deterrence of China in Japan	2010	Ongoing	Japan	Deterrence
64	U.S. Deterrence of Russia in Europe/Operation Atlantic Resolve	2014	Ongoing	Latvia, Lithuania, Estonia, Poland, Hungary, Romania, Bulgaria, Germany, Belgium, Netherlands, Italy, United Kingdom, Norway	Deterrence
65	Combined Joint Task Force Against the Islamic State (Operation Inherent Resolve)	2014	Ongoing	Iraq, Syria, Turkey, Jordan, Kuwait	Combat, Stabilization

Table A.2
Historical Force Requirement Dataset: Variable Definitions

Number	Variable Name	Description
1	intervention_id	Integer used as unique identifier for each intervention
2	intervention_name	Common name for the intervention, often incorporates country location and activity type
3	start_year	Integer denoting year intervention began
4	end_year	Integer denoting year intervention ended
5	duration_months	Duration of intervention calculated in months
6	country_1	Name of the primary country/countries where the intervention occurred
7	country_2	Name of the secondary country/countries where the intervention occurred, if applicable; these countries generally include the locations of temporary U.S. footprints at main operating bases or forward operating bases during major combat and stability operations (e.g., thousands of U.S. forces at K-2 and Manas Air Force Base in Uzbekistan and Kyrgyzstan, respectively, during the early years of the U.S. occupation of Afghanistan, post-2001)
8	activity_1	Integer describing the primary activity in which U.S. ground forces were engaged during the intervention
9	activity_2	Integer describing the secondary activity in which U.S. ground forces were engaged during the intervention
10	activity_3	Integer describing the tertiary activity in which U.S. ground forces were engaged during the intervention
11	intervention_typology	Integer denoting the involvement of conventional combat/counterinsurgency, deterrence, and/or stability operations activities during the intervention
12	troop_size_minimum	Integer denoting the lower bound annual estimate (i.e., nadir) for the number of U.S. ground troops involved in the intervention
13	troop_size_maximum	Integer denoting the upper bound annual estimate (i.e., peak) for the number of U.S. ground troops involved in the intervention
14	troop_size_average	Integer denoting the "average" annual estimate for the number of U.S. ground troops involved in the intervention
15	heavy_force_ratio	Integer denoting the "average" annual estimate of "heavy" U.S. ground troops involved in the intervention as a percentage of the total U.S. ground troops
16	globalist_era_binary	Binary variable (0,1) denoting whether intervention began in the pre- or postglobalist era (i.e., before or after 1940)
17	war_type_binary	Binary variable (0,1) denoting whether interstate or intrastate armed conflict occurred (if the intervention involved U.S. armed combat)

Table A.2—continued

Number	Variable Name	Description
18	adversary_strength_CINC_proxy	Integer denoting the host country's Composite Index of National Capability (CINC) score as a proxy for adversary strength in cases of interstate war
19	adversary_strength_GDPPC_proxy	Integer denoting the host country's GDP per capita as a proxy for adversary strength in cases of intrastate war
20	adversary_strength_binary	Binary variable (0,1) denoting the strength of interstate adversary (based on CINC score) or intrastate adversary (based on GDP per capita), as applicable depending on war type
21	threat_index_max_value	Integer denoting the host country's maximum "Threat Index" score during a deterrence intervention (measured on a scale of 0.00 to 6.00)
22	threat_index_binary	Binary variable (0,1) denoting whether the host country's maximum "Threat Index" score during a deterrence intervention exceeded 4.00 (on a scale of 0.00 to 6.00)
21	US_ally_defense_treaty_binary	Binary variable (0,1) denoting whether the United States and host country were formal defense treaty or alliance partners during a deterrence intervention
22	US_adversarial_war_year_prior	Binary variable (0,1) denoting whether a stabilization intervention occurred following an adversarial war between the United States and the host country, as a proxy for consent level
23	population_size	Integer denoting the host country's average population over the during a stabilization operation
24	population_size_binary	Binary variable (0,1) denoting whether the host country's average population over the during a stabilization operation exceeded 10 million

Input Variable Descriptions and Coding Notes

Below, we describe in greater detail the coding rules, sources, and any anomalies encountered for each of the input variables defined in Table A.2:

- **Start/end year.** This discrete variable is coded as the first/last year that the number of ground troops deployed in a combat, deterrence, and/or stabilization role equaled or exceeded the threshold of 100 "person-years." In a handful of cases, the start/end years in this dataset do not therefore correspond to those in previous iterations of RUGID because of these three definitional components relating to activity types, force types, and build-up/drawdown tails. First, if ground forces ceased to pursue either combat, deterrence, or stabilization activities during a deployment, the case in the historical force requirement dataset is coded as terminating with the end of those activity types. For instance, in this dataset, the U.S. Deterrence/Training/Advisory Mission in Honduras is coded as concluding in 1992—the last year that U.S. ground troops served a Cold War deterrence function—even though U.S. ground troops in excess of 100 per year have remained in-country beyond 1992 operating in other functional capacities, such as training and advisory roles. Second, if ground forces departed a host country, but air and/or naval forces continued the mission in their absence, then the case is coded as terminated with the withdrawal of Army/Marine Corps troops. For example, the U.S. Deterrence Mission in Taiwan is coded as concluding in 1979, the last year that U.S. ground troops were deployed to the island, even though the U.S. Seventh Fleet has continued to serve a naval deterrence function since then. Finally, consistent with RUGID, in many cases U.S. ground forces may have been deployed to a country before/after the start/end year identified in this dataset, but numbered fewer than 100 troops annually; these troop build-up/drawdown "tails" are excluded from the dataset.

- **Duration.** To the degree of accuracy possible, this discrete variable measures the number of months between the start and end dates of the intervention, as defined by the satisfaction of the three components identified above (deployment size greater than 100 person-years, activity type, and troop type). Partial months are rounded up (e.g., 3 months and 12 days would be coded as 4 months). In cases in which only start/end years are known (rather than precise start/end days or months), the duration is calculated from January 1 of the start year through December 31 of the end year. Some interventions—namely long-term deterrence missions—thus risk duration overestimation by as many as 11 months. For perspective on the degree of bias this may introduce in deterrence missions, the average duration of the 21 deterrence cases in this dataset is 281 months, and the median duration is 276 months. Generally, however, a greater degree of precision is possible for measuring the duration of combat and stabilization missions. Lastly, it should be noted that for ongoing interventions, the duration has been calculated through May 2018, the month this research was completed.
- **Activity Type.** All cases have been assigned a primary, secondary, and/or tertiary activity that reflects RAND's analysis of the most important function(s) of the ground force component involved in each intervention based on a careful reading of the case. The activity type taxonomy developed by RAND Arroyo Center includes seven possible categorical values that capture the full range of major activities that ground interventions perform:[5]
    1. advisory/training of foreign internal defense forces (FID)
    2. counterinsurgency (COIN)
    3. combat/conventional warfare
    4. deterrence
    5. humanitarian assistance/disaster relief (HA/DR)

[5] Intentionally absent from this activity type taxonomy are categories for noncombatant evacuation operations and general logistics, support, and communications. Also note, in the second iteration of RUGID that included air and naval interventions, three additional intervention types were added to this taxonomy: interdiction (including blockades and no-fly zones), lift and transport, and intelligence and reconnaissance.

6. security
7. stability operations.[6]

- In some cases, only one or two activity types may have been relevant to a given case, while in others more than three could arguably apply. For instance, in the case of Vietnam (1962–1975), we have coded the primary activity type as "combat," the secondary type as "counterinsurgency," and the tertiary type as "stabilization"; one might reasonably argue, however, that the advisory/training function of deployed U.S. forces was equally or more important than one of these other activities. These coding decisions have therefore been subjected to a rigorous adjudication process involving multiple rounds of coding reviews and debates among RAND analysts and external experts and defense officials to ensure the highest possible degree of consensus and transparency.
- **Intervention Typology.** This nonexclusive, discrete variable is coded "1" if the case, at least in part, involved combat or counterinsurgency activities; "2" if the case, at least in part, involved deterrence activities; and/or "3" if the case, at least in part, involved stabilization activities. As noted above, any intervention that did not contain combat/counterinsurgency, deterrence, and/or stability operations among the intervention's activity types have been excluded from the case universe of this study.
- **Intervention Size (Minimum/Maximum/Average).** For every intervention, we have coded the high/low/average number of ground forces deployed both at the country-year level and at the aggregate case level. These data have been collected from numerous sources, such as official histories published by the U.S. Army's Center of Military History; academic journals and studies; statistical reference publications such as *The Military Balance*; military graduate theses published by the various U.S. war colleges; and other existing databases, namely the DMDC historical time-series publications on *Worldwide Manpower Distribution by Geographical Area*.

---

[6] Kavanagh et al. (2017) define each activity type as follows:

> **1 (Advisory/FID):** Interventions involving U.S. military advisors or trainers. The focus of these interventions is typically on preparing host nation personnel to operate on their own. **2 (COIN):** Interventions involving counterinsurgency activities, which, according to JP 3-24 includes "comprehensive civilian and military efforts designed to simultaneously defeat and contain insurgency and address its root causes" (pg. iii). **3 (Combat/Conventional Warfare):** Interventions involving traditional military operations and fighting, characterized by large formations of organized military forces on both sides. **4 (Deterrence):** Interventions involving activities intended to dissuade an adversary from taking an action not desired by the United States. This may also include intimidation interventions aimed at the same purpose. **5 (HA/DR):** Interventions involving humanitarian and relief operations, including responses to natural disasters and conflict. **6 (Security):** Interventions involving protection of U.S. assets or personnel during periods of threat or unrest. **7 (Stability Operations):** Interventions involving operations to stabilize or maintain peace in postconflict situations. This may include operations following coups or other situations causing unrest among the civilian population.

Of course, collecting historical data at this level of detail presents multiple coding challenges. First, in several interventions (particularly in the early 20th century), annual troop data could not be readily found for every year of the deployment; in these cases of missing values, we have necessarily had to extrapolate best estimates from known country-years.

Second, the number of U.S. boots on the ground often varied substantially within a given year as troops entered/exited the country, or our research uncovered different annual troop level estimates from different sources. As a general coding rule, when the minimum and maximum estimates were found to be relatively close, we have used the average of the two as the "best estimate" at the country-year level; however, if the minimum and maximum varied widely in a given year due to a rapid deployment or withdrawal of troops (e.g., World War II), we have used the peak number as the "best estimate" at the country-year level.

Third, in several instances, two interventions overlapped in a given country in a given year (e.g., Post–World War II Occupation of Germany, 1945–1952, and U.S. Cold War Deterrence Posture in Europe, 1946–1989). To avoid double-counting U.S. troop deployments in these country-years, we have endeavored to assign an appropriate number to each respective mission. For instance, during the concurrent stabilization and deterrence missions in West Germany between 1946–1955, we judged that prior to the intensification of the Cold War in late 1950, the majority of troops in West Germany were primarily devoted to stabilization activities; during the years 1946–1950, plausibly a single armored division may have been focused on deterrence, suggesting a 4:1 stabilization to deterrence ratio of forces. After the outbreak of the Korean War, however, force levels in West Germany subsequently increased steadily from 1951–1955; all of these incremental troop increases have therefore been assigned at the country-year level to the Cold War Deterrence in Europe case.

Fourth, as a general rule, the "average" or "typical" troop deployment size aggregated to the overall case level has been calculated by taking the simple arithmetic mean of the country-year level "best estimates." In a handful of cases, however, exceptions have been made to this methodology, namely due to wide variance in troop levels year-by-year in major combat operations (e.g., Vietnam, 1962–1975). In these cases, we have taken the mean of *monthly* troop deployments instead of *annual* best estimates.

Fifth, users of our troop data should understand that, to the greatest extent possible, we have attempted to count only ground forces deployed (i.e., U.S. Army and Marine Corps service members), though this has sometimes been problematic. To this end, DMDC's data, which break out annual force deployments by geographic location and service branch, were particularly valuable in isolating Army and Marine Corps personnel in many cases, particularly during long-term deterrence interventions.[7]

[7] For historical data files on U.S. military personnel deployed overseas by country and year, see DMDC, "DoD Personnel, Workforce Reports, & Publications," undated.

(Consequently, the ground force troop numbers contained in this dataset for some long-term deterrence missions—such as the U.S. Cold War Deterrence Posture in Morocco, 1951–1977, which involved nearly all Naval and Air Force personnel—may significantly understate the *total* number of U.S. military personnel involved.) On the other hand, for some major combat operations it was impossible within the scope of this research effort to disaggregate service members from other branches, such as Navy SEALs, who may have been deployed alongside Army and Marine Corps ground forces.

Finally, it should be noted that, for transparency's sake, all of these coding challenges and anomalies have been carefully documented in the country-year level coding notes.

- **Globalist era.** This binary dummy variable indicates whether the intervention occurred before or after the onset of the "globalist era," defined as beginning in 1940 ("0" = pre-globalist era; "1" = globalist era). In coding this variable, we encountered one anomalous case (Cold War Deterrence in Panama, 1915–1989) that spanned both the pre-globalist and globalist eras. This case has been coded pre-globalist based on its start date.
- **Heavy force ratio.** For every case in the globalist era, we have estimated the average annual ratio of "heavy" forces deployed as a percentage of the average annual number of total ground forces deployed. Building on recent research conducted within RAND Arroyo Center,[8] we adopted a typology that defined heavy ground forces as those that included armored units, armored cavalry units, mechanized units, fires (artillery) units, and combat attack aviation units. Excluded from this definition were "light" forces such as light infantry/cavalry, airborne, air mobile/air assault, and some special forces units as well as support, logistics, engineering, medical units, etc.

To calculate the overall heavy force ratio, we first conducted research to determine the type and number of heavy ground troop units deployed in each country in each year (i.e., the annual Order of Battle (ORBAT)). This analysis was conducted primarily at the battalion level. For instance, for every army infantry division deployed in conflict, we isolated and counted as heavy the organic artillery and tank battalions rather than counting the entire division as light. As one exception to this general rule, in a handful of small interventions in which total forces numbered less than a battalion equivalent, our analysis attempted to be even more granular and attempted to identify heavy units at the company level. In identifying the annual ORBATs for every case, we relied heavily on Army and Marine Corps official histories. These resources were supplemented by other secondary sources, such as the International Institute of

[8] Bryan Frederick, Stephen Watts, Matthew Lane, Abby Doll, Ashley L. Rhoades, and Meagan L. Smith, *Understanding the Deterrent Impact of U.S. Overseas Forces*, Santa Monica, Calif.: RAND Corporation, RR-2533-A, 2020.

Strategic Studies' annual publication *The Military Balance*, as well as other think tank reports and academic publications from the war colleges. In a handful of years, particularly in long-term deterrence cases, we encountered gaps in available data. In these cases, assumptions were made about deployment composition based on the most alike closest year. Notes within the dataset provide greater detail on these assumptions made to fill out missing years.

Second, after determining the type and number of heavy units deployed annually in each intervention, we calculated the number of personnel in heavy units by applying guidelines from historical tables of organization and equipment (TO&Es). Relevant TO&Es were drawn from a number of U.S. military and government publications, but two in particular were most useful: (1) John B. Wilson, *Maneuver and Firepower: The Evolution of Divisions and Separate Brigades*, Washington, D.C.: Center of Military History, U.S. Army, 1998; and (2) Congressional Budget Office, *The U.S. Military's Force Structure: A Primer*, Washington, D.C., July 2016.

Third, we calculated an average heavy force ratio for each year by dividing the estimated number of personnel in heavy units (again, primarily at the battalion level) by the estimated number of total ground troops. Finally, we determined an aggregate estimated heavy force ratio at the intervention level by calculating the average of each annual ratio.

- **War Type.** For all ground interventions involving conventional or counterinsurgent warfare, the case has been assigned a binary code indicating whether it was an instance of interstate ("0") or intrastate ("1") conflict. These coding assignments have been exported from two authoritative datasets: The Correlates of War Project's "COW War Data, 1816–2007 (v4.0)" and UCDP/PRIO's "Armed Conflict Dataset, 1946–2016 (v.17.2)."[9] It should be noted that a few cases did not fit neatly into the classic definitions of either interstate or intrastate war. For instance, the U.S. military invasions of Grenada (1983) and Panama (1989–1990) have been coded as interstate wars (or more precisely, as interstate *conflicts*), even though total battle deaths in these cases did not exceed the threshold of 1,000 that is frequently used to define instances of war in the academic literature on armed conflict. Also, two early 20th century cases that occurred within the context of the Mexican Revolution (Pershing's Expedition to capture Pancho Villa, 1916–1917, and the U.S.-Mexico Border Wars, 1918–1919) are coded as intrastate wars even though the American component of the conflict might more precisely be defined as *extrastate* war. Finally, one case in our dataset (Vietnam War, 1962–1975) is coded as an instance of both interstate and intrastate war;[10] in this case,

[9] Sarkees and Wayman, 2010; Gleditsch et al., 2002; Allansson, Melander, and Themner, 2017.

[10] We note that two other cases might also reasonably be coded as instances of both interstate and intrastate armed conflict. First, during the U.S. invasion of Afghanistan (October–December 2001), U.S. forces joined

the United States was simultaneously fighting against both state and nonstate actors (i.e., the North Vietnamese Army and the Vietcong).

- **Adversary Strength During *Interstate* Conflicts.** As a proxy measure for the relative strength of adversaries during interstate conflicts, we used each government's "Composite Index of National Capability (CINC)" score, as published in the Correlates of War Project's "National Material Capabilities (v5.0)" dataset.[11] As an index of state "power," CINC/NMC scores are based on several component variables: total population, urban population, military personnel, military expenditures, primary energy consumption, and iron and steel production.[12] The CINC score, which ranges from 0 to 1, "aggregates the six individual measured components of national material capabilities into a single value per state-year. . . . [It] reflects an average of a state's share of the system total of each element of capabilities in each year, weighting each component equally."[13] In the present forecasting demand dataset, the adversary's numeric CINC score was first coded in the start year of interstate war or armed conflict. If the United States was fighting multiple governments in the same interstate war (e.g., World War II), then the highest state CINC score has been used. Second, the numeric variable was then translated into a derivative binary variable whereby "0" indicates a relatively negligible or low interstate adversary strength (if the CINC score rounds to 0.001 or lower), and "1" indicates a relatively substantial or high interstate adversary strength (if the CINC score rounds to 0.002 or higher). There was no precise basis in either theory or empirical work for establishing a particular breakpoint or threshold between substantial and minor adversaries. The breakpoint used,

---

the Northern Alliance in their ongoing intrastate conflict to overthrow the Taliban. However, because, from the United States' perspective, this was a conflict fought against the Taliban government, we code it as only an interstate war. Second, the Korean War might also reasonably be coded as both an interstate and intrastate armed conflict. We have not done so in this dataset, however, for two reasons. First, the CoW and UCDP/PRIO datasets code this case only as an instance of interstate war. Second, while U.S. forces did provide the South Korean government military assistance—particularly, in terms of training, advising, and close air support—in fighting the leftist insurgency in the southern areas of the peninsula after the frontlines had stabilized around the 38th parallel, the majority of counterinsurgent operations were conducted by the South Korean armed forces themselves. For a more detailed account of U.S. involvement in counterinsurgency operations during the Korean War, see Mark J. Reardon, "Chasing a Chameleon: The U.S. Army Counterinsurgency Experience in Korea, 1945–1952," in Richard Davis, ed., *The U.S. Army and Irregular Warfare, 1775–2007: Selected Papers from the 2007 Conference of Army Historians*, Washington, D.C.: Center of Military History, U.S. Army, 2008, pp. 213–228.

[11] For the articles of record on COW's National Material Capabilities (NMC) dataset, see J. David Singer, Stuart Bremer, and John Stuckey, "Capability Distribution, Uncertainty, and Major Power War, 1820–1965," in Bruce Russett, ed., *Peace, War, and Numbers*, Beverly Hills, Calif.: Sage, 1972, pp. 19–48, as well as Singer, 1987.

[12] For a more detailed discussion on the coding methodology of the CINC, see the codebook: J. Michael Greig and Andrew J. Enterline, "Correlates of War Project: National Material Capabilities (NMC) Data Documentation, Version 5.0 (Period Covered: 1816–2012)," 2017.

[13] Greig and Enterline, 2017.

however, generally reflects contemporaneous estimates of the conventional capabilities of U.S. adversaries.

- **Adversary strength during *intrastate* conflicts.** As a proxy measure for the relative strength of adversaries during intrastate conflicts, we used the GDP per capita for the host country in the first year of the conflict.[14] Because of the difficulty of obtaining reliable, comparative GDP data on economic growth and income levels across long periods of time and national economies, we have utilized the "Maddison Project Database 2018" hosted by the University of Groningen, which contains real GDP per capita (in 2011 dollars) for most countries in the world dating back to the 19th century.[15] Second, for each case in which U.S. ground forces were engaged in an intrastate conflict, the host country's GDP per capita has been translated into a four-tier variable categorizing the country's income level (i.e., as a proxy for government resources and strength relative to nonstate armed groups). To do so, we have utilized the guidelines of the World Bank's 2011 economic groupings; as part of its determination of operational lending categories, the World Bank divides economies into four income groupings (low, low-middle, upper-middle, and high).[16] Finally, we have converted each country's income category into a derivative binary variable for intrastate adversary strength, whereby upper-middle and high income countries are coded as "0" (indicating that the intrastate adversary was likely relatively weak compared with the government), and low and low-middle income countries are coded as "1" (indicating that the intrastate adversary was likely relatively strong compared with the government).
- **Threat Index.** For every deterrence intervention, we have first coded a numeric value ranging from 0.00 to 6.00 indicating the "maximum threat index" value. This index is adapted from prior RAND Arroyo Center research on the drivers of deterrent interventions.[17] The index incorporates measures of whether the host is

[14] As explained in Chapter Two, this proxy measure is not entirely satisfactory. It has been used in the social science literature as a broad estimate of the capabilities of counterinsurgents (i.e., the states fighting insurgent groups). It thus provides some analytic leverage on the relative strength of insurgents (because most strong states are able to defeat or at least substantially degrade rebel forces). Unfortunately, we were not aware of any more precise measure of relative insurgent strength that could be projected 20 or more years into the future with any degree of confidence, as was required for the forecasting models in this research. Consequently, we were forced to rely on this indirect and imprecise proxy.

[15] Bolt et al., 2018.

[16] Technically, the World Bank's economic groupings are based on gross national income per capita (not GDP per capita), measured in U.S. dollars, converted from local currency using the World Bank Atlas method. The World Bank's income thresholds for these categories change annually; we have used the 2011 benchmarks because the Maddison Project's data are measured in real GDP per capita in 2011 dollars. In 2011, the World Bank's category thresholds were: (1) low: less than or equal to $1,025; (2) low-middle: $1,026–$4,035; (3) upper-middle: $4,036–$12,475; and (4) high: greater than or equal to $12,476.

[17] Kavanagh et al 2017.

a target of a higher salience territorial claim, has a history of militarized interstate disputes with its neighbors, and has neighbors that have notably higher military capabilities and/or a lack of joint democracy. Second, we translated the maximum threat index score into a derivative, binary dummy variable whereby index scores equal to or less than 4.00 were coded as "0" indicating a relatively low threat level, and scores greater than 4.00 were coded as "1" indicating a relatively high threat level. Again, there was no existing theoretical or empirical basis for establishing a particular breakpoint or threshold between higher and lower levels of threat. The threshold selected, however, generally sorts cases into what are commonly understood as higher- and lower-threat adversaries, although certainly some specific cases can be debated.

- **U.S. Defense Treaty/Ally.** For every deterrence intervention, we coded this binary dummy variable to indicate whether the country in which U.S. forces were deployed had a formal defense treaty alliance with the United States—including mutual defense treaties, nonaggression treaties, neutrality pacts, and ententes—for the majority of the duration of the intervention, i.e., for more than 50 percent of the country years ("0"= no alliance; "1" = yes alliance). The data for this variable have been extracted from the Correlates of War Project's "Formal Alliances Data Set (v4.1)," which covers the period 1816–2012.[18] Notably, in five long-term deterrence cases, formal treaty alliances did not exist at the beginning of the intervention, but they subsequently came into effect:
  - *U.S. Cold War Deterrence in Panama (1915–1989).* The first entente existed from 1936–1945, and then a formal defense treaty (and later nonaggression pact) was signed in 1945, which has remained in effect to the present.
  - *U.S. Cold War Deterrence in Europe (1946–1989).* An entente existed between the United States and both France and the United Kingdom from 1921 to 1931. Subsequently, a formal treaty and nonaggression pact did not begin until the creation of NATO in 1949.
  - *U.S. Cold War Deterrence in Taiwan (1950–1979).* A defense treaty and nonaggression pact began in 1954 and lasted until the removal of U.S. ground troops from Taiwan in 1979.
  - *U.S. Cold War Deterrence in Asia/Pacific (1946–1989).* Defense treaties with Japan and the Philippines were signed in 1951 and have remained in effect to the present.
  - *U.S. Cold War Deterrence in Iran (1953–1978).* An entente existed between the United States and Iran beginning in 1959 and lasted through the overthrow of the Shah in 1979.

[18] Gibler, 2009.

Lastly, it should be noted that the CoW Alliance Data Set (v4.1) incorrectly omits the Baltic States as NATO allies following their accession to the alliance in 2004. As an exception to the rules outlined above, Operation Atlantic Resolve/Deterrence of Russian Aggression in Europe (2014–present), is thus coded "1" as for this variable.

- **U.S. adversarial war in year prior.** For every stabilization mission, we have coded this binary dummy variable ("0" = no, "1" = yes) to indicate whether the United States engaged in an adversarial war with the host nation at some point in the 12 months prior to the beginning of stabilization operations. It should be noted that in two cases (Bosnia, 1995 and Kosovo, 1999), we have coded these stabilization interventions as "1" indicating an adversarial war in the year prior, even though the war involved air strikes, not ground forces.
- **Population size.** For every stabilization mission, we have first coded the average annual population of the host nation. These data are drawn from the "Maddison Project Database 2018" hosted by the University of Groningen, which contains annual population data for most countries in the world dating back to the 19th century.[19] We have then translated these raw numbers into a binary code whereby 0 = a relatively small population, and 1 = a relatively large population. We have used an annual average population of 10 million as the breakpoint between these categories. This threshold distinguishes between cases that have been frequently discussed as relatively less demanding due to their small population sizes (e.g., Bosnia and Kosovo) and those that are much more demanding (e.g., Afghanistan and Iraq).[20]

Coding Results of the Historical Force Requirement Dataset

Tables A.3–A.5 contain the detailed coding results of the historical force requirement dataset. These tables are presented at the individual case level and are organized according to intervention type: combat missions (Table A.3), deterrence missions (Table A.4), and stabilization missions (Table A.5).

---

[19] Bolt et al., 2018.

[20] Using the force-to-population ratio of 20 forces per 1,000 inhabitants proposed by James Quinlivan and formalized for several years in U.S. Army doctrine, a country with a population of 10 million would require 200,000 forces to stabilize—a number higher than any U.S. stabilization mission other than the post–World War II occupation of Germany and the Vietnam War. See Quinlivan, 1995/96.

Table A.3
Historical Force Requirement Coding Results: Combat Interventions

	Historical Requirements			Operating Environment				
						Adversary Strength		
Intervention	Ground Troop Size (average)	Duration (months)	Heavy Force Ratio (est.)	Geostrategic Era	Conflict Type	NMC (Interstate Proxy)	GDP per capita (Intrastate Proxy)	Adversary Type
Spanish-American War (1898)	43,500	8	n/a	Pre-globalist	Interstate	0.017	n/a	Major
Boxer Rebellion (1900)	2,350	5	n/a	Pre-globalist	Interstate	0.120	n/a	Major
U.S. Invasion of Mexico: Pershing's Expedition (1916–1917)	11,000	11	n/a	Pre-globalist	Intrastate	n/a	$1,607	Major
World War I (1917–1918)	624,000	19	n/a	Pre-globalist	Interstate	0.158	n/a	Major
U.S.-Mexican Border Wars (II) (1918–1919)	19,000	24	n/a	Pre-globalist	Intrastate	n/a	$1,741	Major
American Expeditionary Forces in Vladivostok and Archangel (1918–1920)	9,500	21	n/a	Pre-globalist	Intrastate	n/a	$1,327	Major
World War II (Asian/Pacific Theater) (1941–1945)	1,055,000	44	12%	Globalist	Interstate	0.067	n/a	Major
World War II (European/ Mediterranean/N. African Theater) (1941–1945)	1,053,000	40	20%	Globalist	Interstate	0.176	n/a	Major
Korean War (1950–1953)	243,000	37	11%	Globalist	Interstate	0.118	n/a	Major
Vietnam War (1962–1975)	253,000	160	9%	Globalist	Interstate, Intrastate	0.004	$1,197	Major
U.S. Invasion of Grenada (1983)	8,000	2	12%	Globalist	Interstate	0.00001	n/a	Minor
U.S. Invasion of Panama (1989–1990)	27,000	2	12%	Globalist	Interstate	0.0003	n/a	Minor

Table A.3—continued

	Historical Requirements			Operating Environment				
						Adversary Strength		
Intervention	Ground Troop Size (average)	Duration (months)	Heavy Force Ratio (est.)	Geostrategic Era	Conflict Type	NMC (Interstate Proxy)	GDP per capita (Intrastate Proxy)	Adversary Type
Persian Gulf War (1991)	343,000	2	47%	Globalist	Interstate	0.008	n/a	Major
U.S. Invasion of Afghanistan (2001)	1,900	3	0%	Globalist	Interstate	0.001	$692	Minor
U.S. Occupation of Afghanistan (2001–)	34,000	204*	12%	Globalist	Intrastate	n/a	$692	Major
U.S. Invasion of Iraq (Operation Iraqi Freedom) (2003)	150,000	2	20%	Globalist	Interstate	0.007	n/a	Major
U.S. Occupation of Iraq (2003–2011)	124,000	103	22%	Globalist	Intrastate	n/a	$2,898	Major
CJTF Against Islamic State (2014–)	3,000	48*	11%	Globalist	Intrastate	n/a	$12,889	Minor

NOTES: (1) Heavy force ratios could not be reliably calculated for pre-1940 interventions due to lack of unit type data. (2) *Denotes ongoing interventions; duration calculated through May 2018.

Table A.4
Historical Force Requirement Coding Results: Deterrence Interventions

	Historical Requirements			Operating Environment				
					Level of Threat		Level of Commitment	
Intervention	Ground Troop Size (average)	Duration (months)	Heavy Force Ratio (est.)	Geostrategic Era	Threat Index Score	Threat Level	Defense Treaty Ally	Commitment
U.S. Cold War Deterrence Posture in the Panama Canal Zone (1915–1989)	12,000	892	n/a	Pre-globalist	2.00	Lower	Yes	Higher
U.S. Cold War Deterrence Posture in Saudi Arabia (1945–1989)	150	540	0%	Globalist	2.23	Lower	No	Lower
U.S. Occupation of South Korea (Post–World War II) (1945–1949)	42,000	46	13%	Globalist	4.00	Lower	No	Lower
U.S. Cold War Deterrence Posture in Europe (1946–1989)	194,000	528	37%	Globalist	5.95	Higher	Yes	Higher
U.S. Cold War Deterrence Posture in Libya (1948–1970)	200	276	0%	Globalist	2.00	Lower	No	Lower
Berlin Airlift (Operation Vittles) (1948–1949)	n/a	15	n/a	Globalist	4.00	Lower	No	Lower
U.S. Cold War Deterrence Posture in Ethiopia/Eritrea (1950–1976)	1,000	276	0%	Globalist	3.75	Lower	No	Lower
U.S. Cold War Deterrence Posture in Taiwan (1950–1979)	1,020	360	0%	Globalist	6.00	Higher	Yes	Higher
U.S. Cold War Deterrence Posture in Morocco (1951–1977)	300	324	0%	Globalist	3.48	Lower	No	Lower
U.S. Cold War Deterrence Posture in Asia/Pacific (1946–1989)	44,000	528	12%	Globalist	4.82	Higher	Yes	Higher
U.S. Cold War Deterrence Posture in Iran (1953–1978)	400	300	0%	Globalist	5.69	Higher	Yes	Higher

Table A.4—continued

	Historical Requirements			Operating Environment				
				Level of Threat			Level of Commitment	
Intervention	Ground Troop Size (average)	Duration (months)	Heavy Force Ratio (est.)	Geostrategic Era	Threat Index Score	Threat Level	Defense Treaty Ally	Commitment
U.S. Cold War/Post–Cold War Deterrence Posture in South Korea (1953–)	34,500	780*	25%	Globalist	6.00	Higher	Yes	Higher
U.S. Show of Force During Laos Crisis (1962)	4,250	3	0%	Globalist	4.00	Lower	No	Lower
Multinational Force and Observers in Sinai (MFO) (1982–)	900	432*	0%	Globalist	3.79	Lower	No	Lower
U.S. Show of Force/Cold War Deterrence Posture in Honduras (1983–1992)	1,100	108	0%	Globalist	1.56	Lower	Yes	Higher
U.S. Peacekeeping Force in Kuwait (1991)	4,300	10	0%	Globalist	6.00	Higher	No	Lower
U.S. Deterrence Posture in Turkey/ Northern Iraq (1991–2003)	400	143	0%	Globalist	6.00	Higher	Yes	Higher
U.S. Deterrence Force in the Persian Gulf (1992–2003)	2,600	127	0%	Globalist	5.50	Higher	No	Lower
Multinational Deterrence Force in Macedonia (1993–1999)	450	78	0%	Globalist	3.71	Lower	No	Lower
U.S. Deterrence of China in Japan (2010–)	19,200	96*	22%	Globalist	3.71	Lower	Yes	Higher
U.S. Deterrence of Russia in Europe (2014–)	34,000	48*	44%	Globalist	4.00	Lower	Yes	Higher

NOTES: (1) Heavy force ratios could not be reliably calculated for pre-1940 interventions due to lack of unit type data. (2) *Denotes ongoing interventions; duration calculated through May 2018. (3) Reliable troop size data could not be obtained for ground forces assigned to Berlin Airlift/ Operation Vittles as opposed to those assigned to overlapping post–World War II stabilization and deterrence operations in Germany.

Table A.5
Historical Force Requirement Coding Results: Stabilization Interventions

				Operating Environment				
	Historical Requirements				Level of Consent		Population Size	
Intervention	Ground Troop Size (average)	Duration (months)	Heavy Force Ratio (est.)	Geostrategic Era	U.S. Adversarial War Year Prior	Consent Coding	Average Population (thousands)	Population Coding
Boxer Rebellion (1900)	2,350	5	n/a	Pre-globalist	No	More	400,000	Larger
U.S. Occupation of Panama (1903–1915)	500	135	n/a	Pre-globalist	No	More	323	Smaller
Cuban Pacification Intervention (1906–1909)	4,600	26	n/a	Pre-globalist	No	More	2,065	Smaller
U.S. Peacekeeping Force in Cuba (1912)	8,000	2	n/a	Pre-globalist	No	More	2,358	Smaller
Marines Landing During Nicaraguan Revolution (1912–1925)	300	168	n/a	Pre-globalist	No	More	621	Smaller
U.S. Occupation of Veracruz (1914)	4,700	7	n/a	Pre-globalist	No	More	14,960	Larger
U.S. Occupation of Haiti (1915–1934)	1,500	229	n/a	Pre-globalist	No	More	2,101	Smaller
U.S. Occupation of Dominican Republic (1916–1924)	2,300	100	n/a	Pre-globalist	No	More	N/A	Smaller
Allied Occupation of the Rhineland Post–World War I (1918–1923)	67,000	49	n/a	Pre-globalist	Yes	Less	62,076	Larger
U.S. Occupation of Nicaragua (1926–1933)	4,250	80	n/a	Pre-globalist	No	More	681	Smaller
"China Marines" Deployment (1927–1941)	3,750	177	n/a	Pre-globalist	No	More	502,053	Larger
U.S. Occupation of Germany (Post–World War II) (1945–1955)	247,000	120	18%	Globalist	Yes	Less	68,097	Larger

Table A.5—continued

Intervention	Historical Requirements				Operating Environment			
					Level of Consent		Population Size	
	Ground Troop Size (average)	Duration (months)	Heavy Force Ratio (est.)	Geostrategic Era	U.S. Adversarial War Year Prior	Consent Coding	Average Population (thousands)	Population Coding
Allied Occupation of Austria (Post–World War II) (1945–1955)	17,400	125	4%	Globalist	Yes	Less	6,935	Smaller
U.S. Occupation of Japan (Post–World War II) (1945–1952)	144,000	80	6%	Globalist	Yes	Less	81,137	Larger
U.S. Occupation of South Korea (Post–World War II I) (1945–1949)	42,000	46	13%	Globalist	No	More	19,481	Larger
Northern China Marines (Post–World War II) (1945–1949)	14,000	44	6%	Globalist	No	More	538,259	Larger
U.S. Military Government in S. Korea (1953–1957)	66,000	52	10%	Globalist	No	More	21,703	Larger
Lebanon Crisis of 1958 (1958)	14,000	3	13%	Globalist	No	More	1,692	Smaller
Vietnam War (1962–1975)	253,000	160	9%	Globalist	Yes	Less	40,967	Larger
U.S. Occupation of Dominican Republic (1965–1966)	23,000	17	5%	Globalist	No	More	3,866	Smaller
Lebanese Civil War (1982–1984)	1,800	20	24%	Globalist	No	More	3,065	Smaller
U.S. Invasion of Grenada (1983)	8,000	2	12%	Globalist	Yes	Less	96	Smaller
U.S. Invasion of Panama (1989–1990)	27,000	2	12%	Globalist	Yes	Less	2,370	Smaller
U.S. Peacekeeping Force in Panama (1990–1994)	7,600	56	0%	Globalist	Yes	Less	2,490	Smaller

Table A.5—continued

					Operating Environment			
	Historical Requirements				Level of Consent		Population Size	
Intervention	Ground Troop Size (average)	Duration (months)	Heavy Force Ratio (est.)	Geostrategic Era	U.S. Adversarial War Year Prior	Consent Coding	Average Population (thousands)	Population Coding
U.S. Peacekeeping Force in Kuwait (1991)	4,300	10	0%	Globalist	No	More	2,088	Smaller
Multinational Peacekeeping Force in Bosnia (1995–2008)	4,100	156	28%	Globalist	Yes	Less	3,865	Smaller
Multinational Peacekeeping Force in Somalia (1992–1995)	10,500	27	8%	Globalist	No	More	7,574	Smaller
U.S. Peacekeeping Force in Haiti (1994–1996)	10,100	18	12%	Globalist	No	More	7,464	Smaller
Multinational Peacekeeping Force in Kosovo (1999–)	2,100	228*	22%	Globalist	Yes	Less	1,754	Smaller
U.S. Occupation of Afghanistan (2001–)	34,000	204*	12%	Globalist	Yes	Less	28,250	Larger
Joint Task Force Horn of Africa (2002–)	2,000	192*	0%	Globalist	No	More	739	Smaller
U.S. Occupation of Iraq (2003–2011)	124,000	103	22%	Globalist	Yes	Less	29,625	Larger
U.S. Peacekeeping Force in Liberia (2003)	250	3	0%	Globalist	No	More	2,983	Smaller
Multinational Peacekeeping Force in Haiti (2004)	1,900	3	0%	Globalist	No	More	8,900	Smaller
CJTF Against the Islamic State (2014–)	5,000	48*	11%	Globalist	No	More	56,883	Larger

NOTES: (1) Heavy force ratios could not be reliably calculated for pre-1940 interventions due to lack of unit type data. (2) *Denotes ongoing interventions; duration calculated through May 2018.

APPENDIX B

Forecasting Model Assumptions, Components, and Processes

This appendix serves to provide greater detail concerning the forecasting model detailed throughout this report. First, this appendix provides details about the data sources used to operationalize the variables in our statistical models of armed conflict and U.S. ground interventions, as well as the full results of the statistical, logit models summarized previously in Chapter Three. The tables in this appendix are presented in the same order as the discussion in Chapter Three.

This appendix then provides a more detailed description of the inner workings of our forecasting model. It describes in detail the processes that our model uses to forecast trends in armed conflict and interventions in the 2017–2040 period. It also describes the assumptions and technical modeling decisions that we made when building the forecasting model.

Operationalizing Statistical Models of Intrastate Armed Conflict

Table B.1 and its associated notes provide details on the operationalization and the data sources used for the variables employed in the intrastate conflict models, previously discussed in Chapter Three.

Table B.1
Key Factor Concepts and Metrics Affecting Intrastate Armed Conflict Onset and Cessation

Key Factor Name	Key Factor Metric	Intrastate Conflict Onset Model	Intrastate Conflict Cessation Model
Economic development	The state's GDP per capita (per 1,000 people)[a,b]	X	X
Political representation	Whether the state is an anocracy[a,c]	X	X
	Whether the state has experienced a significant regime transition in the prior five years[a,d]	X	
Ethnic discrimination	Whether the state political apparatus discriminates against a significant portion of the population[a,e]	X	X
Societal opportunity	Whether the percentage of the population between the ages of 15 and 29 exceeds 45% of the total population[a,f]	X	X
	The state's population size[a,g]	X	X
Ongoing and recent intrastate conflicts	The number of ongoing intrastate armed conflicts among each state's regional neighbors[a,h]	X	
	The number of previous intrastate armed conflicts experienced by the state[a,i]	X	
Geostrategic environment	Whether the Cold War is ongoing	X	X
Regional and temporal interdependencies	Whether the state is in the Middle East, Eurasia, or East Asia	X	X
	The number of years since the previous intrastate armed conflict in the state	X	
	The number of years that an active intrastate armed conflict in the state remains ongoing		X

[a] Denotes variables that we lagged by one year to better measure the directionality of effects (e.g., the effect of GDP per capita on conflict) and to account for any adverse effects that may be caused by intrastate armed conflicts on our key factors.

[b] International Futures Database, version 7.31, undated; The Maddison Project (Jutta Bolt and Jan Luiten van Zanden, "The Maddison Project: Collaborative Research on Historical National Accounts," *Economic History Review*, Vol. 67, No. 3, 2014, pp. 627–651); The World Bank, "World Development Indicators," 2012.

[c] In keeping with commonly used standards in the academic literature, we identified anocracies as states possessing a Polity score of between –6 and +6 on the scale that ranges from –10 to +10 (Marshall, Gurr, and Jaggers, 2017).

[d] We identified a significant regime transition as a change in the state's Polity score of +/– 3 or more from the previous year (Marshall, Gurr, and Jaggers, 2017).

Table B.1—continued

[e] We identified highly discriminatory states as those states that discriminate against at least 20% of their populace in the Ethnic Power Relations dataset, or the top 10% of discriminatory states (Cederman, Wimmer, and Min, 2010; Wimmer, Cederman, and Min, 2009.)

[f] This metric represents a youth bulge in the state. Demographic data used to construct this measure from the International Futures Database, version 7.31, undated.

[g] International Futures Database, version 7.31, undated.

[h] Specifically, we use a weighted average spatial lag to measure the level of intrastate armed conflict around each state, which increases as more proximate states experience intrastate conflicts. Data on intrastate conflicts used to construct this measure from the UCDP/PRIO Armed Conflict Database (Gleditsch et al., 2002; Allansson, Melander, and Themner, 2017).

[i] Data from the UCDP/PRIO Armed Conflict Database (Gleditsch et al., 2002; Allansson, Melander, and Themner, 2017).

Results of Statistical Models of Intrastate Armed Conflict

Table B.2
Effects of Key Factors on Intrastate Armed Conflict Onset and Cessation

Key Factor Metric	Effect on Intrastate Armed Conflict Onsets	Effect on Intrastate Armed Conflict Cessation
GDP per capita (per 1,000 people)	−0.042*** (0.014)	0.038** (0.017)
Anocracy	0.584*** (0.167)	−0.264 (0.202)
Significant regime transition	0.277* (0.173)	
Ethnic discrimination	0.363** (0.169)	−0.318* (0.187)
Youth bulge	0.405** (0.176)	−0.402* (0.235)
State population size	0.241*** (0.070)	−0.413*** (0.074)
Neighborhood/regional intrastate conflicts	0.699** (0.315)	
Number of previous intrastate conflicts	0.252*** (0.065)	
Cold War	0.326* (0.194)	−0.616*** (0.209)
Middle East	0.515*** (0.194)	−0.470* (0.289)
Eurasia	0.702** (0.326)	−0.591 (0.377)
East/Southeast Asia	−0.592* (0.344)	
Number of years since last intrastate armed conflict	−0.068** (0.035)	
Number of years of ongoing intrastate armed conflict		−0.162*** (0.051)
Number of observations	6,235	1,330
Model Pseudo R^2	0.1524	0.1495

NOTE: Robust standard errors clustered by country in parentheses. *** $p < 0.01$, ** $p < 0.05$, * $p < 0.1$. Model of intrastate armed conflict onset also includes cubic polynomials of time since last intrastate armed conflict (not shown). Model of intrastate armed conflict cessation also includes cubic polynomials of years of ongoing intrastate armed conflict (not shown).

Figure B.1
Intrastate Armed Conflict Onset Statistical Model Performance

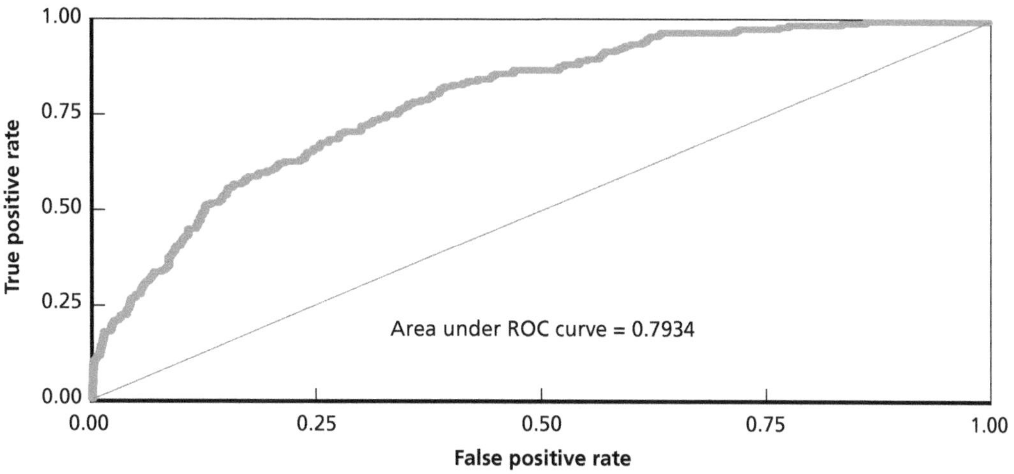

NOTE: Figure shows the receiver operating characteristic (ROC) curve and related area under the curve (AUC) associated with our statistical model of intrastate armed conflict onset. The ROC curve shows the ability of the statistical model to accurately identify actual intrastate armed conflicts in the data (the true positive rate) while limiting the number of observations incorrectly identified as intrastate armed conflicts (the false positive rate). The diagonal line of Figure B.1 delimits a model that correctly identifies intrastate armed conflict onsets no better than random chance. Higher AUC scores indicate statistical models that perform increasingly better than random chance in classifying the data.

Figure B.2
Intrastate Armed Conflict Cessation Statistical Model Performance

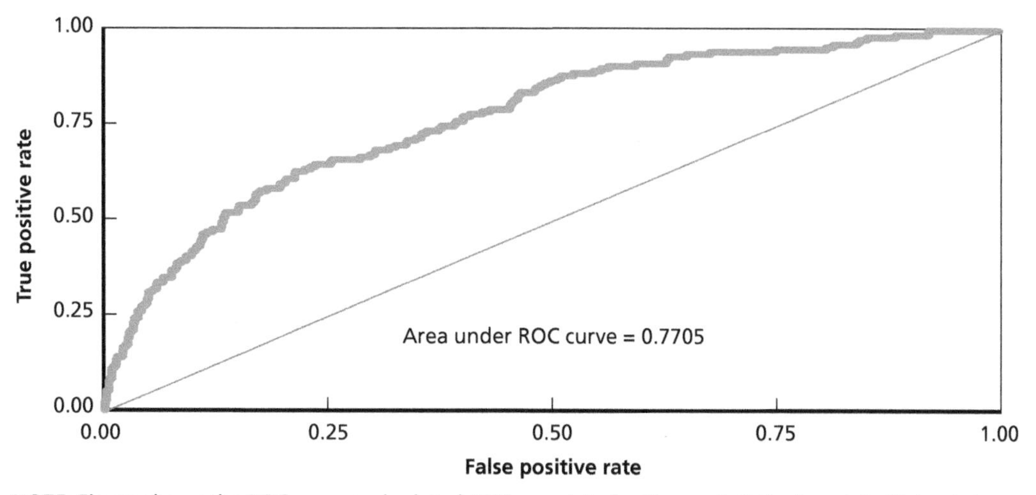

NOTE: Figure shows the ROC curve and related AUC associated with our statistical model of intrastate armed conflict cessation. The ROC curve shows the ability of the statistical model to accurately identify actual intrastate armed conflict cessations in the data (the true positive rate) while limiting the number of observations incorrectly identified as intrastate armed conflict cessations (the false positive rate). The diagonal line of Figure B.2 delimits a model that correctly identifies intrastate armed conflict cessations no better than random chance. Higher AUC scores indicate statistical models that perform increasingly better than random chance in classifying the data.

Operationalizing Statistical Models of Interstate War

Table B.3 and its associated notes provide details on the operationalization and the data sources used for the variables employed in the interstate war models, previously discussed in Chapter Three.

Table B.3
Key Factor Concepts and Metrics Affecting Interstate War Onset and Cessation

Key Factor Name	Key Factor Metric	Interstate War Onset Model	Interstate War Cessation Model
Degree of regional hegemony	Ratio of capabilities between first and second-most powerful states in each region[b]	X	X
	Power transition—whether the regional capabilities ratio crossed a 2:1 threshold in the previous five years[b]	X	X
	Number of U.S. heavy ground forces forward deployed in each region[a,c]	X	X
Balance of military capabilities	Ratio of military capabilities between both states in a dyad[d]	X	X
	Whether both states in a dyad fall under a nuclear umbrella[e]	X	
Territorial contestation	Whether the states in a dyad contest a territorial claim of medium or high salience[f]	X	X
	Whether the states in a dyad are contiguous by a land border	X	X
Economic interdependence	The minimum ratio of bilateral trade to GDP in the dyad[a,g]	X	X
	Whether both states in a dyad belong to the same or different trading blocs[h]	X	
Political congruence	Whether both states in a dyad are established democracies[i]	X	
Strength of international norms	Percentage of states in each region that have ratified multiple multilateral treaties requiring the pacific settlement of international disputes[j]	X	X
Temporal interdependencies	The number of years since the previous interstate armed conflict in the state	X	
	The number of years that an active interstate armed conflict in the state remains ongoing		X

[a] Denotes variables that we lagged by one year to better measure the directionality of effects (e.g., the effect of GDP per capita on conflict) and to account for any adverse effects that may be caused by intrastate armed conflicts on our key factors. Watts et al., 2017a.

[b] Watts et al., 2017a.

Table B.3—continued

[c] Specifically, we use a weighted average spatial lag to measure the number of U.S. heavy forces around each state, weighted by the number of U.S. forces in the region and the relative distance military forces can travel in a day, which increases as the number of U.S. forces around each state increases. To ensure that U.S. troops in nearby states never have a larger effect than troops in a given country, we add a value of 200 miles, or the largest assumed distance that can be traveled by U.S. forces in a given day, to our proximity calculations. Data on U.S. forward presence from the Defense Manpower Data Center. Data on U.S. forward presence troop types from RUGID (Kavanagh et al., 2017). Data on distance between states from the cShapes Dataset. Defense Manpower Data Center, *Historical Report—Military Only (aggregated data 1950–current),* Alexandria, Va., 2016; Kavanagh et al., 2017; Nils B. Weidmann, Doreen Kuse, and Kristian Skrede Gleditsch, "The Geography of the International System: the cShapes Dataset," *International Interactions*, Vol. 36, No. 1, 2010; O'Mahony et al., 2017.

[d] Singer, 1987.

[e] Data on nuclear possession from Erik Gartzke and Matthew Kroenig, "A Strategic Approach to Nuclear Proliferation," *Journal of Conflict Resolution,* Vol. 53, No. 2, 2009. Data on which states fall under a nuclear umbrella provided by another state from International Law and Policy Institute, "The Nuclear Umbrella States," ILPI Nuclear Weapons Project Nutshell Paper No. 5, 2012.

[f] The threshold for identifying a medium- or high-salience territorial claim is a score of 6 or higher on the 12-point scale used in the Issue Correlates of War Territorial Claims Dataset. Bryan A. Frederick, Paul R. Hensel, and Christopher Macaulay, "The Issue Correlates of War Territorial Claims Data, 1816–2001," *Journal of Peace Research,* Vol. 54, No. 1, 2017.

[g] Bilateral trade data from Katherine Barbieri and Omar Keshk, *Correlates of War Project Trade Data Set Codebook*, Version 3.0, 2012. GDP data from International Futures Database, version 7.31, undated; The Maddison Project (Bolt and van Zanden, 2014).

[h] Watts et al., 2017a.

[i] Both states in a dyad possess a Polity score of between –6 and +6 on the scale that ranges from –10 to +10 (Marshall, Gurr, and Jaggers, 2017).

[j] Paul R. Hensel, Multilateral Treaties of Pacific Settlement (MTOPS) Data Set, Version 1.4, 2005.

Results of Statistical Models of Interstate War

Table B.4
Effects of Key Factors on Interstate War Onset and Cessation

Key Factor Metric	Effect on Interstate War Onsets	Effect on Interstate War Cessation
Regional hegemony ratio	−0.201*** (0.082)	1.351* (0.732)
Power transition	0.905*** (0.213)	1.435 (0.991)
U.S. heavy forces forward presence	−0.236*** (0.059)	0.062 (0.189)
Dyadic balance of capabilities	−0.655 (0.746)	5.467 (3.447)
Nuclear umbrella	−1.187* (0.713)	
Medium or high-salience territorial claim	1.607*** (0.229)	0.546 (0.939)
Land border	−0.228 (0.241)	1.917* (1.159)
Bilateral trade/GDP in dyad	−1.490 (1.284)	−1.605 (2.566)
States in different trading blocs	1.198*** (0.344)	
Dyadic democracy	−2.364*** (0.586)	
Prevalence of regional norms	−0.655** (0.316)	2.200 (2.340)
Number of years since last interstate armed conflict	0.002 (0.016)	
Number of years of ongoing interstate armed conflict		0.689 (1.253)
Number of observations	43,313	72
Model Pseudo R^2	0.2132	0.2644

NOTE: Robust standard errors clustered by dyad in parentheses. *** $p < 0.01$, ** $p < 0.05$, * $p < 0.1$. Model of interstate war onset also includes cubic polynomials of time since last interstate armed conflict (not shown). Model of interstate war cessation also includes cubic polynomials of years of ongoing interstate armed conflict (not shown).

Figure B.3
Interstate War Onset Statistical Model Performance

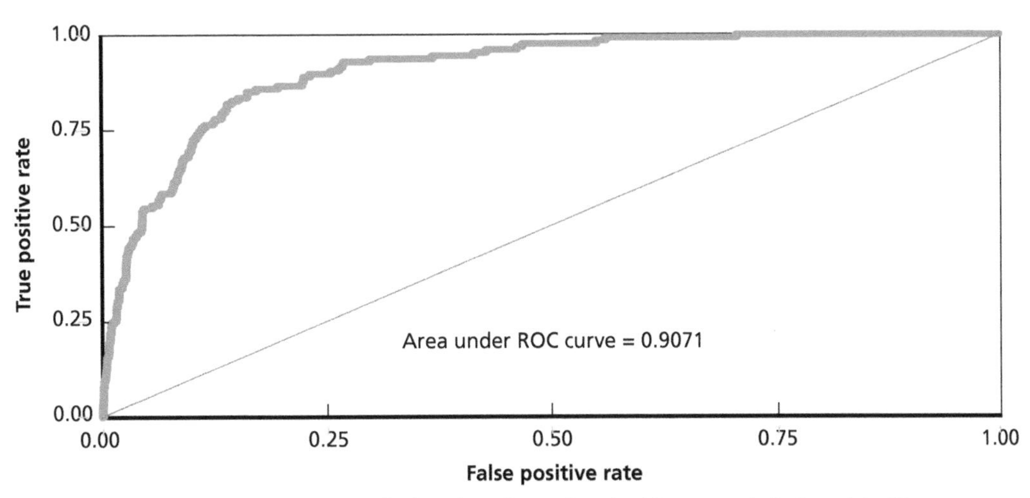

NOTE: Figure shows the ROC curve and related AUC associated with our statistical model of interstate war onset. The ROC curve shows the ability of the statistical model to accurately identify actual interstate war onsets in the data (the true positive rate) while limiting the number of observations incorrectly identified as interstate wars (the false positive rate). The diagonal line of Figure B.3 delimits a model that correctly identifies interstate war onsets no better than random chance. Higher AUC scores indicate statistical models that perform increasingly better than random chance in classifying the data.

Figure B.4
Interstate War Cessation Statistical Model Performance

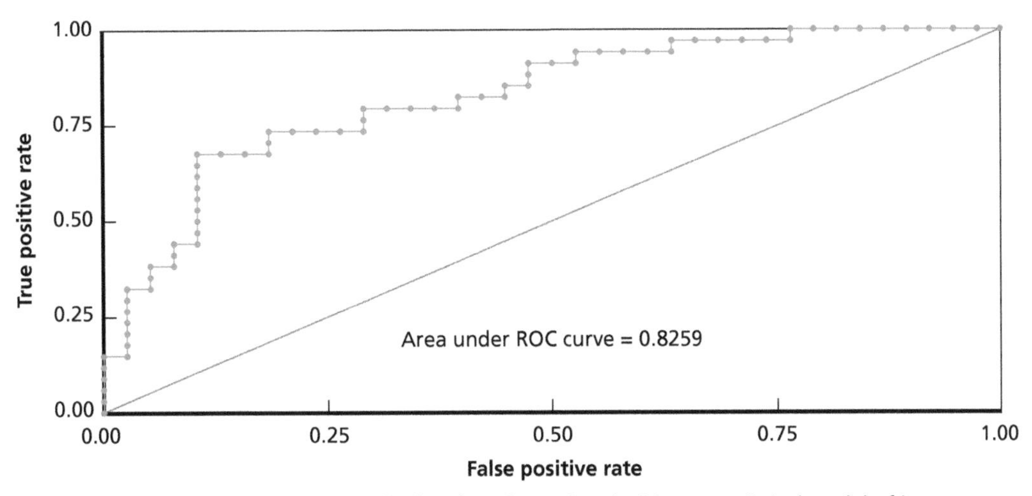

NOTE: Figure shows the ROC curve and related AUC associated with our statistical model of interstate war cessation. The ROC curve shows the ability of the statistical model to accurately identify actual interstate war cessations in the data (the true positive rate) while limiting the number of observations incorrectly identified as interstate war cessations (the false positive rate). The diagonal line of Figure B.4 delimits a model that correctly identifies interstate war cessations no better than random chance. Higher AUC scores indicate statistical models that perform increasingly better than random chance in classifying the data.

Operationalizing Statistical Models of U.S. Ground Interventions

Table B.5 and its associated notes provide details on the operationalization and the data sources used for the variables employed in the U.S. ground interventions models, previously discussed in Chapter Three.

Table B.5
Key Factor Concepts and Metrics Affecting U.S. Ground Intervention Onset and Cessation

Key Factor Name	Key Factor Metric	Deterrence Intervention Onset	Deterrence Intervention Cessation	Armed Conflict Intervention Onset	Stabilization Intervention Onset	Stabilization Intervention Cessation
U.S. economic outlook	U.S. GDP growth[a,b]	X	X			
U.S. military capabilities	U.S. aggregate military capabilities[a,c]	X	X			
	Number of ongoing U.S. ground interventions[a,d]			X		
Partner-state economic and strategic resources	Partner state GDP per capita (per 1,000 people)[a,e]	X	X	X	X	X
	Partner state oil production[a,f]	X	X	X		
	Partner state population size[a,g]				X	
Partner-state political system	Whether the state is an anocracy[a,h]				X	
	Level of partner state democracy[a,i]			X		
Partner states under threat	Whether the partner state is the target of a high-value territorial claim by an adversary state[a,j]	X	X			
U.S.–partner state strategic relationship	U.S. alliance with partner state[a,k]	X	X		X	X
	Distance between a partner state and the United States[a,l]				X	
	Whether the United States was involved in a prior armed conflict intervention in the state[m]				X	

Table B.5—continued

Key Factor Name	Key Factor Metric	Deterrence Intervention Onset	Deterrence Intervention Cessation	Armed Conflict Intervention Onset	Stabilization Intervention Onset	Stabilization Intervention Cessation
Geostrategic environment	The number of ongoing armed conflicts among each state's regional neighbors[n]			X		
Regional and temporal interdependencies	Whether the state is in Europe	X	X			
	Whether the state is in sub-Saharan Africa				X	
	The number of years since the previous intervention (of each type) in the state	X		X	X	
	The number of years that an intervention in the state remains ongoing		X			

[a] Denotes variables that we lagged by one year to better measure the directionality of effects (e.g. the effect of GDP per capita on conflict) and to account for any adverse effects that may be caused by intrastate armed conflicts on our key factors.

[b] International Futures Database, version 7.31, undated; The Maddison Project (Bolt and van Zanden, 2014; World Bank, 2012).

[c] Singer, 1987; International Futures Database, version 7.31, undated.

[d] Data from RUGID (Kavanagh et al., 2017).

[e] International Futures Database, version 7.31, undated; The Maddison Project (Bolt and van Zanden, 2014); World Bank, 2012.

[f] International Futures Database, version 7.31, undated.

[g] International Futures Database, version 7.31, undated.

[h] In keeping with commonly used standards in the academic literature, we identified anocracies as states possessing a POLITY score of between –6 and +6 on the scale that ranges from –10 to +10 (Marshall, Gurr, and Jaggers, 2017).

[i] Marshall, Gurr, and Jaggers, 2017.

[j] Frederick, Hensel, and Macaulay, 2017.

[k] Douglas M. Gibler, Correlates of War Formal Alliances Dataset (v4.1), 2014.

[l] RAND calculations using the cShapes dataset (Weidmann, Kuse, and Gleditsch, 2010).

[m] Data from RIGID (Kavanagh et al., 2017).

[n] Specifically, we use a weighted average spatial lag to measure the level of armed conflict around each state, which increases as more proximate states experience conflicts. Data on conflicts used to construct this measure from the UCDP/PRIO Armed Conflict Database and the Correlates of War Interstate War Dataset (Sarkees and Wayman, 2010; Gleditsch et al., 2002; Allansson, Melander, and Themner, 2017).

Results of Statistical Models of U.S. Ground Interventions

Table B.6
Effects of Key Factors on U.S. Ground Intervention Onset and Cessation

Key Factor Metric	Deterrence Intervention Onset	Deterrence Intervention Cessation	Armed Conflict Intervention Onset	Stabilization Intervention Onset	Stabilization Intervention Cessation
U.S. GDP growth	–16.288* (6.787)	18.562*** (4.938)			
U.S. aggregate military capabilities	13.997*** (5.083)	1.406 (4.576)			
Number of ongoing U.S. ground interventions			–0.187 (0.152)		
Partner state GDP per capita (per 1,000 people)	0.001 (0.005)	0.025 (0.201)	0.147** (0.070)	–0.229*** (0.090)	0.483 (0.438)
Partner state oil production	0.041* (0.022)	–0.017 (0.034)	–0.043 (0.060)		
Partner state population size				–1.204*** (0.482)	
Whether the state is an anocracy				0.899 (0.660)	
Level of partner state democracy			–0.256** (0.102)		
Whether the partner state is the target of a high-value territorial claim by an adversary state	0.063 (0.277)	–0.414 (0.422)			
U.S. alliance with partner state	2.678*** (0.606)	–0.211 (0.635)		0.960 (1.182)	4.619* (2.709)
Distance between a partner state and the United States				–0.813*** (0.204)	
Whether the United States was involved in a prior armed conflict intervention in the state				4.681*** (0.878)	-5.364** (2.279)
The number of ongoing armed conflicts among each state's regional neighbors			0.292 (0.242)		
Whether the state is in Europe	1.888*** (0.391)	0.015 (0.559)			
Whether the state is in sub-Saharan Africa				–0.903 (1.021)	
The number of years since the previous intervention (of each type) in the state	–0.729*** (0.223)		–0.144 (0.272)		

Table B.6—continued

Key Factor Metric	Deterrence Intervention Onset	Deterrence Intervention Cessation	Armed Conflict Intervention Onset	Stabilization Intervention Onset	Stabilization Intervention Cessation
The number of years that an intervention in the state remains ongoing		−0.183 (0.154)			
Number of Observations	8,138	917	1,264	1,218	36
Pseudo R^2	0.3843	0.1133	0.2114	0.3418	0.3269

NOTE: Robust standard errors clustered by state in parentheses. *** $p < 0.01$, ** $p < 0.05$, * $p < 0.1$. Models of deterrence intervention onset and armed conflict intervention onset also include cubic polynomials of time since last intervention (not shown). Models of deterrence intervention cessation and armed conflict intervention cessation also include cubic polynomials of years of ongoing intervention (not shown).

Figure B.5
Deterrence Intervention Onset Statistical Model Performance

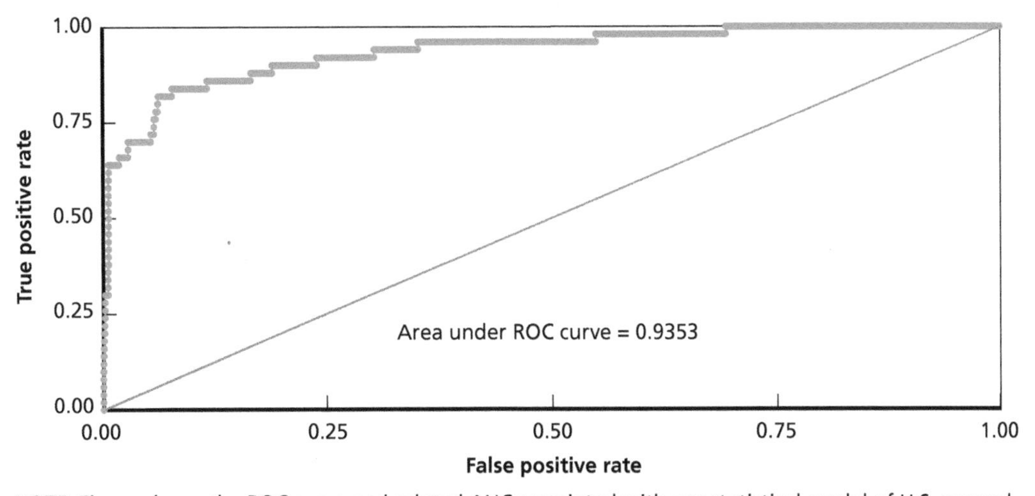

NOTE: Figure shows the ROC curve and related AUC associated with our statistical model of U.S. ground deterrence intervention onset. The ROC curve shows the ability of the statistical model to accurately identify actual deterrence interventions in the data (the true positive rate) while limiting the number of observations incorrectly identified as deterrence interventions (the false positive rate). The diagonal line of Figure B.5 delimits a model that correctly identifies deterrence interventions no better than random chance. Higher AUC scores indicate statistical models that perform increasingly better than random chance in classifying the data.

Figure B.6
Deterrence Intervention Cessation Statistical Model Performance

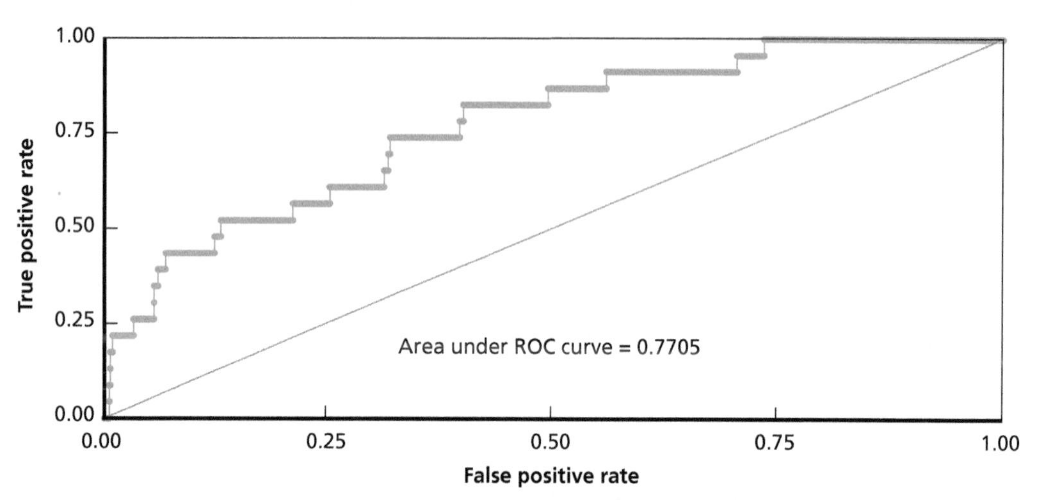

NOTE: Figure shows the ROC curve and related AUC associated with our statistical model of U.S. ground deterrence intervention cessation. The ROC curve shows the ability of the statistical model to accurately identify actual deterrence intervention cessations in the data (the true positive rate) while limiting the number of observations incorrectly identified as deterrence intervention cessations (the false positive rate). The diagonal line of Figure B.6 delimits a model that correctly identifies deterrence intervention cessations no better than random chance. Higher AUC scores indicate statistical models that perform increasingly better than random chance in classifying the data.

Figure B.7
Armed Conflict Intervention Onset Statistical Model Performance

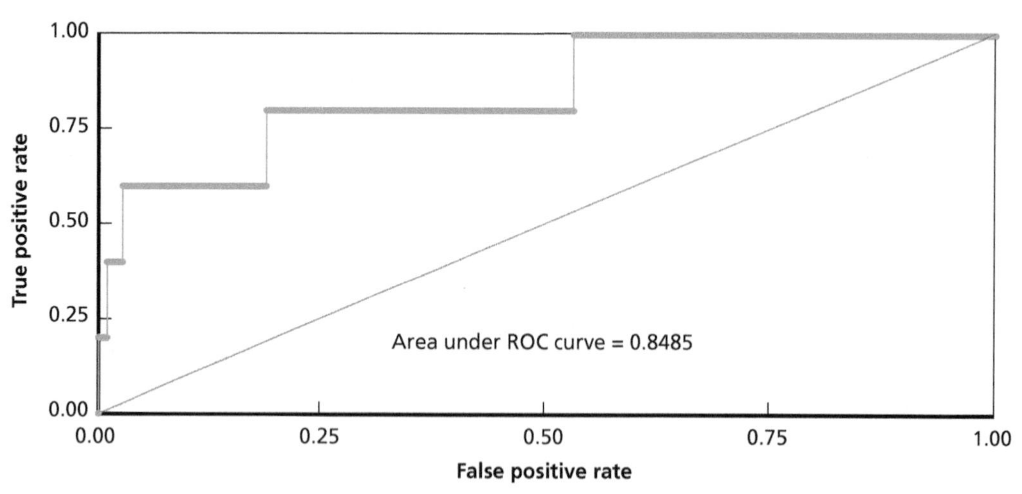

NOTE: Figure shows the ROC curve and related AUC associated with our statistical model of U.S. ground armed conflict intervention onset. The ROC curve shows the ability of the statistical model to accurately identify actual armed conflict interventions in the data (the true positive rate) while limiting the number of observations incorrectly identified as armed conflict interventions (the false positive rate). The diagonal line of Figure B.7 delimits a model that correctly identifies armed conflict interventions no better than random chance. Higher AUC scores indicate statistical models that perform increasingly better than random chance in classifying the data.

Figure B.8
Stabilization Intervention Onset Statistical Model Performance

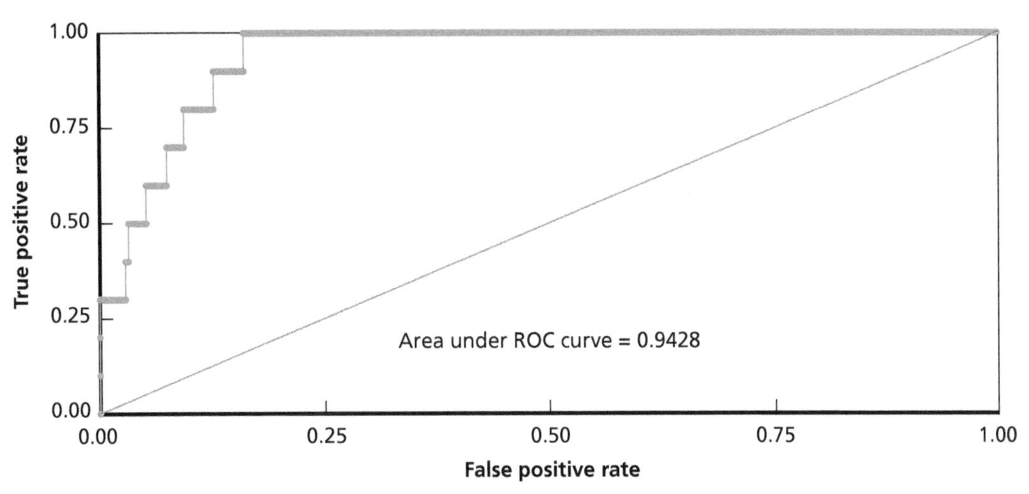

NOTE: Figure shows the ROC curve and related AUC associated with our statistical model of U.S. ground stabilization intervention onset. The ROC curve shows the ability of the statistical model to accurately identify actual stabilization interventions in the data (the true positive rate) while limiting the number of observations incorrectly identified as stabilization interventions (the false positive rate). The diagonal line of Figure B.8 delimits a model that correctly identifies stabilization interventions no better than random chance. Higher AUC scores indicate statistical models that perform increasingly better than random chance in classifying the data.

Figure B.9
Stabilization Intervention Cessation Statistical Model Performance

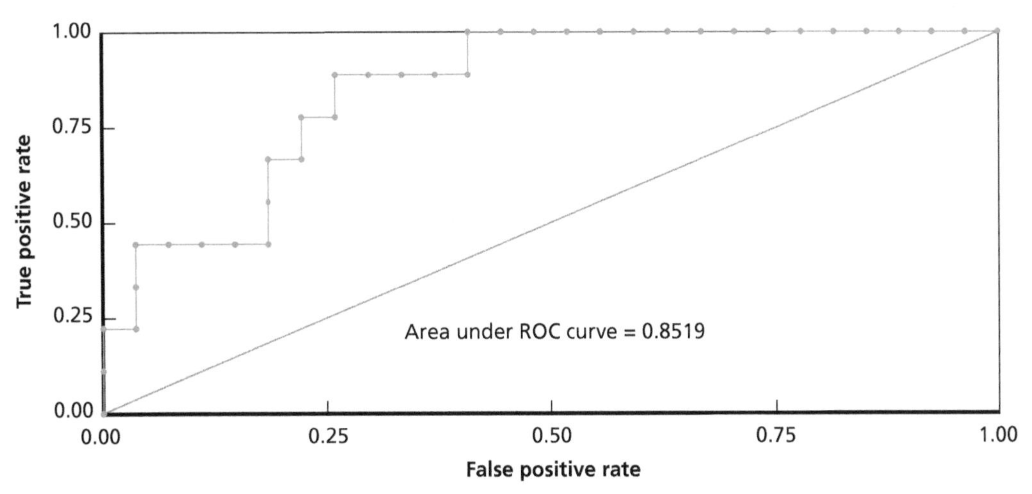

NOTE: Figure shows the ROC curve and related AUC associated with our statistical model of U.S. ground stabilization intervention cessation. The ROC curve shows the ability of the statistical model to accurately identify actual stabilization intervention cessations in the data (the true positive rate) while limiting the number of observations incorrectly identified as stabilization intervention cessations (the false positive rate). The diagonal line of Figure B.9 delimits a model that correctly identifies stabilization intervention cessations no better than random chance. Higher AUC scores indicate statistical models that perform increasingly better than random chance in classifying the data.

Detailed Description of Forecasting Model Processes

As discussed in Chapter Three, our forecasting model includes four main components: detailing the future strategic environment, forecasting intrastate and interstate armed conflicts, forecasting U.S. ground interventions, and adjudicating the force requirements of those forecasted interventions. This section provides step-by-step details about how the latter three of those components are processed by our forecasting model to develop our annual forecasts. Details on the future strategic environment component, which provides the data on the key factors used in the other components for the years 2017–2040, are already discussed at length in Chapters Three and Four, with the exception of certain variables that were not projected directly by International Futures but for which we instead developed projection models based on prior academic work and variable inputs from International Futures. In total, there were four such models, covering the future incidence of interstate territorial claims, nuclear weapon capability, state military power or capabilities, and international normative strength.[1]

Returning to the overall modeling process, each of the steps detailed in the remainder of this appendix illustrate the process by which our forecasting model produces one year of forecasts.[2] For simplicity, the discussion in this appendix focuses on producing annual forecasts for one state/dyad, but, in reality, the steps outlined below occur simultaneously across all states/dyads in our data.

The process detailed below is then repeated to produce forecasts of armed conflict and interventions for the next year, and so forth for the entire 2017–2040 period. To increase the robustness of our forecasts, we then iterate that entire process 500 times, and base our forecasts on the average predictions across those iterations.[3] A single iteration of our model therefore involves the full simulation of each model component annually for each year from 2017 to 2040. Each subsequent iteration then re-simulates the entire 2017 to 2040 period using a different random seed.

---

[1] The territorial claims models incorporated metrics of regime type, distance, bilateral trade, nuclear weapon possession, state age, and regional dummies, building on research in Frederick (2012). The nuclear weapon capabilities models incorporated metrics of GDP, GDP per capita, state age, regime type, military expenditures, and years since 1945, building on models developed by Horowitz (Michael C. Horowitz, *The Diffusion of Military Power: Causes and Consequences for International Politics*, Princeton, N.J.: Princeton University Press, 2010). The normative strength models projected the regional incorporation of pacific dispute settlement language into treaties using regional metrics of GDP per capita, international trade, nuclear weapon possession, and regional dummies, building on research by Hensel (2001). The state military power models were extensive, incorporating metrics of state military spending and size, economic base, and technological sophistication, while taking into account the geographic location of states and their power projection capabilities. These models are described in detail in Watts et al., 2017a, pp. 244–248.

[2] The code to run our forecasting model actually involves opening and saving many separate data files to perform various calculations. We do not explicitly state here when the model switches between different data files to perform various processes.

[3] Specifically, we run 500 iterations of our baseline model and each alternative future scenario. For replicability, we use the same set of random seeds (1–500) for all of our models.

Forecasting Intrastate Armed Conflicts

Our forecasting model starts by predicting levels of intrastate armed conflict in the current year in a given state.[4] First, the model identifies whether the state experienced an ongoing intrastate conflict that did not end in the previous year. Because the conflict did not end in the previous year, the model marks the state as being in conflict in the current year. The model also updates the number of years that the ongoing conflict in the state has been ongoing.[5] Alternatively, if the model identifies that an ongoing intrastate armed conflict in the state ended in the previous year, then it begins counting years of peace.[6] The number of years in which the state has been in either conflict or peace are important explanatory variables in our statistical models.

Having adjudicated all ongoing conflicts from the previous year, the model then determines whether the state experiences the onset of a new intrastate armed conflict in the current year.[7] To do this, the statistical model of intrastate armed conflict onset is fit, then generates a predicted probability of intrastate conflict onset based on the values of our key factors for the state in the current year. To help reduce the number of iterations of our model that needed to be run to smooth out random fluctuations, we also rebalanced the predicted probabilities by assigning the potential for conflict in the lowest-probability states to the higher-probability states and setting the potential for conflict in these lower-probability states equal to zero.[8]

To determine whether a new intrastate conflict begins in the state, we then simulate from a Bernoulli distribution, with the probability of conflict set to the predicted probability generated by our statistical model and rebalanced as described above.[9] In

---

[4] While conflicts, representing opportunities for intervention, had to be forecast before the interventions themselves, the choice to begin with intrastate conflict over interstate wars was arbitrary, and the choice does not affect the projections produced.

[5] Technically, the model assigns a value of $\text{conflictyears}_{t-1}+1$ for the current year.

[6] In this case, the model assigns a value of peaceyears = 1 for the current year.

[7] Our model only performs this step if the model did not already mark the state as experiencing an ongoing intrastate armed conflict.

[8] Specifically, we sum the collected probability of conflict in all those states with values below the mean predicted probability value, and then increase the probability of conflict of each state above the mean predicted probability value equally such that the total number of conflicts predicted by the model remains the same. However, the specific conflicts that do occur can now only fall among states whose initial predicted probability was above the mean. After repeated experiments, we found this step to be necessary because the number of iterations our model required for the projections from the simulated runs to converge to state-level projections approximating the predicted probabilities from our regressions was in the thousands, exceeding the computing time available. Rebalancing the probabilities in this manner allowed us to shorten those requirements will still retaining appropriate and credible state-level projections.

[9] The Bernoulli distribution is a simple binomial probability distribution that takes the form $\Pr(X = 1) = p$. In words, the probability that an observation takes on a value of 1 is p, and the probability that an observation takes on a value of 0 is $1 - p$, where p is some simple probability. In this case, p is the probability, generated by our logit models, that a specific observation takes on a value of 1. For instance, an observation with a predicted

this way, the state is assigned either a "1" or "0" for "conflict" or "no conflict," with states having higher predicted probabilities more likely to draw a "1" than states with a lower predicted probability. If the state is assigned a "1," then the model starts counting the duration of the conflict; otherwise the model continues counting years of peace.[10]

The model then determines whether any of the ongoing intrastate armed conflicts end in the current year, including those conflicts that just started in the current year. To do this, the statistical model of intrastate armed conflict cessation is fit, whose components are detailed above in Table B.2, and then generates a predicted probability of intrastate conflict cessation based on the values of our key factors for the state in the current year. To determine whether an ongoing intrastate conflict ends in the state, we then simulate from a Bernoulli distribution, with the probability of the conflict ending set to the predicted probability generated by our statistical model and rebalanced toward higher-probability states as described above. In this way, the state is assigned either a "1" or "0" for "conflict ends" or "conflict does not end," with states having higher predicted probabilities more likely to draw a "1" than states with a lower predicted probability.

That three-step process completes our model's forecasting of new, ongoing, and terminated intrastate armed conflicts in the current year. However, because one of our key factors, the level of ongoing intrastate conflicts around each state, is dynamic and based on our model's annual forecasts, our model must also update our key factor metric for ongoing conflicts in the state's neighborhood. To do this, our model then creates a weighted average spatial lag of ongoing intrastate armed conflicts around each state. This weighted average spatial lag takes the general form

$$W_{ijt} = \sum_{j=1}^{n} \left(\frac{\dfrac{1}{distance_{ijt}}}{\sum \dfrac{1}{distance_{ijt}}} \times IntrastateConflict_{jt} \right).$$

In words, the weighted average spatial lag is a continuous measure that captures the inverse minimum distance between a state i and all other states experiencing ongoing intrastate armed conflicts j in a given year t, normalized by the sum of the inverse minimum distance between state i and all other states in the international system j for

probability of 0.60 generated from our statistical model of intrastate conflict onset has a 60 percent chance of being a 1 (i.e., conflict in that year), whereas another observation with a predicted probability of 0.20 has only a 20 percent chance of being a 1. This same principle holds for all other uses of the Bernoulli distribution in our forecasting model, with the caveat that the event of interest (conflict onset, conflict cessation, intervention onset, etc.) changes in each instance as discussed in the description below.

[10] If the state experiences an intrastate armed conflict onset in the current year, then the model sets conflictyears = 1. If the state does not experience the onset of an intrastate conflict in the current year, then the model assigns a value of $peaceyears_{t-1}+1$ for the current year.

the given year *t*. This measure ranges from 0 to 1 and measures the relative proximity and scale of ongoing intrastate armed conflicts around each state, with higher values indicating that more proximate states are experiencing ongoing intrastate conflicts. We calculate this measure for each state in the system, allowing our model to dynamically account for possible contagion effects from forecasted intrastate conflicts in the international system.

Forecasting Interstate Wars

After forecasting all new, ongoing, and ended intrastate conflicts in a given year, our forecasting model forecasts all interstate wars in dyads in that year in a process that is roughly analogous to our process for forecasting intrastate conflicts. First, the model identifies the dyad that experienced an ongoing interstate war that did not end in the previous year. Because the war did not end in the previous year, the model marks the dyad as being at war in the current year. The model also updates the number of years that the war in the dyad has been ongoing.[11] Alternatively, if the model identifies that an ongoing interstate war in the dyad ended in the previous year, then it begins counting years of peace.[12]

Having adjudicated all ongoing wars from the previous year, the model then determines whether the dyad experiences the onset of a new interstate war in the current year.[13] To do this, the statistical model of interstate war onset is fit, then generates a predicted probability of interstate war onset based on the values of our key factors for the dyad in the current year. To determine whether a new interstate war begins between the states in the dyad, we then simulate from a Bernoulli distribution, with the probability of conflict set to the predicted probability generated by our statistical model, after being rebalanced toward higher-probability dyads in the same manner described above with regards to the intrastate conflict model. In this way, the dyad is assigned either a "1" or "0" for "war" or "no war," with dyads having higher predicted probabilities more likely to draw a "1" than dyads with a lower predicted probability. If the dyad is assigned a "1," then the model starts counting the duration of the war; otherwise, the model continues counting years of peace.[14]

The model then determines whether any of the ongoing interstate wars end in the current year, including those wars that just started in the current year. To do this, the statistical model of interstate war cessation is fit, then generates a predicted prob-

[11] Technically, the model assigns a value of $\text{conflictyears}_{t-1}+1$ for the current year.

[12] In this case, the model assigns a value of peaceyears = 1 for the current year.

[13] Our model only performs this step if the model did not already mark the dyad as experiencing an ongoing interstate armed conflict.

[14] If the dyad experiences an interstate armed conflict onset in the current year, then the model sets conflictyears = 1. If the dyad does not experience the onset of an interstate war in the current year, then the model assigns a value of $\text{peaceyears}_{t-1}+1$ for the current year.

ability of interstate war cessation based on the values of our key factors for the dyad in the current year. To determine whether an ongoing interstate war ends in the dyad, we then simulate from a Bernoulli distribution, with the probability of the war ending set to the predicted probability generated by our statistical model, after being rebalanced toward higher-probability dyads in the same manner described above with regards to the intrastate conflict model. In this way, the dyad is assigned either a "1" or "0" for "war ends" or "war does not end," with dyads having higher predicted probabilities more likely to draw a "1" than dyads with a lower predicted probability.

That three-step process completes our model's forecasting of new, ongoing, and ended interstate wars among dyads in the current year.

Combining Intrastate Conflict and Interstate War Projections

Having completed the given year's forecasts of intrastate conflicts and interstate wars, our model then prepares those forecasts for use in forecasting U.S. ground interventions. Because our forecasts of interstate wars occur at the dyadic level, but our forecasts of U.S. ground interventions occur at the state level, our model transforms those dyad-year forecasts into country-year forecasts, marking each state in a given year by whether it is involved in an interstate war. Our model then merges our forecasts of interstate wars and intrastate armed conflicts into a single data frame.

In that combined data frame, our model marks each state as being in "conflict"—meaning that the state experiences either an intrastate armed conflict or an interstate war in a given year. This combined "conflict" marker is crucial to our ability to accurately forecast different types of interventions. States, for instance, may experience a one-year interstate war immediately followed by several years of intrastate conflict, providing more opportunities for the United States to undertake an armed conflict intervention. Alternatively, because in our model stabilization missions only occur in postconflict states, the model will only allow for stabilization missions when all ongoing intrastate conflicts and interstate wars in a state have ceased. Once our model assesses whether the state is in any kind of conflict in a given year, it also updates the postconflict window, the five years after ongoing conflicts in the state have ceased, for use in forecasting U.S. ground my stabilization missions.

Forecasting U.S. Ground Interventions

Once our model has forecast all intrastate and interstate armed conflicts in a given year, the model turns to forecasting U.S. ground interventions into those conflicts. The processes for forecasting our three types of U.S. ground interventions—deterrence missions, interventions into ongoing armed conflicts, and postconflict stabilization missions—are roughly analogous to the processes outlined for forecasting armed conflicts, but with some important changes.

Forecasting U.S. Ground Force Deterrence Interventions

The first type of intervention forecast by our model is for U.S. ground force deterrence missions.[15] First, our model identifies whether the state experienced an ongoing deterrence mission that did not end in the previous year. Because the deterrence mission did not end in the previous year, the model marks the state as hosting a U.S. deterrence mission in the current year. The model also updates the number of years that the deterrence mission in the state has been ongoing.[16] Alternatively, if the model identifies that an ongoing deterrence mission in the state ended in the previous year, then it begins counting years of nondeterrence.[17]

Having adjudicated all ongoing deterrence missions in different states from the previous year, the model then determines whether the state experiences the onset of a new deterrence mission in the current year.[18] To do this, the statistical model of U.S. deterrence mission onset is fit, then generates a predicted probability of a deterrence intervention onset based on the values of our key factors for the state in the current year. The universe of potential cases, in contrast to the other intervention types, includes all states in the international system, apart from the United States itself. To determine whether a new deterrence mission begins in the state, we then simulate from a Bernoulli distribution, with the probability of a new deterrence mission set to the predicted probability generated by our statistical model, after being rebalanced toward higher-probability states in the same manner described above with regards to the intrastate conflict model. In this way, the state is assigned either a "1" or "0" for "deterrence mission" or "no deterrence mission," with states having higher predicted probabilities more likely to draw a "1" than states with a lower predicted probability. If the state is assigned a "1," then the model starts counting the duration of the deterrence mission; otherwise the model continues counting years of nonintervention.[19]

The model then determines whether any of the ongoing U.S. deterrence interventions end in the current year, including those interventions that just started in the current year. To do this, the statistical model of deterrence mission cessation is fit,

---

[15] Unlike the order of our forecasts for intrastate and interstate armed conflicts, the order of our forecasts for U.S. ground interventions is not entirely arbitrary. Deterrence missions, which occur only in years of peace, can be theoretically forecast any time after armed conflicts are forecast. However, because some armed conflict interventions automatically become postconflict stabilization missions in certain circumstances, interventions into ongoing armed conflicts must be forecast before postconflict stabilization missions.

[16] Technically, the model assigns a value of $\text{deterrenceyears}_{t-1}+1$ for the current year.

[17] In this case, the model assigns a value of nondeterrenceyears = 1 for the current year. Nondeterrence years are analogous to the peaceyears variables in our conflict models.

[18] Our model only performs this step if the model did not already mark the state as experiencing an ongoing armed conflict intervention.

[19] If the dyad experiences an interstate armed conflict onset in the current year, then the model sets armedinterventionyears = 1. If the dyad does not experience the onset of an armed conflict intervention in the current year, then the model assigns a value of $\text{non-armedinterventionyears}_{t-1}+1$ for the current year.

then generates a predicted probability of deterrence mission cessation based on the values of our key factors for the state in the current year. To determine whether an ongoing deterrence mission ends in the state, we then use a draw from a Bernoulli distribution, with the probability of the deterrence mission ending set to the predicted probability generated by our statistical model, after being rebalanced toward higher-probability states in the same manner described above with regards to the intrastate conflict model. In this way, the state is assigned either a "1" or "0" for "deterrence mission ends" or "deterrence mission does not end," with states having higher predicted probabilities more likely to draw a "1" than states with a lower predicted probability. Alternatively, if the state experiences the onset of an interstate armed conflict in the current year, then our model automatically assesses any ongoing deterrence mission in that state as ending in the current year.[20]

Forecasting U.S. Ground Force Interventions into Ongoing Armed Conflicts

Having forecast U.S. deterrence missions in the given year, our model then turns to forecasting U.S. ground interventions into going intrastate conflicts and interstate wars. First, our model identifies whether the state experienced an ongoing armed conflict intervention that did not end in the previous year. Because the armed conflict intervention did not end in the previous year, the model marks the state as hosting a U.S. armed conflict intervention in the current year. The model also assigns all states forecasted as a belligerent in an interstate war versus the United States in the given year as experiencing a U.S. ground force armed conflict intervention. The model also updates the number of years that the armed conflict intervention in the state has been ongoing.[21]

Having adjudicated all ongoing armed conflict interventions from the previous year, the model then determines whether the state experiences the onset of a new armed conflict intervention in the current year.[22] If, in the interstate war projections, a state began an interstate war versus the United States in the current year, then the model automatically assesses that state as experiencing the onset of an armed conflict intervention in the current year. For all other states in ongoing conflicts, the statistical model U.S. armed conflict intervention onset is fit, then generates a predicted probability of an armed conflict intervention onset based on the values of our key factors for the state in the current year. Importantly, this process only occurs for states experiencing ongoing armed intrastate conflicts or interstate wars. To determine whether a new armed conflict intervention begins in the state, we then use a draw from a Ber-

---

[20] However, states that experience the onset of intrastate armed conflicts during ongoing deterrence missions do not automatically see those deterrence missions end.

[21] Technically, the model assigns a value of $armedinterventionyears_{t-1}+1$ for the current year.

[22] Our model only performs this step if the model did not already mark the state as experiencing an ongoing armed conflict intervention.

noulli distribution, with the probability of a new armed conflict intervention set to the predicted probability generated by our statistical model, after being rebalanced toward higher-probability states in the same manner described above with regards to the intrastate conflict model. In this way, the state is assigned either a "1" or "0" for "armed conflict intervention" or "no armed conflict intervention," with states having higher predicted probabilities more likely to draw a "1" than states with a lower predicted probability. If the state is assigned a "1," then the model starts counting the duration of the armed conflict intervention; otherwise the model continues counting years of nonintervention.[23]

Unlike many of our other forecasting processes, our model does not generate predicted probabilities of armed conflict intervention cessation from a statistical model. Rather, we assume that, once U.S. forces are committed to an armed conflict intervention, the United States stays involved until the ending of the conflict, such that the ongoing armed conflict intervention continues until the period of intrastate conflict or interstate war involving the state ends. In the year that the period of armed conflict ends, the U.S. armed conflict intervention then also ends.

Forecasting U.S. Ground Force Postconflict Stabilization Missions

The final type of intervention forecast by our model is postconflict stabilization interventions. Importantly, the universe of opportunities for such interventions is limited to a five-year postconflict window after a period of conflict has ended. As detailed later in this appendix, however, while stabilization missions can only begin in the five years after a period of conflict, they can continue well beyond that five-year postconflict window.

First, our model identifies whether the state experienced an ongoing postconflict stabilization missions that did not end in the previous year. Because the stabilization intervention did not end in the previous year, the model marks the state as hosting a U.S. stabilization intervention in the current year. The model then updates the number of years that the stabilization intervention in the state has been ongoing.[24]

Having adjudicated all ongoing stabilization interventions from the previous year, the model then determines whether the state experiences the onset of a new postconflict stabilization intervention in the current year.[25] This occurs in two parts. First, if the state hosted a U.S. armed conflict intervention in the previous year and the period of conflict in the state ended in the previous year, then our model automatically assesses that armed conflict intervention from the previous year as transitioning to the

---

[23] If the dyad experiences an interstate armed conflict onset in the current year, then the model sets armedinterventionyears = 1. If the dyad does not experience the onset of an armed conflict intervention in the current year, then the model assigns a value of non-armedinterventionyears$_{t-1}$+1 for the current year.

[24] Technically, the model assigns a value of stabilizationyears$_{t-1}$+1 for the current year.

[25] Our model only performs this step if the model did not already mark the state as experiencing an ongoing armed conflict intervention.

state of a postconflict stabilization mission in the current year. In this way, all U.S. ground force armed conflict interventions automatically transition into stabilization interventions once the state enters the postconflict period. That is, we assume that the United States will participate in a stabilization mission following a U.S. intervention into the preceding armed conflict.

For all other states in the postconflict window, the statistical model of U.S. stabilization intervention onset is fit, then generates a predicted probability of a stabilization intervention onset based on the values of our key factors for the state in the current year. Importantly, this process only occurs for states in the five years after a period of armed conflict ends. To determine whether a new stabilization intervention begins in the state, we then use a draw from a Bernoulli distribution, with the probability of a new stabilization intervention set to the predicted probability generated by our statistical model, after being rebalanced toward higher-probability states in the same manner described above with regards to the intrastate conflict model. In this way, the state is assigned either a "1" or "0" for "stabilization intervention" or "no stabilization intervention," with states having higher predicted probabilities more likely to draw a "1" than states with a lower predicted probability. If the state is assigned a "1," then the model starts counting the duration of the stabilization intervention; otherwise the model continues counting years of nonintervention.[26]

Determining the Force Characteristics of Forecasted U.S. Ground Interventions

After our model forecasts these three types of potential interventions in the current year, the final step in the process is to adjudicate the characteristics of the U.S. ground forces required to perform those interventions.

Our model is concerned with two characteristics of U.S. ground intervention forces—the total number of troops required for each intervention and the ratio of heavy versus light ground forces required for each intervention, which we adjudicate in two ways. For interventions ongoing in 2016, the last year of historical data in our model, that are forecast to continue at least into 2017, the model uses the number of troops in the intervention in 2016 as the annual number of troops in the intervention for the forecasted duration of that intervention. That is, an ongoing intervention in 2016 with 50,000 (20,000 heavy) troops that is forecast to continue into 2017 would also use 50,000 (20,000 heavy) troops. If that intervention were then forecast to continue into 2018, then the intervention would also use 50,000 (20,000 heavy) troops in 2018. The assumption that the size of ongoing interventions will not fluctuate is an important assumption in our model.

---

[26] If the dyad experiences an interstate armed conflict onset in the current year, then the model sets stabilizationyears = 1. If the dyad does not experience the onset of a stabilization intervention in the current year, then the model assigns a value of non-stabilizationyears$_{t-1}$+1 for the current year.

For all new forecasted interventions that cannot utilize historical numbers of troops from ongoing interventions in 2016, our model assigns each intervention a total number of troops and the number of heavy versus light ground forces, based on the analyses outlined in Chapter Three and Appendix A. To assign the appropriate forces, our model first places each intervention into a bucket, based on the binning structure outlined in Chapter Three. For instance, one intervention into an ongoing armed conflict might be placed into the bucket "interstate conflict + strong adversary + post-1940 U.S. foreign policy," while another armed conflict intervention might be placed into the bucket "intrastate conflict + weak adversary + post-1940 U.S. foreign policy," based on the characteristics of the intervention. After being assigned such a bucket, our model then assigns to each intervention the average number of troops and average ratio of heavy versus light ground forces likely to be employed based on our analyses of historical U.S. ground interventions.

Deterrent interventions in NATO states are a special case in our forecasts of U.S. ground interventions. NATO interventions are unique in that they, from a strategic standpoint, encompass all NATO states but do not necessarily employ U.S. ground forces in all of those states, and tracking the number of U.S. forces deployed to different NATO states in a given year may be practically difficult and, for the purposes of our model, less important. To accommodate the different status of NATO deterrent interventions, we decouple the presence of an ongoing deterrent intervention in NATO states from the number of U.S. ground forces deployed to those interventions. More specifically, if a deterrent intervention occurs in a given year in any NATO state, then our model automatically assesses a deterrent intervention in all NATO states in that year.[27] If we allow our forecasting model to employ the average number of U.S. ground forces in each of those NATO states, however, then our model would vastly overestimate the number of U.S. ground personnel required to meet those intervention demands. Instead, our model sets the forecasted number of U.S. ground personnel in NATO deployments to zero for all NATO states except Germany, which we treat as the hosting state for the expected number of troops deployed to the intervention overall.[28] In this way, while all NATO states technically experience ongoing interventions, our model only adjudicates one demand of forces for those interventions across all NATO states, and in counting the number of ongoing interventions in a given year, we treat the NATO deterrent intervention as a single intervention.

Identifying these force requirements for interventions technically completes our forecasts of future U.S. ground interventions. However, because one of the key fac-

[27] Performing this step prevents our model from ending ongoing deterrent interventions in some NATO states while continuing them in others, which runs counter to the conceptual aims of NATO deployments.

[28] We chose Germany to hold forces for NATO interventions was made because of its relatively central positioning among the NATO states of western Europe and because Germany has historically been a pivot point of U.S. ground forces operating in Europe.

tors in our interstate armed conflict model, the number of nearby U.S. heavy ground forces, is dynamic and based on our model's annual forecasts of U.S. ground interventions, our model must also update our key factor metric for the numbers of nearby U.S. heavy ground forces. To do this, our model then creates a weighted sum spatial lag of U.S. heavy ground forces around each state, based on the effective strength and travel abilities of U.S. forces in each region. This weighted sum spatial lag takes the general form

$$W_{ijt} = \sum_{j=1}^{n} \left(\frac{EffectiveTravelStrength}{DistanceBetweenStates_{ij} + 200} \times NumberofNonCombatHeavyTroops_{jt} \right).$$

In words, this measure is a modified weighted sum spatial lag, based on inverse weighting of the distance between state i and all other states j, the number of U.S. heavy ground forces in states j in a given year t, and the relative distance U.S. heavy ground forces can travel in a day, based on the total number of U.S. forces in the region. This measure increases as the number of U.S. heavy ground forces in more proximate states around each state i increase. To ensure that U.S heavy forces in nearby states never have a larger effect than U.S. heavy ground forces in the state itself, we add a value of 200 miles, or the largest assumed distance that can be traveled in a day by U.S. forces, to the denominator of our spatial lag equation. Our model calculates this measure for each state in the system, allowing our model to dynamically account for changes in U.S. regional force posture caused by forecasted interventions.

Once this final step is calculated, our forecasting model has completely forecasted a single year of armed conflicts and U.S. ground interventions. This process is then repeated for the following year, inserting the forecasts of armed conflict and U.S. ground interventions from the previous year into the current year's calculations. That sequential process continues until our model has forecast the entire 2017–2040 period. To increase the robustness of our model's forecasts and appropriately account for uncertainty in individual predictions, we then iterate that sequential process of forecasting the 2017–2040 period 500 times and explore average patterns, along with lower and upper bounds, across all iterations.

APPENDIX C

Results of Alternative Future Scenario Forecasting Models

This appendix provides the full set of results of our forecasting model for each of our four alternative future scenarios. The results of each scenario are presented in the same progression as the baseline results presented in Chapter Four.

Forecasting Model Results for Alternative Scenario 1: Global Depression

Figure C.1
Global Depression: Forecasts of Interstate War Onsets, 2017–2040

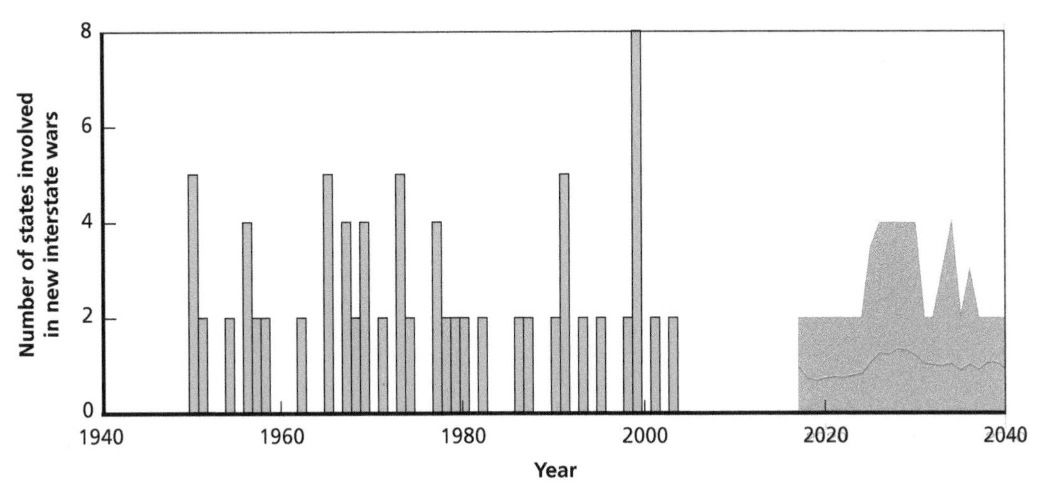

NOTES: The red line denotes the projected mean number of states involved in new interstate wars each year, based on 500 iterations of our forecasting model. The gray shaded area represents the range of forecasts bounded by the 10th and 90th percentiles of state involvement in interstate war onsets each year, based on 500 iterations of our forecasting model.

Figure C.2
Global Depression: Forecasts of Interstate War Occurrence, 2017–2040

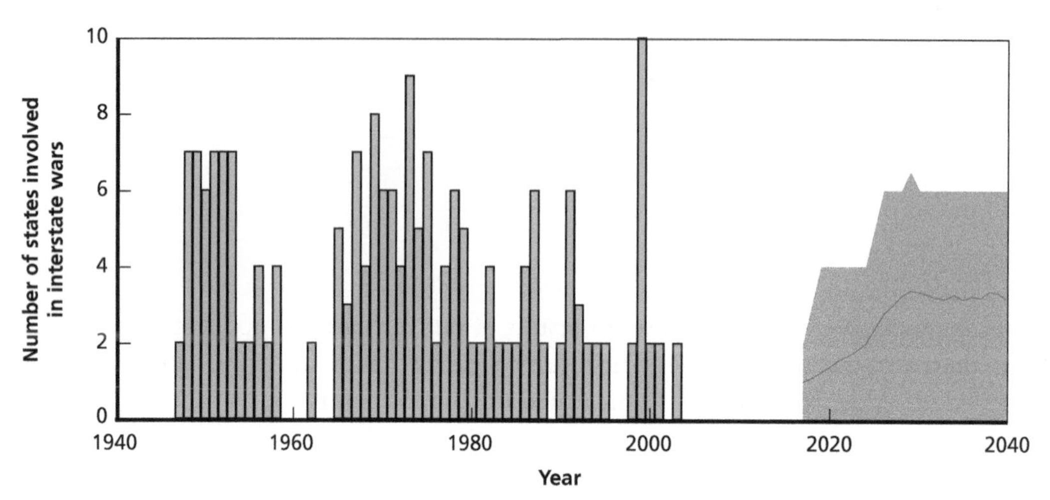

NOTES: The red line denotes the projected mean number of states involved in interstate war each year, based on 500 iterations of our forecasting model. The gray shaded area represents the range of forecasts bounded by the 10th and 90th percentiles of states involved in interstate war each year, based on 500 iterations of our forecasting model.

Figure C.3
Global Depression: Regional Forecasts of Interstate War Occurrence, 2017–2040

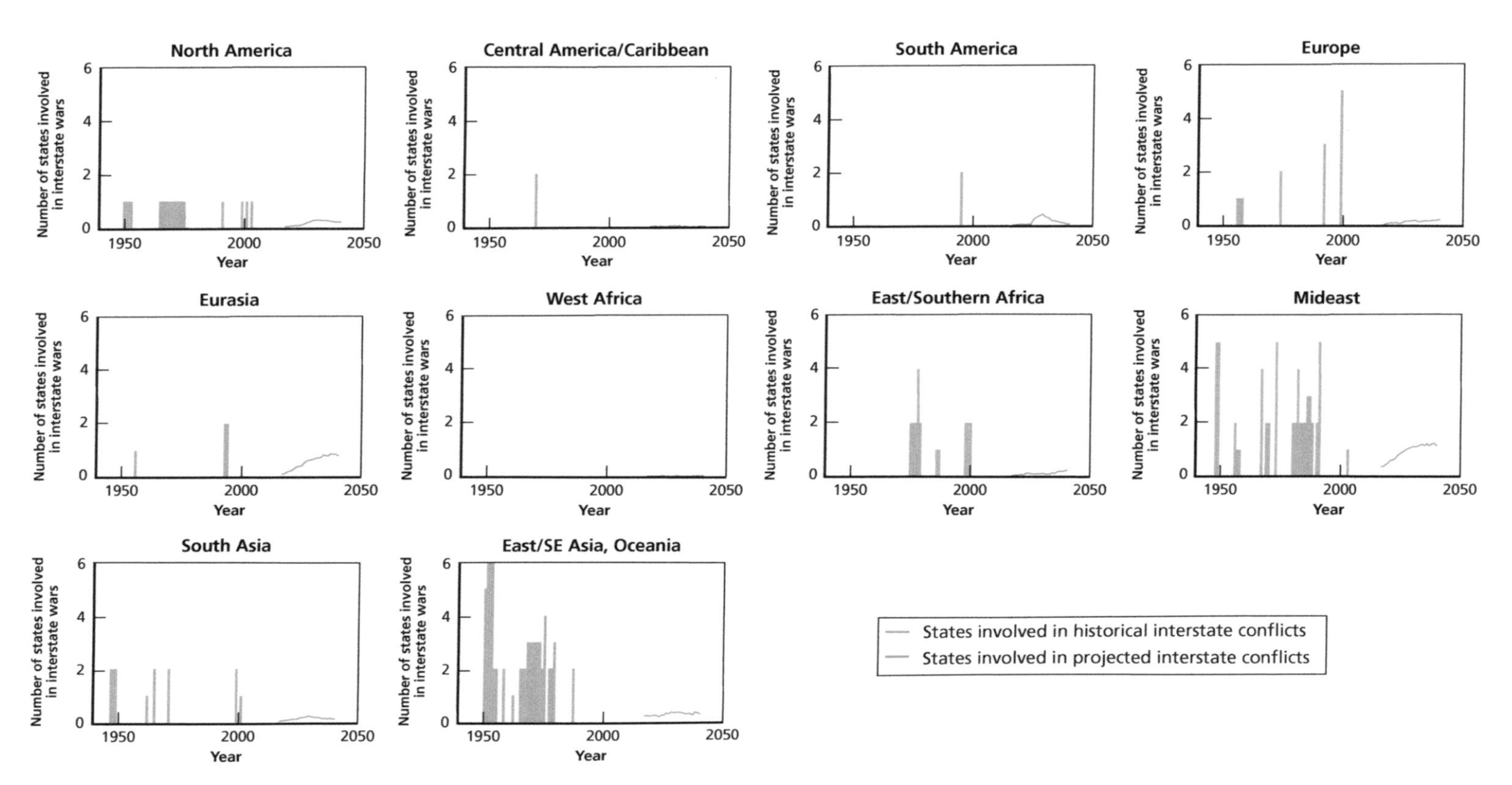

NOTE: The red lines denote the projected mean number of states involved in interstate wars in each region for each year, based on 500 iterations of our forecasting model.

Figure C.4
Global Depression: Forecasts of Intrastate Conflict Onsets, 2017–2040

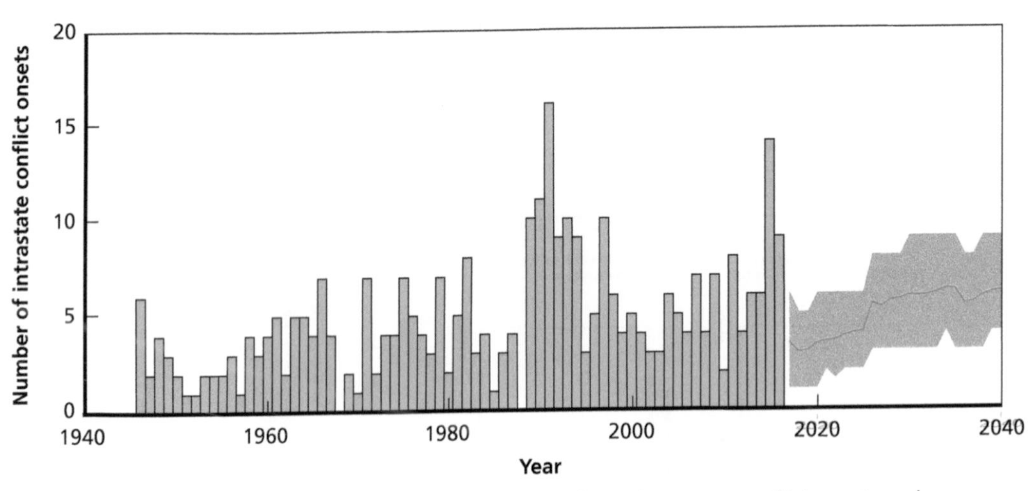

NOTES: The red line denotes the projected mean number of new intrastate conflict onsets each year, based on 500 iterations of our forecasting model. The gray shaded area represents the range of forecasts bounded by the 10th and 90th percentiles of intrastate conflict onsets each year, based on 500 iterations of our forecasting model.

Figure C.5
Global Depression: Forecasts of Intrastate Conflict Occurrence, 2017–2040

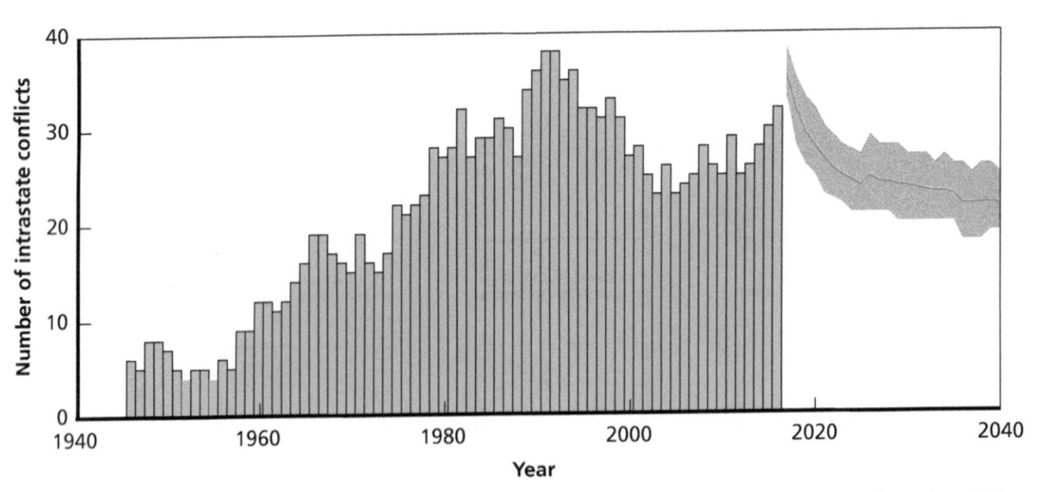

NOTES: The red line denotes the projected mean number of intrastate conflicts each year, based on 500 iterations of our forecasting model. The gray shaded area represents the range of forecasts bounded by the 10th and 90th percentiles of intrastate conflicts each year, based on 500 iterations of our forecasting model.

Figure C.6
Global Depression: Regional Forecasts of Intrastate Conflict Occurrence, 2017–2040

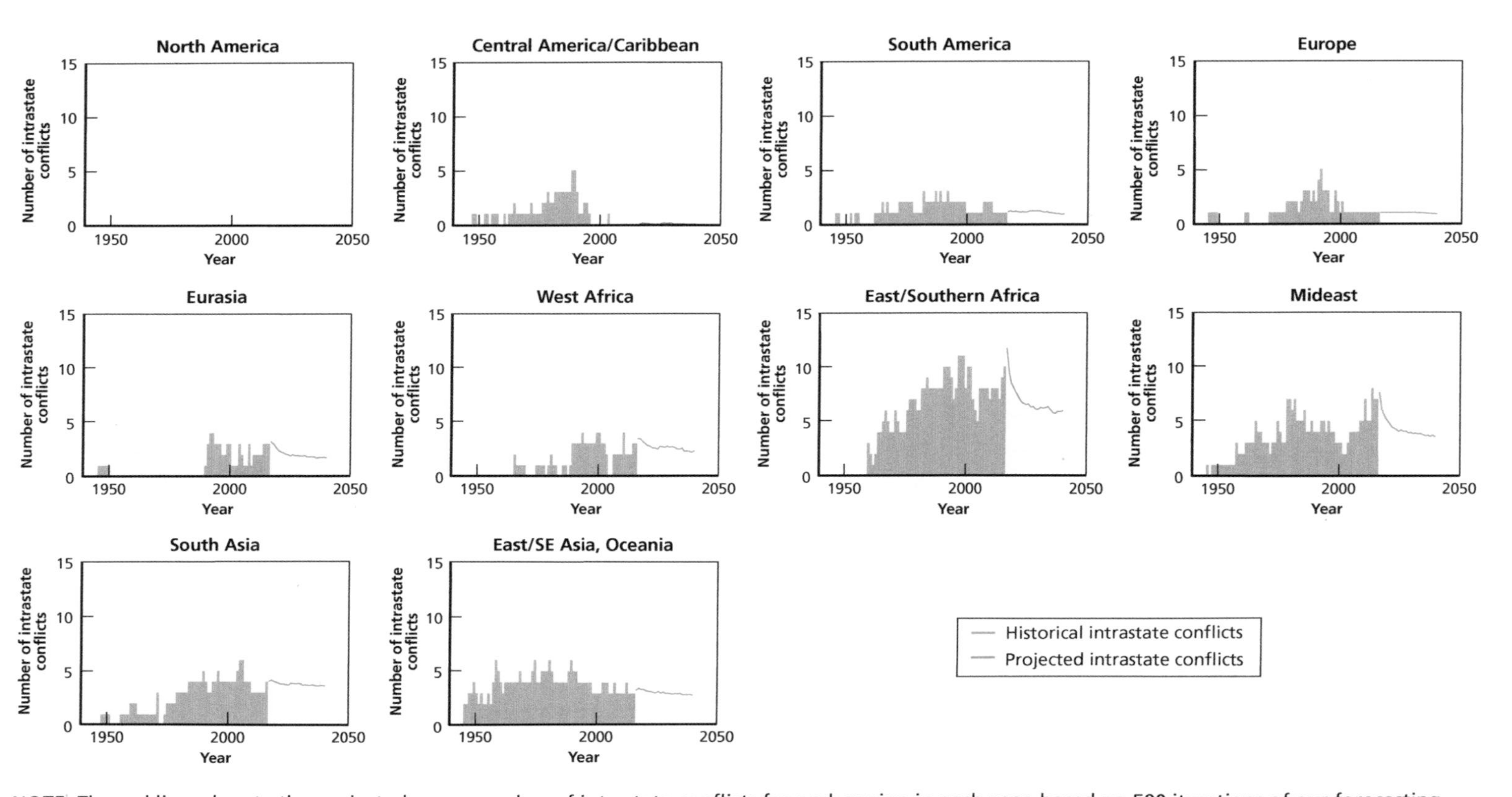

NOTE: The red lines denote the projected mean number of intrastate conflicts for each region in each year, based on 500 iterations of our forecasting model.

Figure C.7
Global Depression: Forecasts of Total U.S. Ground Interventions, 2017–2040

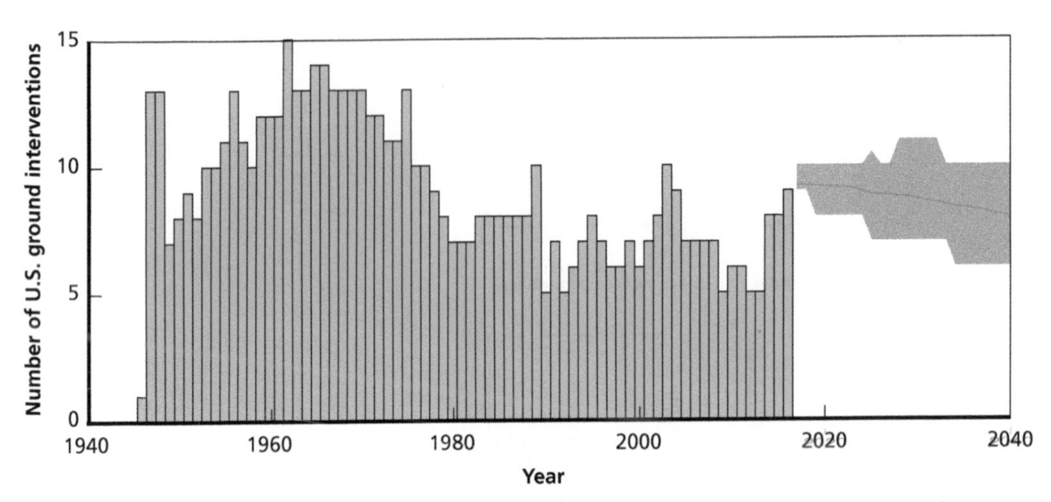

NOTES: The red line denotes the projected mean number of total U.S. ground interventions each year, based on 500 iterations of our forecasting model. The gray shaded area represents the range of forecasts bounded by the 10th and 90th percentiles of U.S. ground interventions each year, based on 500 iterations of our forecasting model.

Figure C.8
Global Depression: Forecasts of Demands for U.S. Ground Intervention Forces, 2017–2040

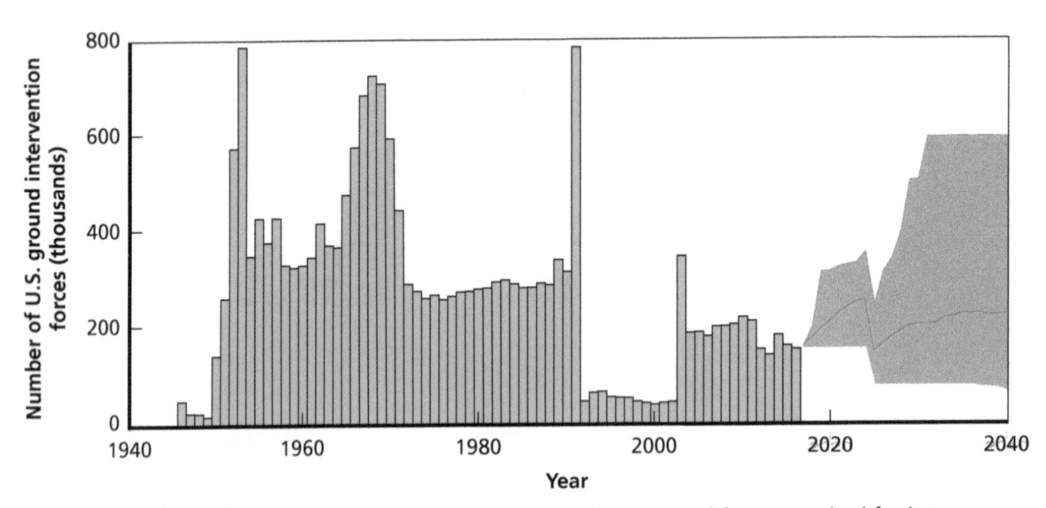

NOTES: The red line denotes the projected mean number of U.S. ground forces required for interventions each year, based on 500 iterations of our forecasting model. The gray shaded area represents the range of forecasts bounded by the 10th and 90th percentiles of U.S. ground forces required for interventions each year, based on 500 iterations of our forecasting model.

Figure C.9
Global Depression: Regional Forecasts of Demands for U.S. Ground Intervention Forces, 2017–2040

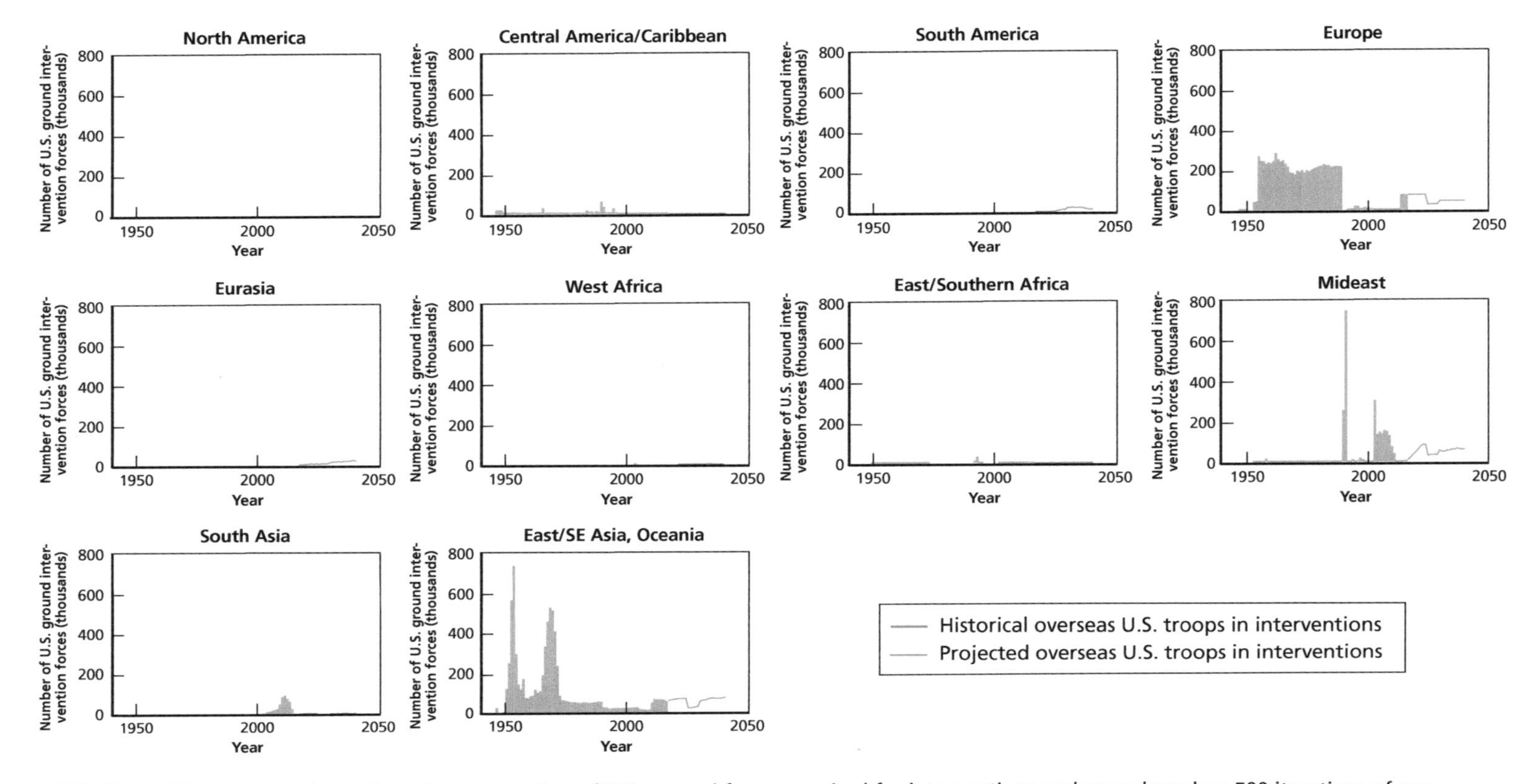

NOTE: The red line denotes the projected mean number of U.S. ground forces required for interventions each year, based on 500 iterations of our forecasting model.

Figure C.10
Global Depression: Forecasts of U.S. Ground Interventions into Armed Conflicts, 2017–2040

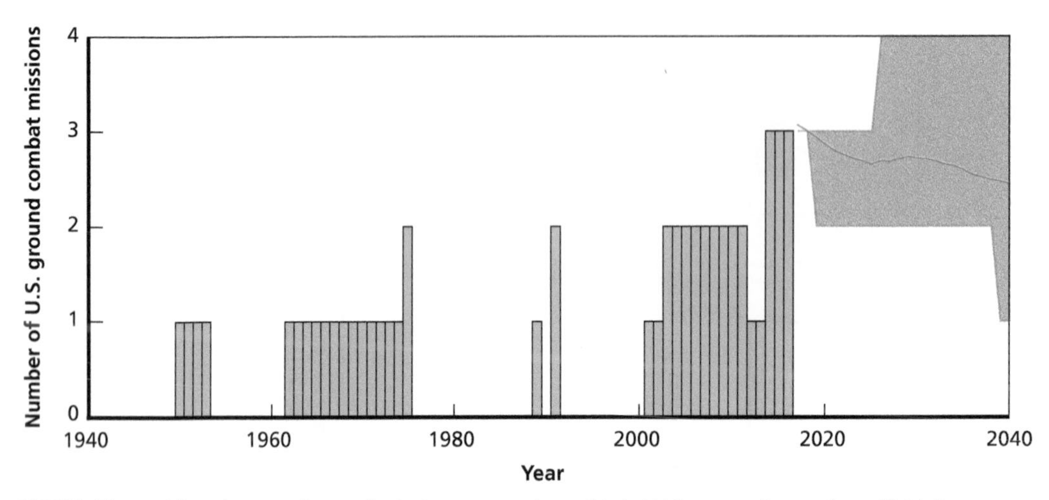

NOTES: The red line denotes the projected mean number of total U.S. ground armed conflict interventions each year, based on 500 iterations of our forecasting model. The gray shaded area represents the range of forecasts bounded by the 10th and 90th percentiles of U.S. ground armed conflict interventions each year, based on 500 iterations of our forecasting model.

Figure C.11
Global Depression: Forecasts of Demands for U.S. Ground Combat Mission Forces, 2017–2040

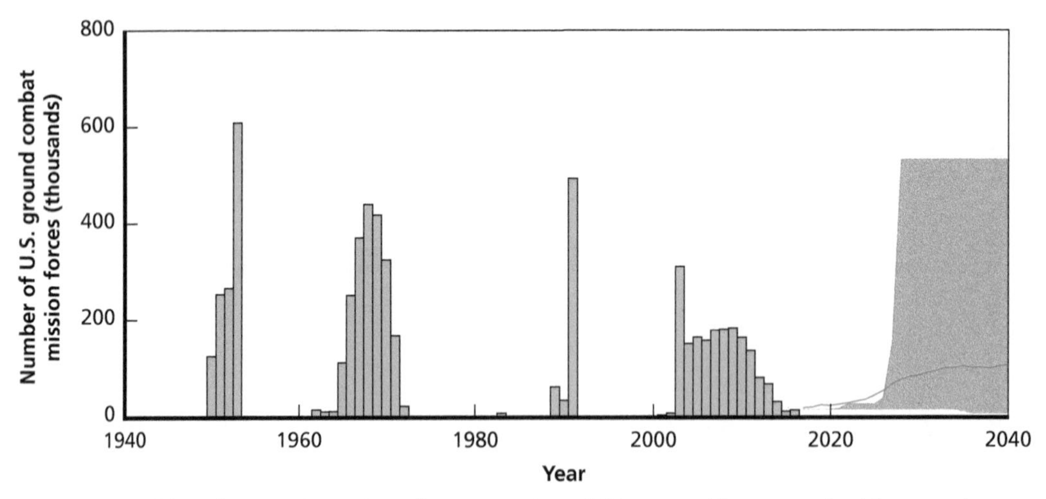

NOTES: The red line denotes the projected mean number of U.S. ground forces required for armed conflict interventions each year, based on 500 iterations of our forecasting model. The gray shaded area represents the range of forecasts bounded by the 10th and 90th percentiles of U.S. ground forces required for armed conflict interventions each year, based on 500 iterations of our forecasting model.

Figure C.12
Global Depression: Forecasts of U.S. Ground Stability Operations, 2017–2040

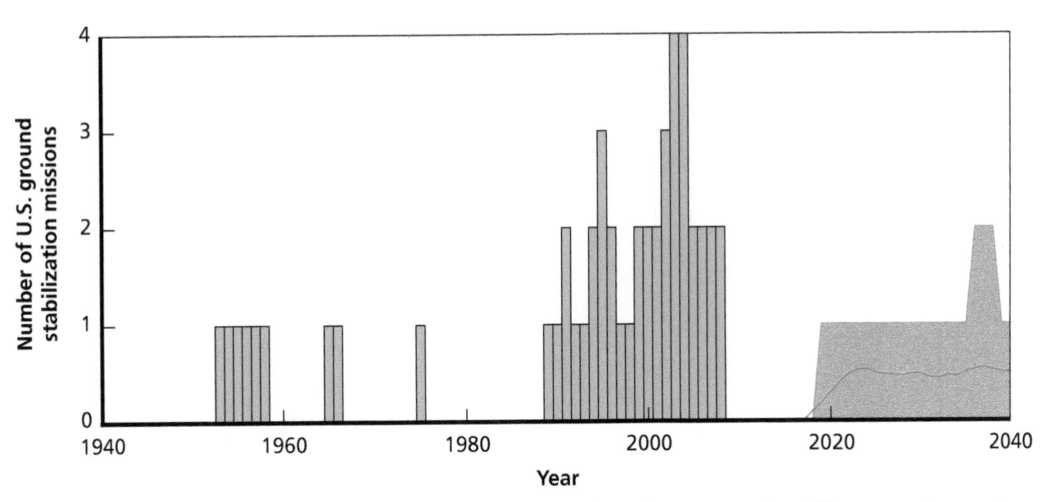

NOTES: The red line denotes the projected mean number of total U.S. ground stability operations each year, based on 500 iterations of our forecasting model. The gray shaded area represents the range of forecasts bounded by the 10th and 90th percentiles of U.S. ground stability operations each year, based on 500 iterations of our forecasting model.

Figure C.13
Global Depression: Forecasts of Demands for U.S. Ground Stability Operations Forces, 2017–2040

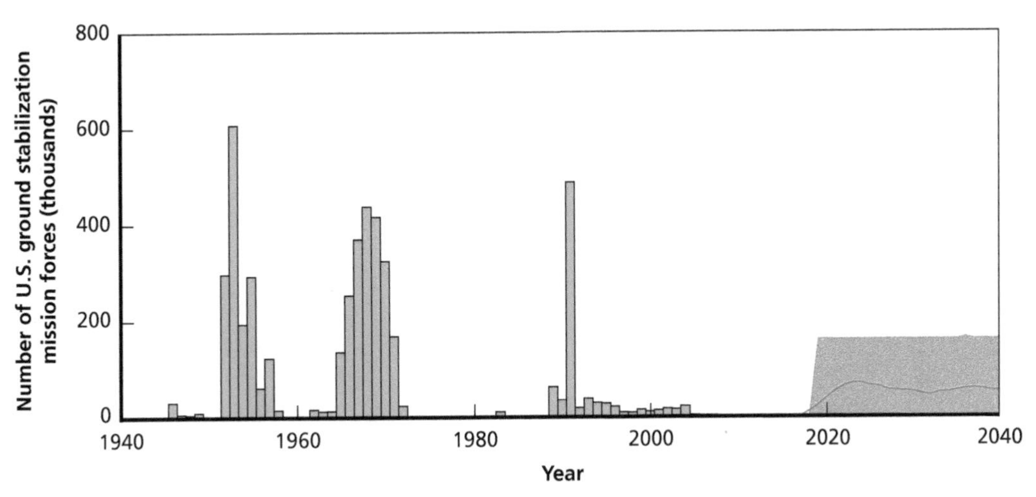

NOTES: The red line denotes the projected mean number of U.S. ground forces required for stability operations each year, based on 500 iterations of our forecasting model. The gray shaded area represents the range of forecasts bounded by the 10th and 90th percentiles of U.S. ground forces required for stability operations each year, based on 500 iterations of our forecasting model.

Figure C.14
Global Depression: Forecasts of U.S. Ground Deterrent Interventions, 2017–2040

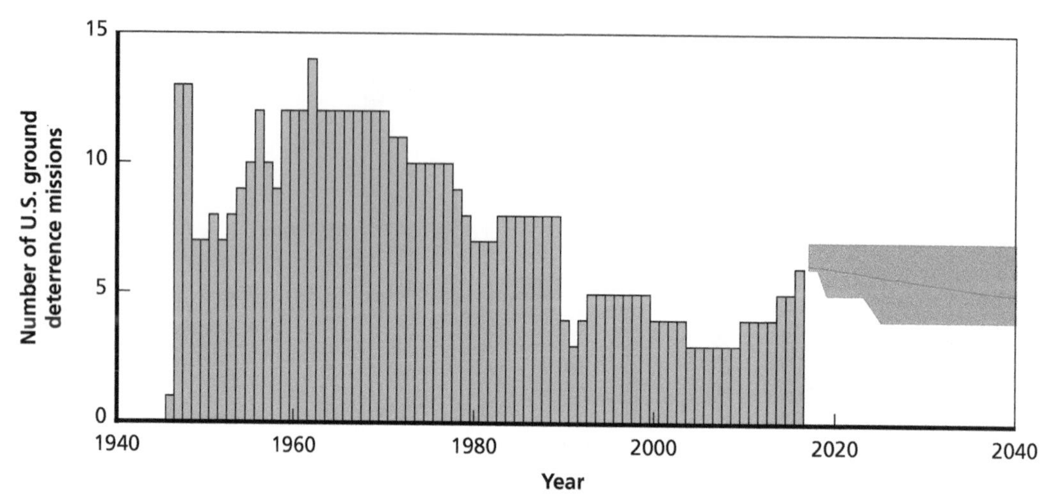

NOTES: The red line denotes the projected mean number of total U.S. ground deterrence missions each year, based on 500 iterations of our forecasting model. The gray shaded area represents the range of forecasts bounded by the 10th and 90th percentiles of U.S. ground deterrence missions each year, based on 500 iterations of our forecasting model.

Figure C.15
Global Depression: Forecasts of Demands for U.S. Ground Deterrent Intervention Forces, 2017–2040

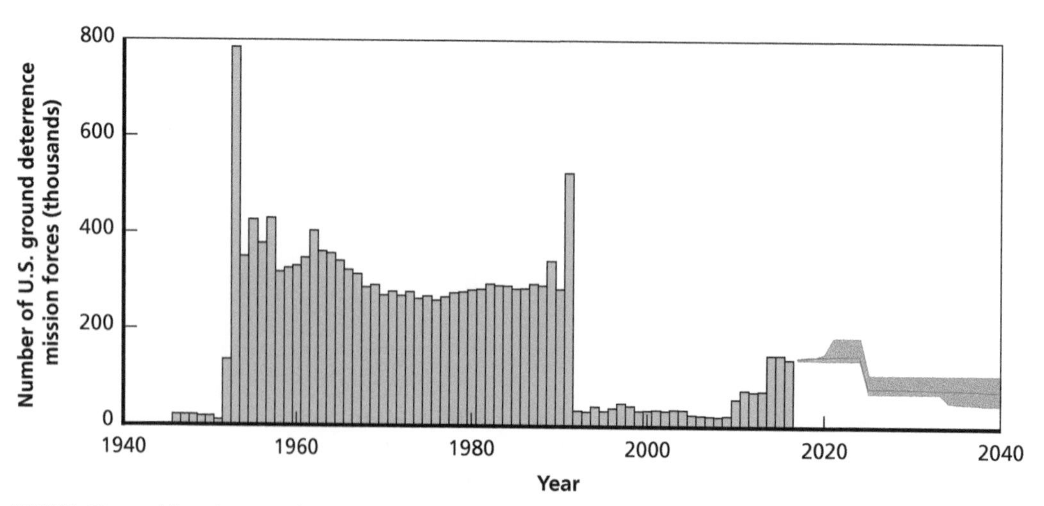

NOTES: The red line denotes the projected mean number of U.S. ground forces required for deterrence missions each year, based on 500 iterations of our forecasting model. The gray shaded area represents the range of forecasts bounded by the 10th and 90th percentiles of U.S. ground forces required for deterrence missions each year, based on 500 iterations of our forecasting model.

Figure C.16
Global Depression: Forecasts of Demands for Heavy U.S. Ground Intervention Forces, 2017–2040

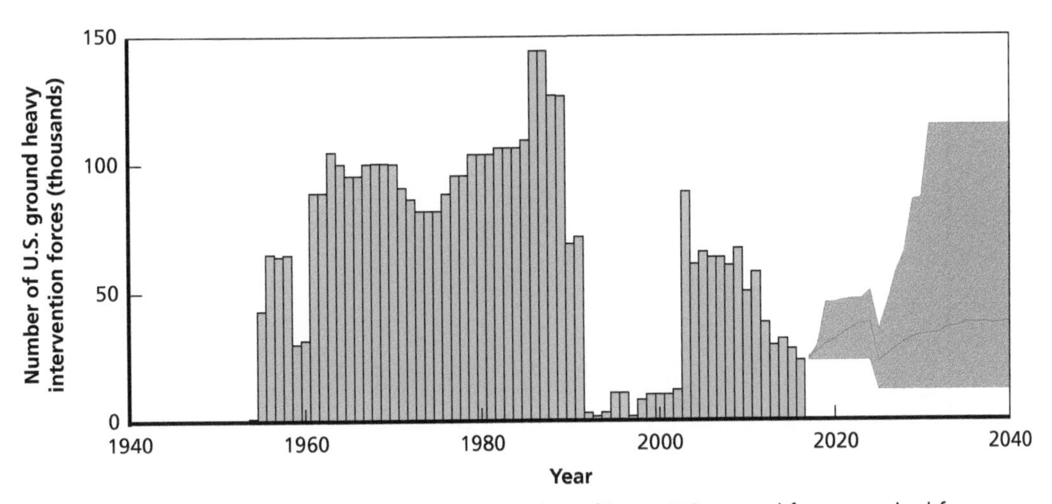

NOTES: The red line denotes the projected mean number of heavy U.S. ground forces required for interventions each year, based on 500 iterations of our forecasting model. The gray shaded area represents the range of forecasts bounded by the 10th and 90th percentiles of heavy U.S. ground forces required for interventions each year, based on 500 iterations of our forecasting model.

Forecasting Model Results for Alternative Scenario 2: Revisionist China

Figure C.17
Revisionist China: Forecasts of Interstate War Onsets, 2017–2040

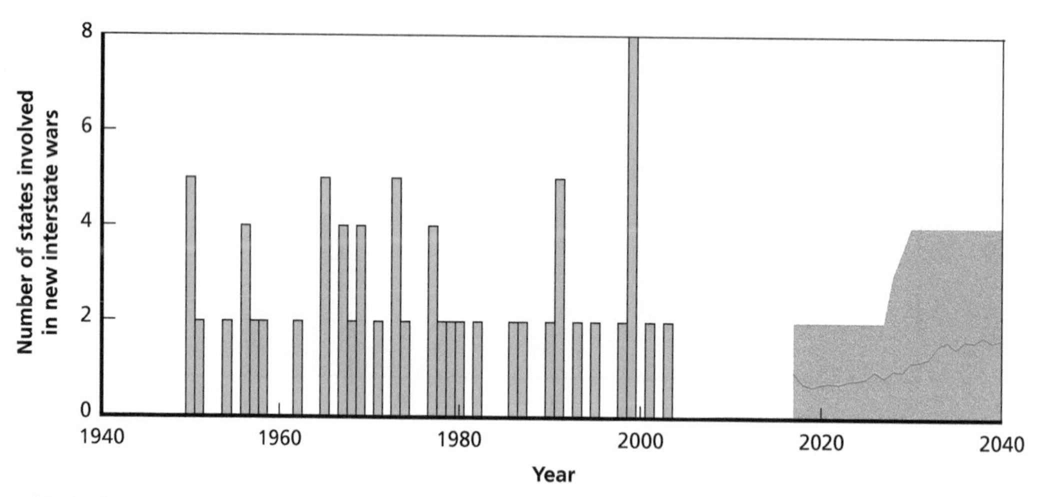

NOTES: The red line denotes the projected mean number of states involved in new interstate wars each year, based on 500 iterations of our forecasting model. The gray shaded area represents the range of forecasts bounded by the 10th and 90th percentiles of state involvement in interstate war onsets each year, based on 500 iterations of our forecasting model.t

Figure C.18
Revisionist China: Forecasts of Interstate War Occurrence, 2017–2040

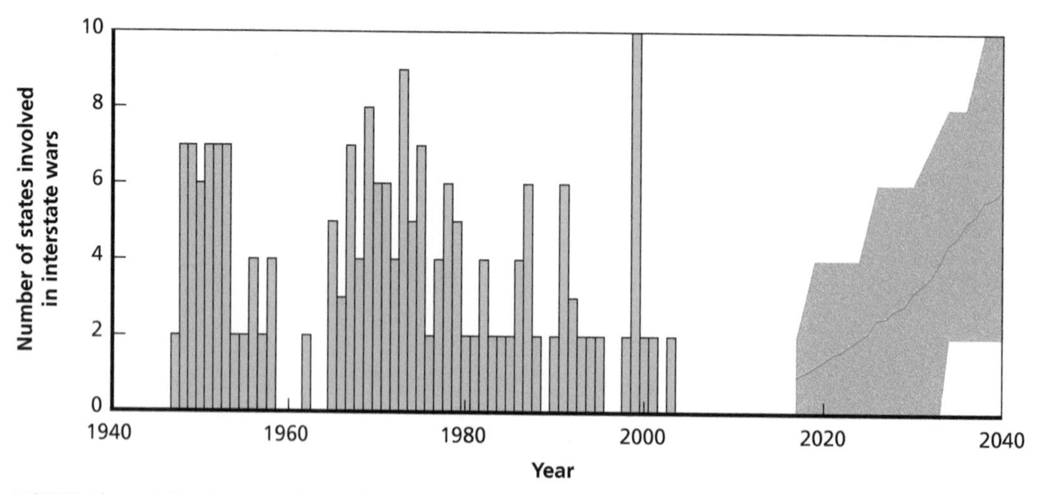

NOTES: The red line denotes the projected mean number of states involved in interstate war each year, based on 500 iterations of our forecasting model. The gray shaded area represents the range of forecasts bounded by the 10th and 90th percentiles of states involved in interstate war each year, based on 500 iterations of our forecasting model.

Figure C.19
Revisionist China: Regional Forecasts of Interstate War Occurrence, 2017–2040

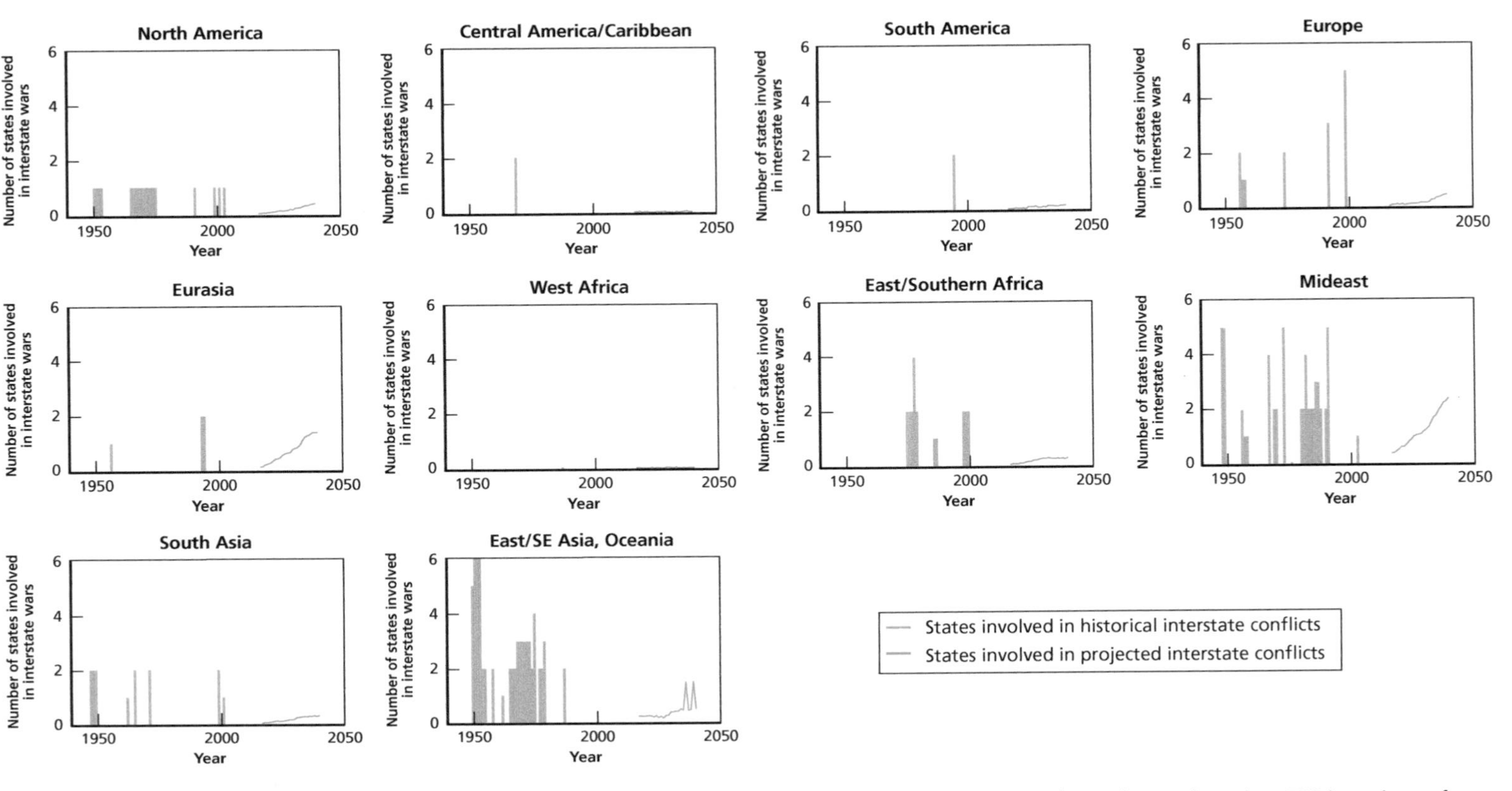

NOTE: The red lines denote the projected mean number of states involved in interstate wars in each region for each year, based on 500 iterations of our forecasting model.

Figure C.20
Revisionist China: Forecasts of Intrastate Conflict Onsets, 2017–2040

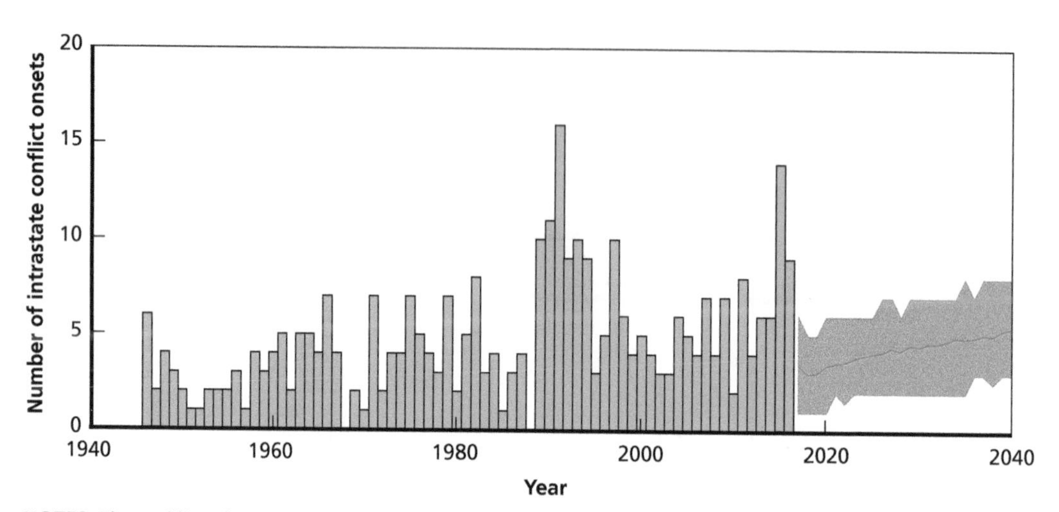

NOTES: The red line denotes the projected mean number of new intrastate conflict onsets each year, based on 500 iterations of our forecasting model. The gray shaded area represents the range of forecasts bounded by the 10th and 90th percentiles of intrastate conflict onsets each year, based on 500 iterations of our forecasting model.

Figure C.21
Revisionist China: Forecasts of Intrastate Conflict Occurrence, 2017–2040

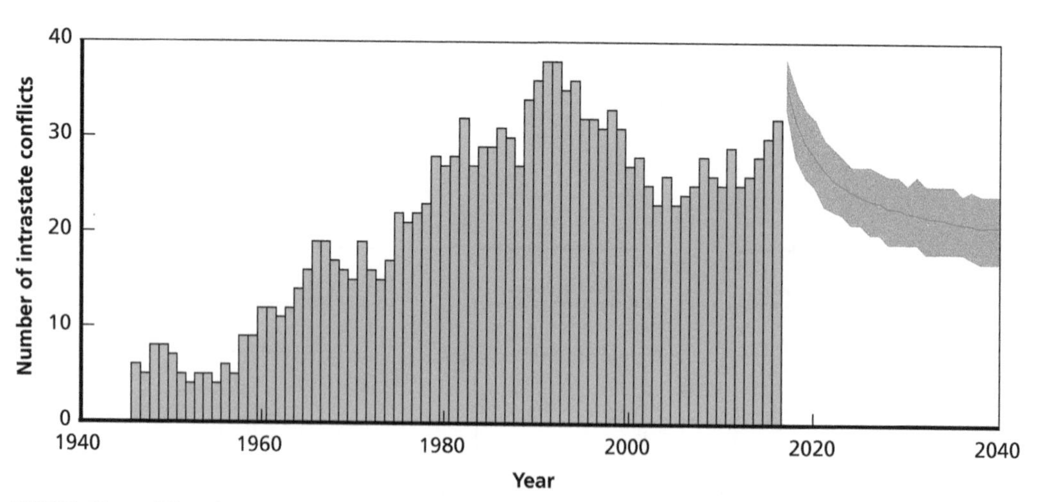

NOTES: The red line denotes the projected mean number of intrastate conflicts each year, based on 500 iterations of our forecasting model. The gray shaded area represents the range of forecasts bounded by the 10th and 90th percentiles of intrastate conflicts each year, based on 500 iterations of our forecasting model.

Figure C.22
Revisionist China: Regional Forecasts of Intrastate Conflict Occurrence, 2017–2040

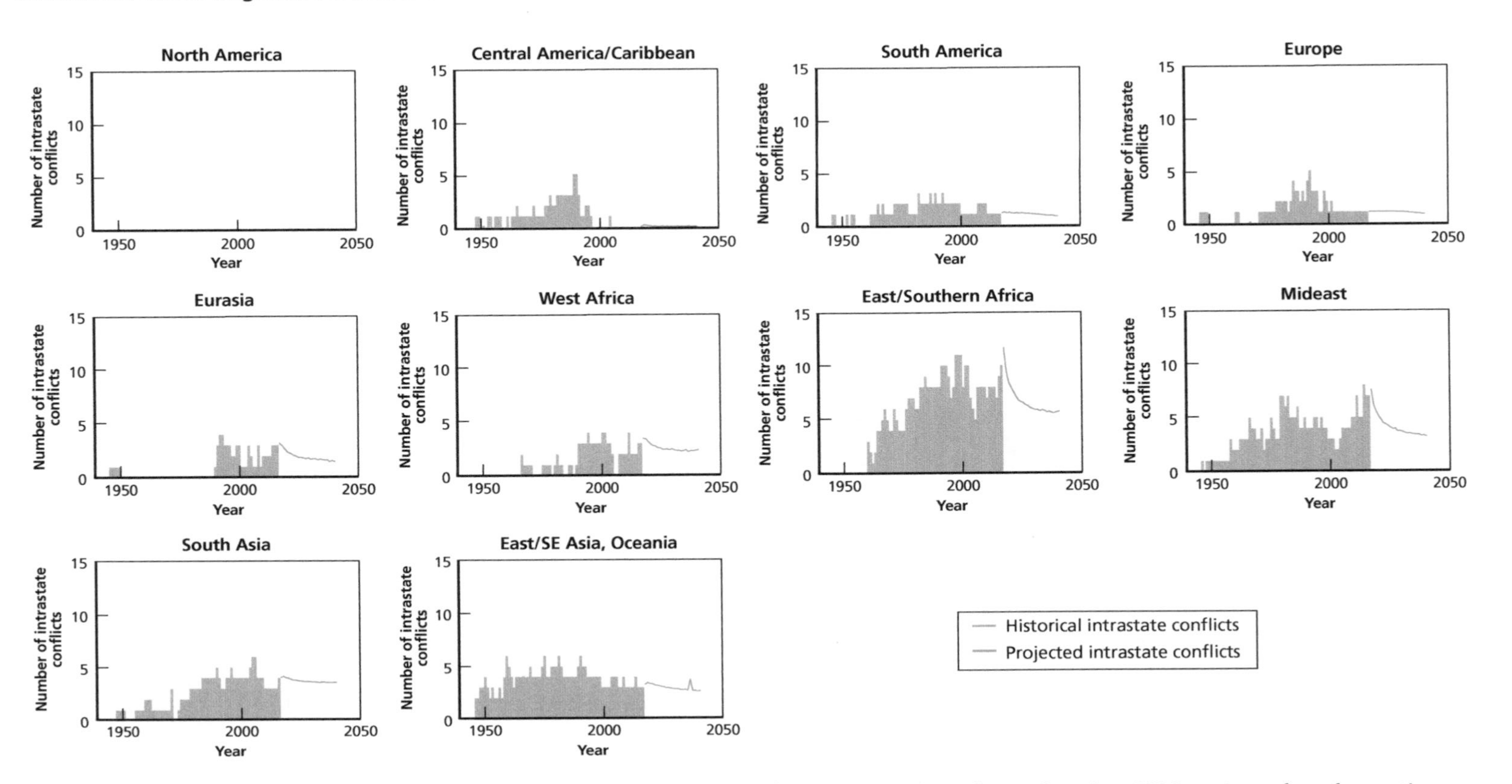

NOTE: The red lines denote the projected mean number of intrastate conflicts for each region in each year, based on 500 iterations of our forecasting model.

Figure C.23
Revisionist China: Forecasts of Total U.S. Ground Interventions, 2017–2040

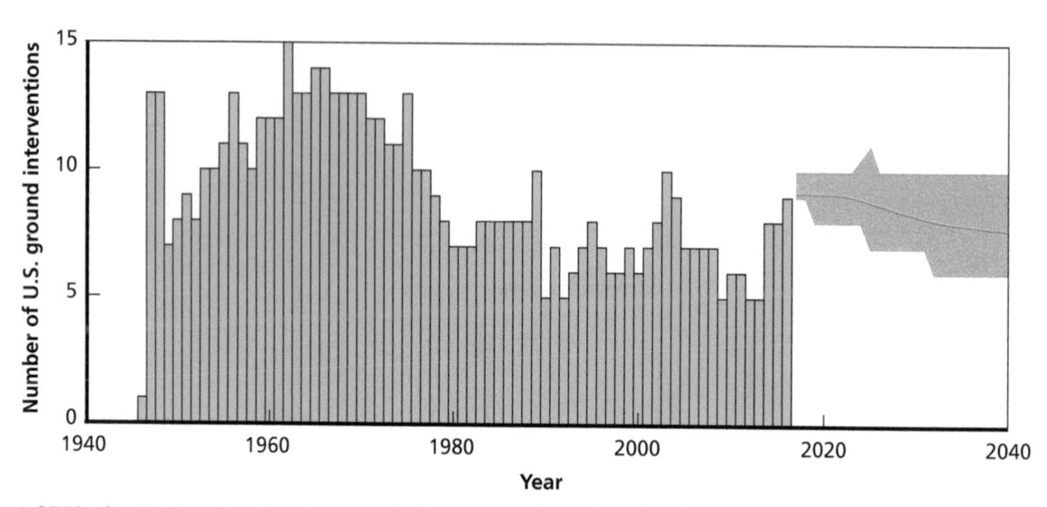

NOTES: The red line denotes the projected mean number of total U.S. ground interventions each year, based on 500 iterations of our forecasting model. The gray shaded area represents the range of forecasts bounded by the 10th and 90th percentiles of U.S. ground interventions each year, based on 500 iterations of our forecasting model.

Figure C.24
Revisionist China: Forecasts of Demands for U.S. Ground Intervention Forces, 2017–2040

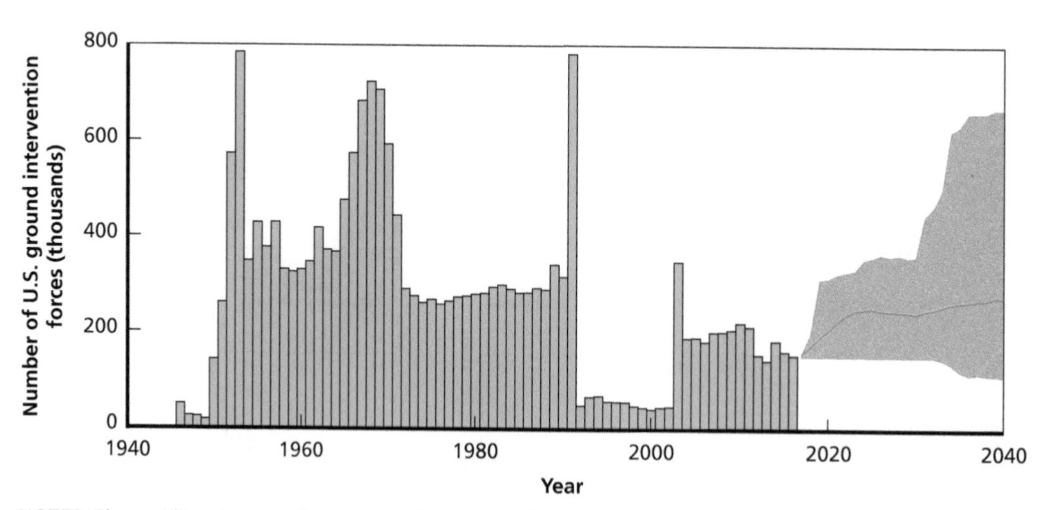

NOTES: The red line denotes the projected mean number of U.S. ground forces required for interventions each year, based on 500 iterations of our forecasting model. The gray shaded area represents the range of forecasts bounded by the 10th and 90th percentiles of U.S. ground forces required for interventions each year, based on 500 iterations of our forecasting model.

Figure C.25
Revisionist China: Regional Forecasts of Demands for U.S. Ground Intervention Forces, 2017–2040

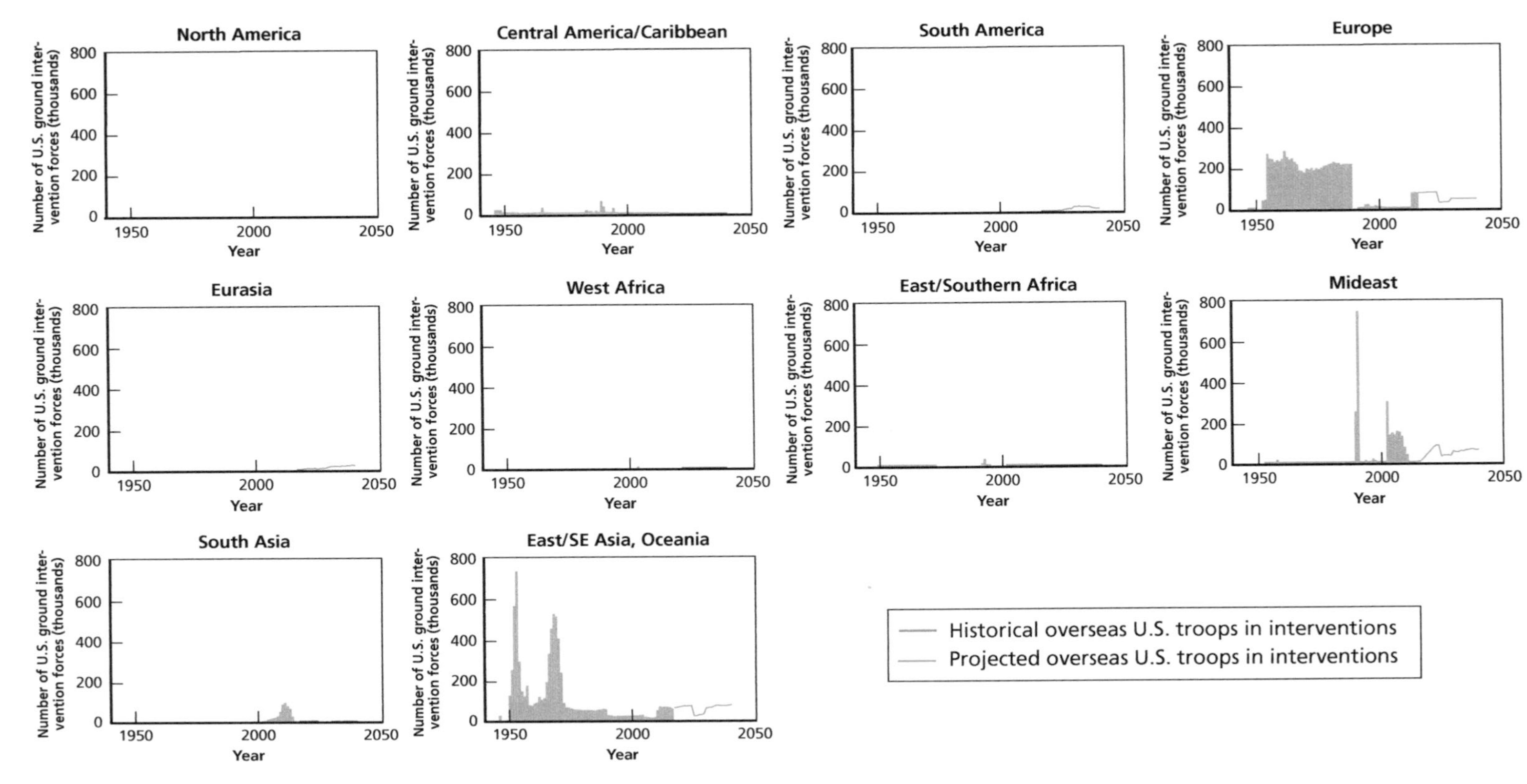

NOTE: The red line denotes the projected mean number of U.S. ground forces required for interventions each year, based on 500 iterations of our forecasting model.

Figure C.26
Revisionist China: Forecasts of U.S. Ground Interventions into Armed Conflicts, 2017–2040

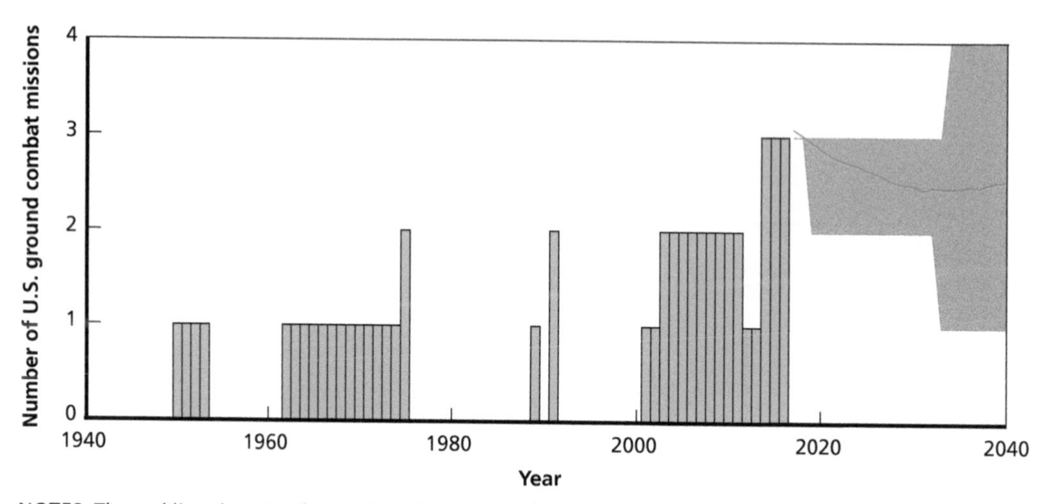

NOTES: The red line denotes the projected mean number of total U.S. ground armed conflict interventions each year, based on 500 iterations of our forecasting model. The gray shaded area represents the range of forecasts bounded by the 10th and 90th percentiles of U.S. ground armed conflict interventions each year, based on 500 iterations of our forecasting model.

Figure C.27
Revisionist China: Forecasts of Demands for U.S. Ground Combat Mission Forces, 2017–2040

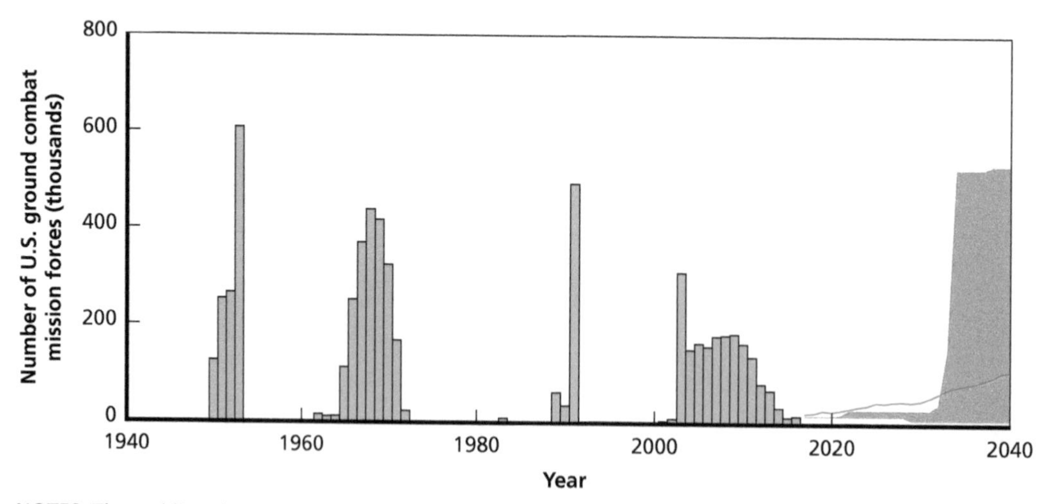

NOTES: The red line denotes the projected mean number of U.S. ground forces required for armed conflict interventions each year, based on 500 iterations of our forecasting model. The gray shaded area represents the range of forecasts bounded by the 10th and 90th percentiles of U.S. ground forces required for armed conflict interventions each year, based on 500 iterations of our forecasting model.

Figure C.28
Revisionist China: Forecasts of U.S. Ground Stability Operations, 2017–2040

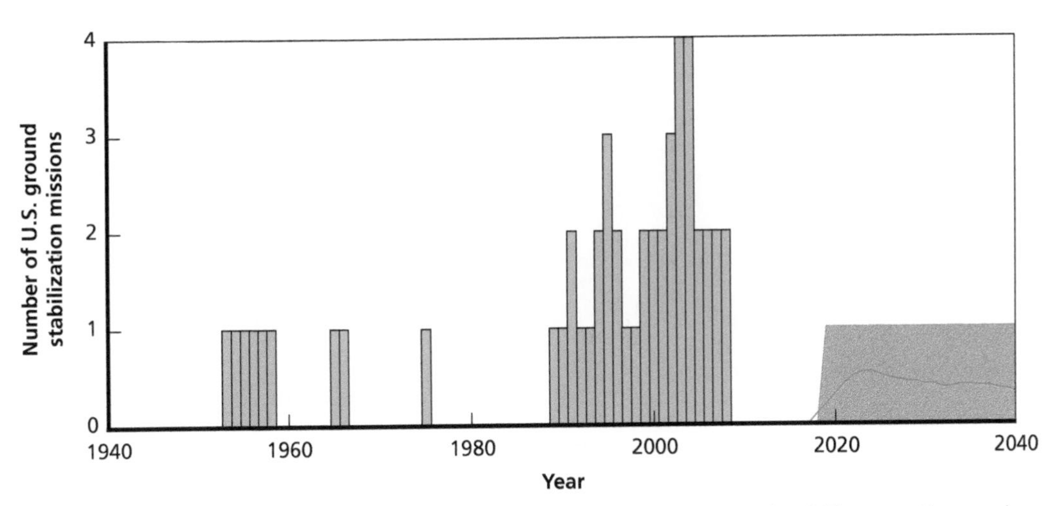

NOTES: The red line denotes the projected mean number of total U.S. ground stability operations each year, based on 500 iterations of our forecasting model. The gray shaded area represents the range of forecasts bounded by the 10th and 90th percentiles of U.S. ground stability operations each year, based on 500 iterations of our forecasting model.

Figure C.29
Revisionist China: Forecasts of Demands for U.S. Ground Stability Operations Forces, 2017–2040

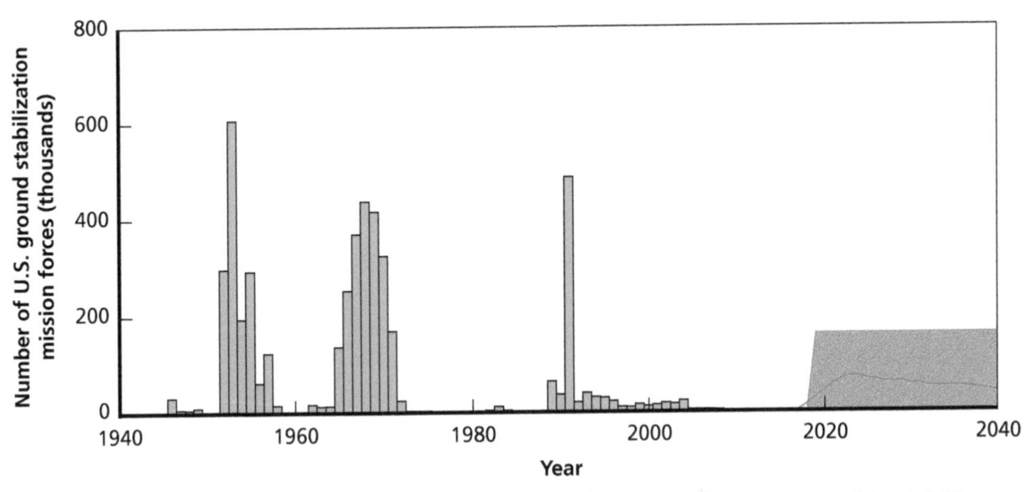

NOTES: The red line denotes the projected mean number of U.S. ground forces required for stability operations each year, based on 500 iterations of our forecasting model. The gray shaded area represents the range of forecasts bounded by the 10th and 90th percentiles of U.S. ground forces required for stability operations each year, based on 500 iterations of our forecasting model.

Figure C.30
Revisionist China: Forecasts of U.S. Ground Deterrent Interventions, 2017–2040

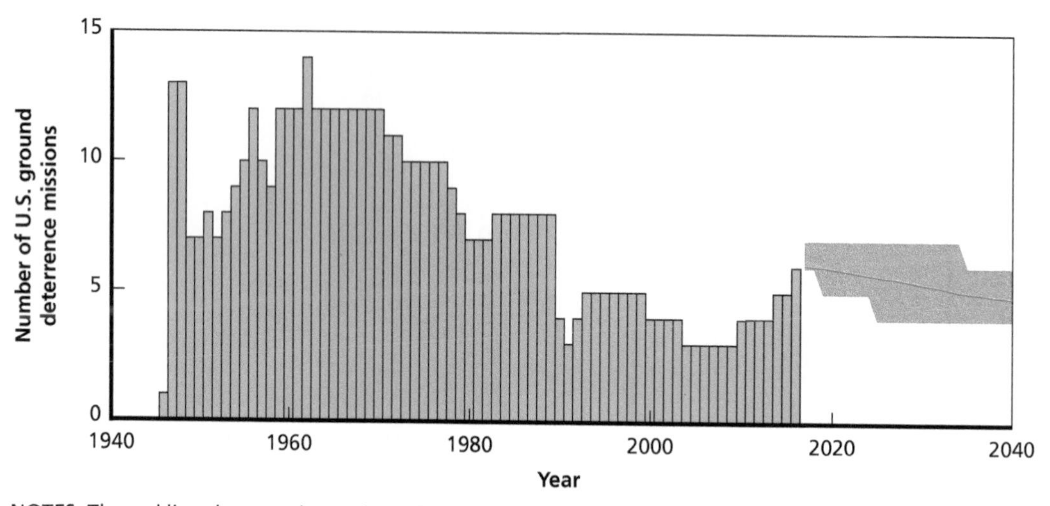

NOTES: The red line denotes the projected mean number of total U.S. ground deterrence missions each year, based on 500 iterations of our forecasting model. The gray shaded area represents the range of forecasts bounded by the 10th and 90th percentiles of U.S. ground deterrence missions each year, based on 500 iterations of our forecasting model.

Figure C.31
Revisionist China: Forecasts of Demands for U.S. Ground Deterrent Intervention Forces, 2017–2040

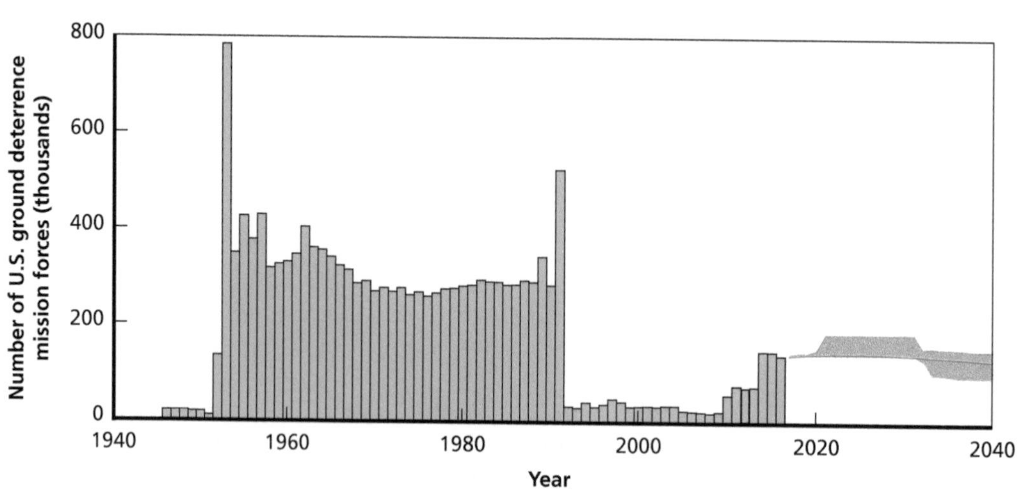

NOTES: The red line denotes the projected mean number of U.S. ground forces required for deterrence missions each year, based on 500 iterations of our forecasting model. The gray shaded area represents the range of forecasts bounded by the 10th and 90th percentiles of U.S. ground forces required for deterrence missions each year, based on 500 iterations of our forecasting model.

Figure C.32
Revisionist China: Forecasts of Demands for Heavy U.S. Ground Intervention Forces, 2017–2040

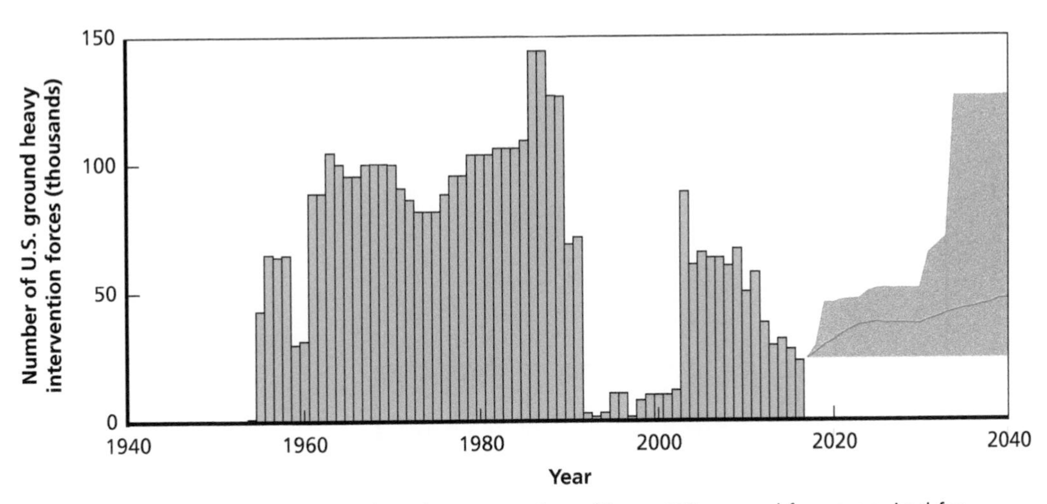

NOTES: The red line denotes the projected mean number of heavy U.S. ground forces required for interventions each year, based on 500 iterations of our forecasting model. The gray shaded area represents the range of forecasts bounded by the 10th and 90th percentiles of heavy U.S. ground forces required for interventions each year, based on 500 iterations of our forecasting model.

Forecasting Model Results for Alternative Scenario 3: Global Pandemic

Figure C.33
Global Pandemic: Forecasts of Interstate War Onsets, 2017–2040

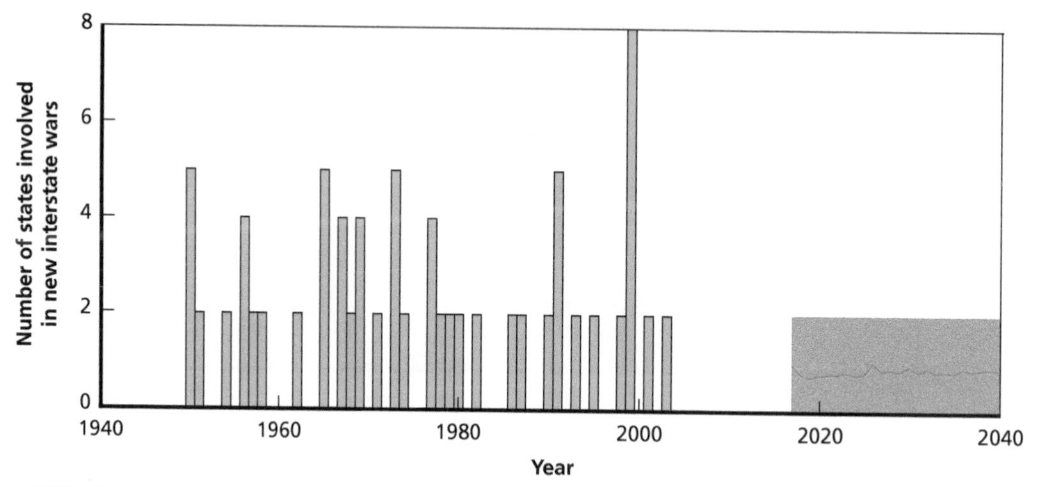

NOTES: The red line denotes the projected mean number of states involved in new interstate wars each year, based on 500 iterations of our forecasting model. The gray shaded area represents the range of forecasts bounded by the 10th and 90th percentiles of state involvement in interstate war onsets each year, based on 500 iterations of our forecasting model.

Figure C.34
Global Pandemic: Forecasts of Interstate War Occurrence, 2017–2040

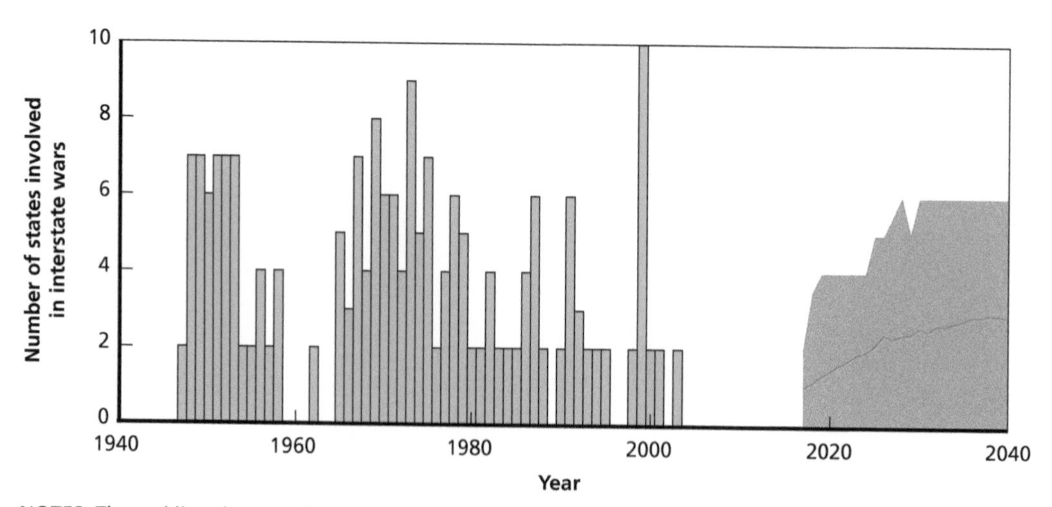

NOTES: The red line denotes the projected mean number of states involved in interstate war each year, based on 500 iterations of our forecasting model. The gray shaded area represents the range of forecasts bounded by the 10th and 90th percentiles of states involved in interstate war each year, based on 500 iterations of our forecasting model.

Figure C.35
Global Pandemic: Regional Forecasts of Interstate War Occurrence, 2017–2040

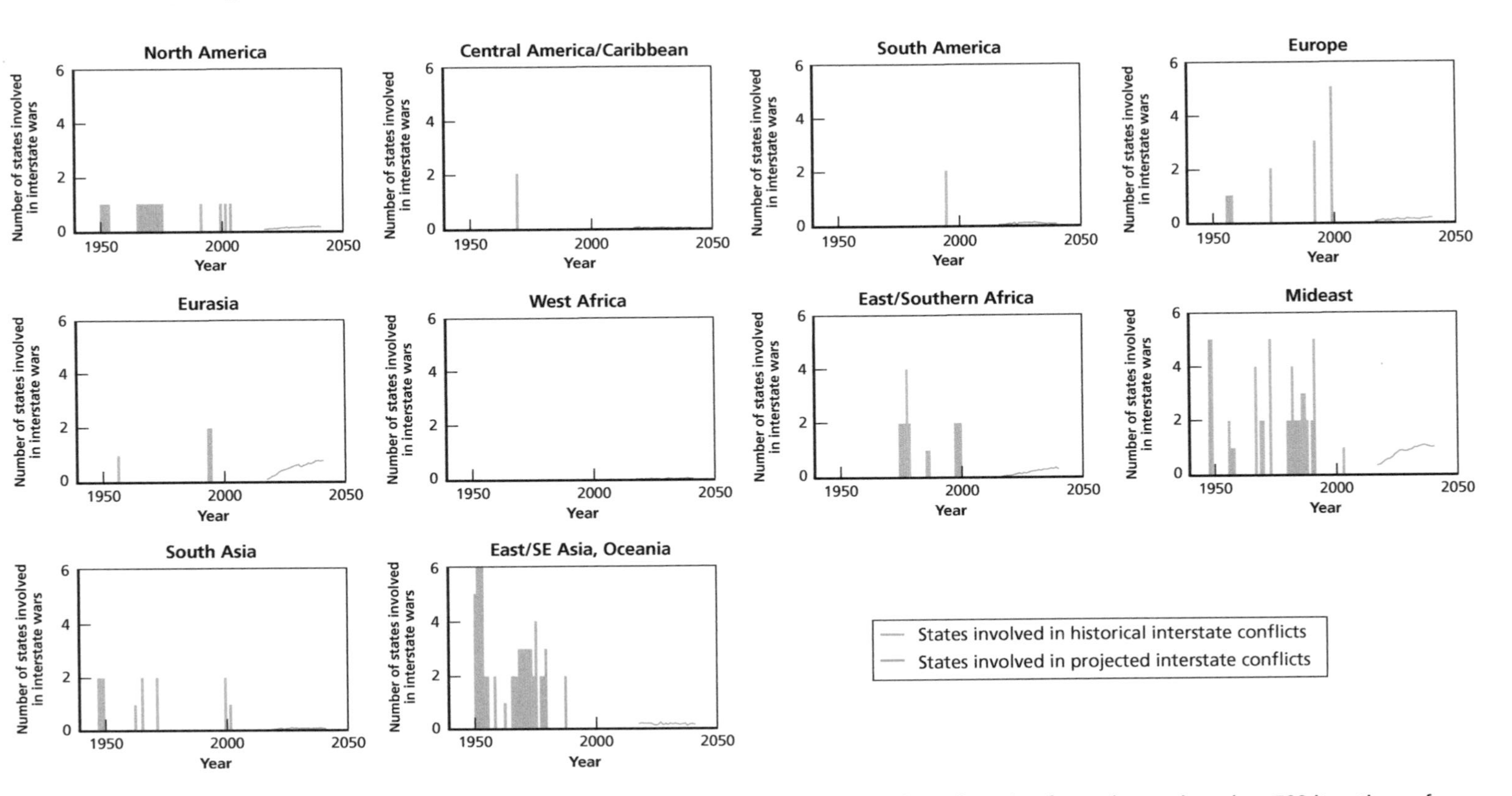

NOTE: The red lines denote the projected mean number of states involved in interstate wars in each region for each year, based on 500 iterations of our forecasting model.

Figure C.36
Global Pandemic: Forecasts of Intrastate Conflict Onsets, 2017–2040

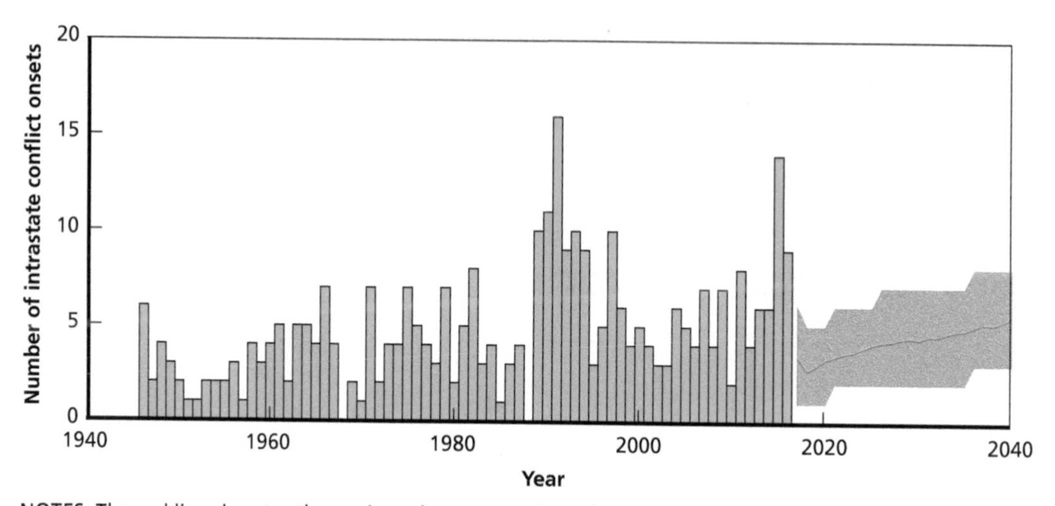

NOTES: The red line denotes the projected mean number of new intrastate conflict onsets each year, based on 500 iterations of our forecasting model. The gray shaded area represents the range of forecasts bounded by the 10th and 90th percentiles of intrastate conflict onsets each year, based on 500 iterations of our forecasting model

Figure C.37
Global Pandemic: Forecasts of Intrastate Conflict Occurrence, 2017–2040

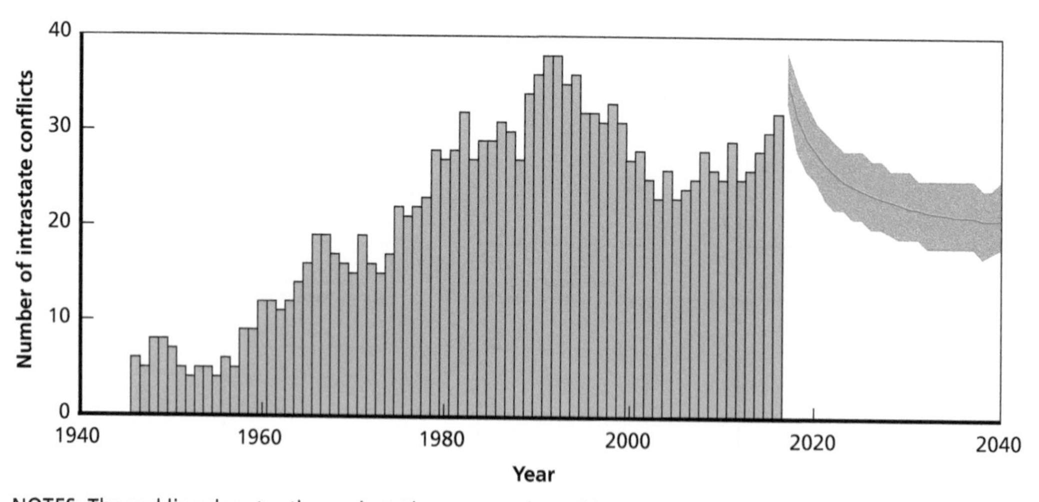

NOTES: The red line denotes the projected mean number of intrastate conflicts each year, based on 500 iterations of our forecasting model. The gray shaded area represents the range of forecasts bounded by the 10th and 90th percentiles of intrastate conflicts each year, based on 500 iterations of our forecasting model.

Figure C.38
Global Pandemic: Regional Forecasts of Intrastate Conflict Occurrence, 2017–2040

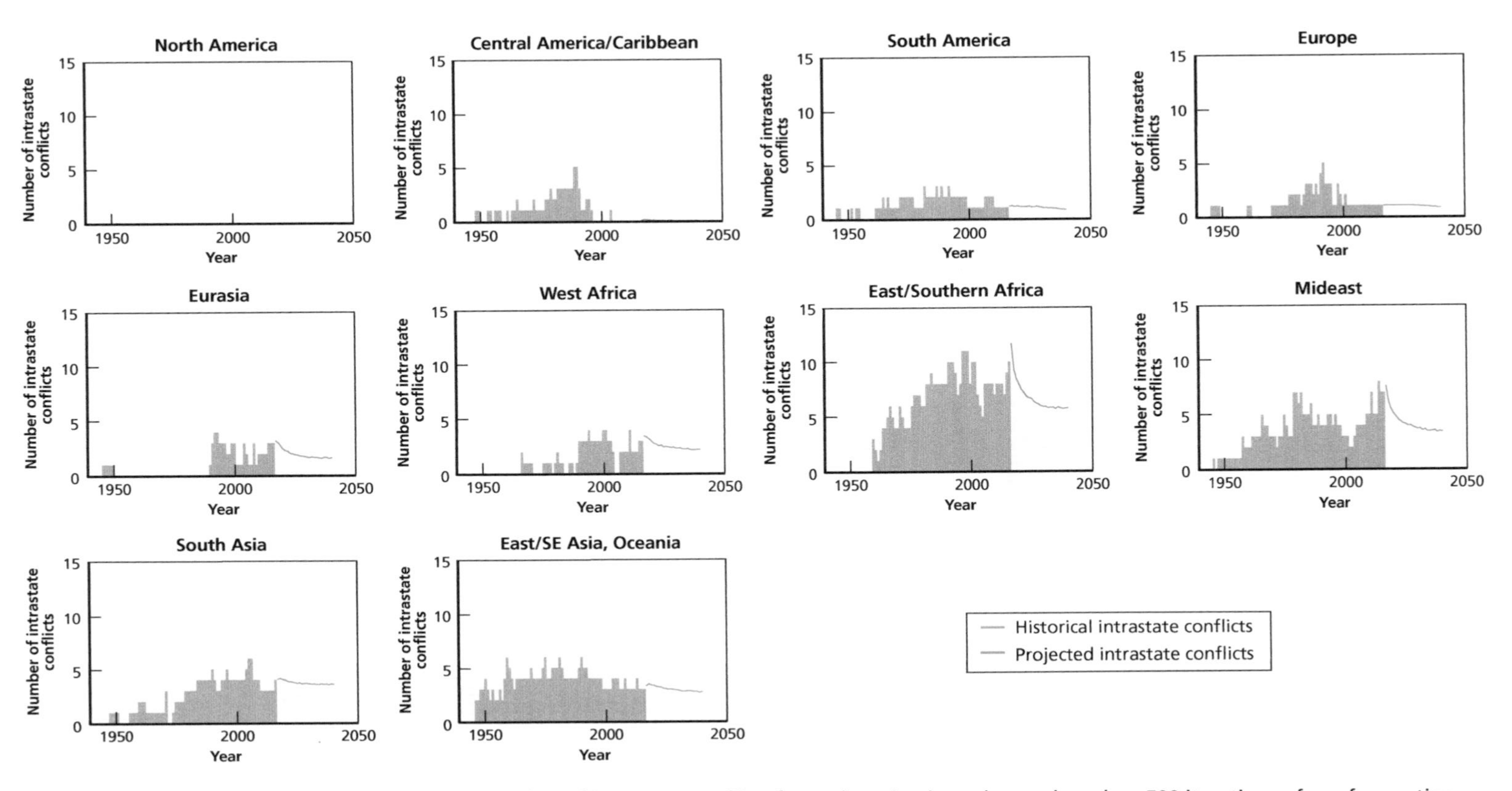

NOTE: The red lines denote the projected mean number of intrastate conflicts for each region in each year, based on 500 iterations of our forecasting model.

Figure C.39
Global Pandemic: Forecasts of Total U.S. Ground Interventions, 2017–2040

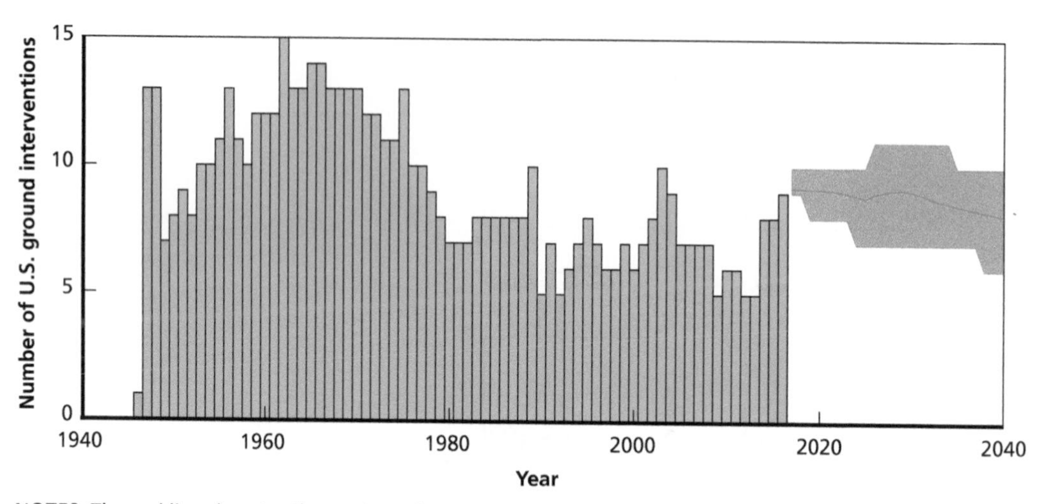

NOTES: The red line denotes the projected mean number of total U.S. ground interventions each year, based on 500 iterations of our forecasting model. The gray shaded area represents the range of forecasts bounded by the 10th and 90th percentiles of U.S. ground interventions each year, based on 500 iterations of our forecasting model.

Figure C.40
Global Pandemic: Forecasts of Demands for U.S. Ground Intervention Forces, 2017–2040

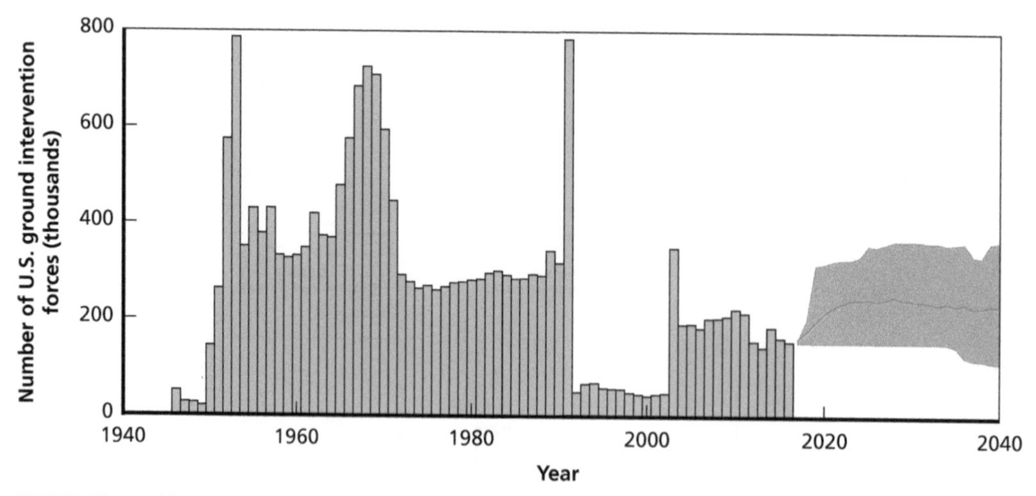

NOTES: The red line denotes the projected mean number of U.S. ground forces required for interventions each year, based on 500 iterations of our forecasting model. The gray shaded area represents the range of forecasts bounded by the 10th and 90th percentiles of U.S. ground forces required for interventions each year, based on 500 iterations of our forecasting model.

Figure C.41
Global Pandemic: Regional Forecasts of Demands for U.S. Ground Intervention Forces, 2017–2040

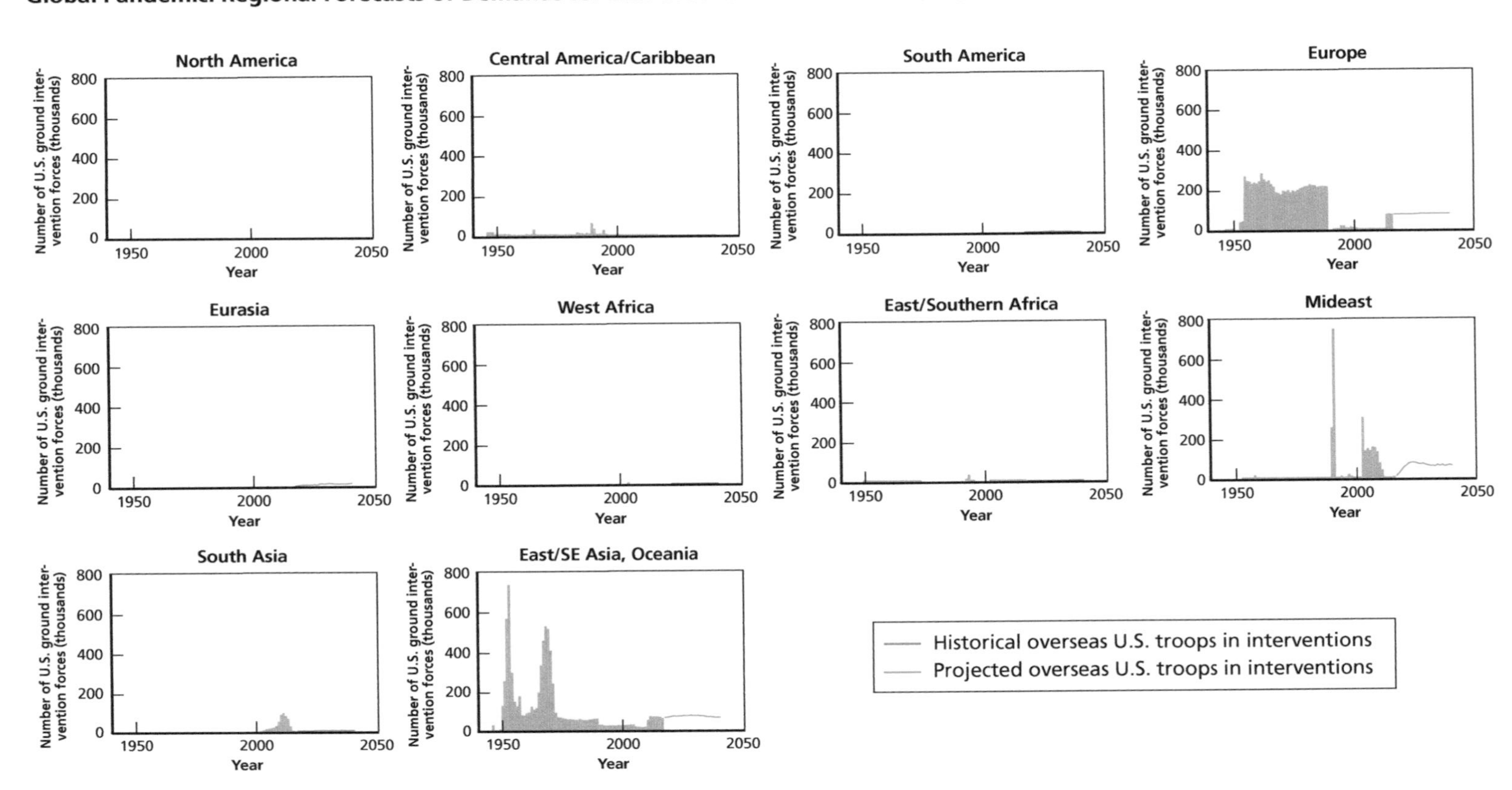

NOTE: The red line denotes the projected mean number of U.S. ground forces required for interventions each year, based on 500 iterations of our forecasting model.

Figure C.42
Global Pandemic: Forecasts of U.S. Ground Interventions into Armed Conflicts, 2017–2040

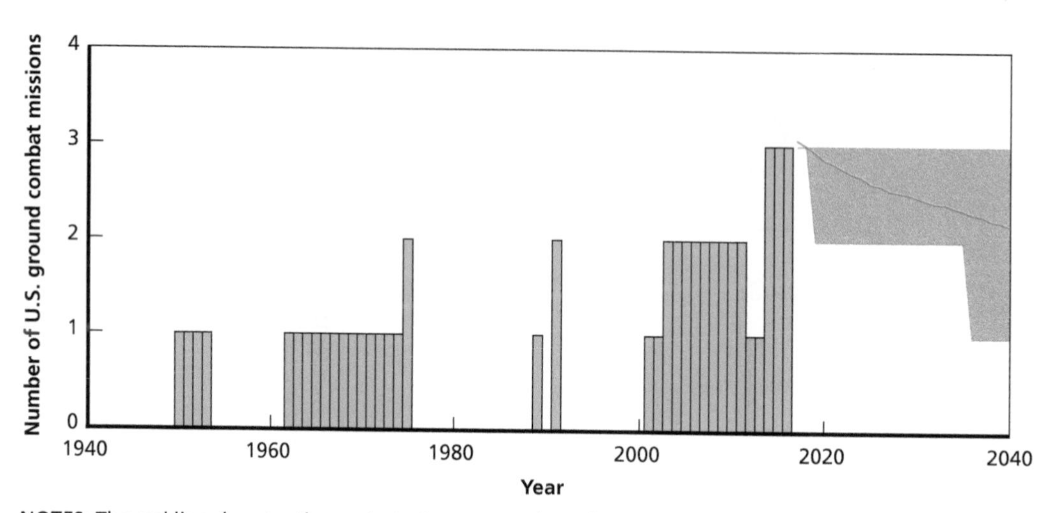

NOTES: The red line denotes the projected mean number of total U.S. ground armed conflict interventions each year, based on 500 iterations of our forecasting model. The gray shaded area represents the range of forecasts bounded by the 10th and 90th percentiles of U.S. ground armed conflict interventions each year, based on 500 iterations of our forecasting model.

Figure C.43
Global Pandemic: Forecasts of Demands for U.S. Ground Combat Mission Forces, 2017–2040

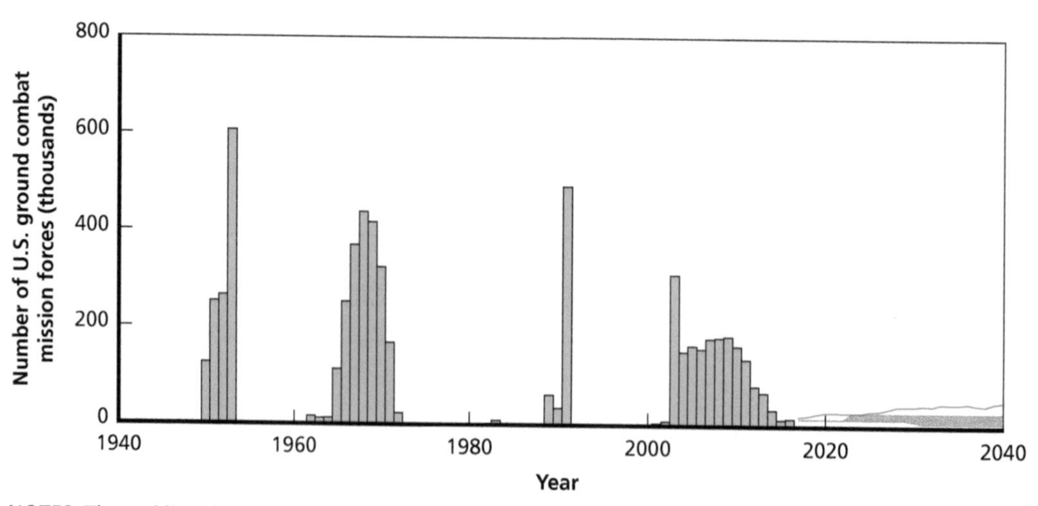

NOTES: The red line denotes the projected mean number of U.S. ground forces required for armed conflict interventions each year, based on 500 iterations of our forecasting model. The gray shaded area represents the range of forecasts bounded by the 10th and 90th percentiles of U.S. ground forces required for armed conflict interventions each year, based on 500 iterations of our forecasting model.

Figure C.44
Global Pandemic: Forecasts of U.S. Ground Stability Operations, 2017–2040

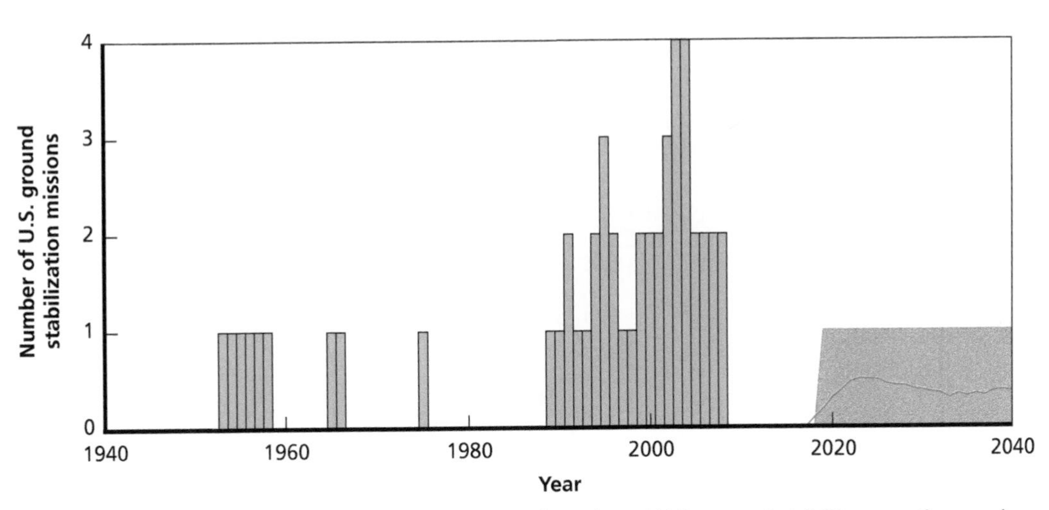

NOTES: The red line denotes the projected mean number of total U.S. ground stability operations each year, based on 500 iterations of our forecasting model. The gray shaded area represents the range of forecasts bounded by the 10th and 90th percentiles of U.S. ground stability operations each year, based on 500 iterations of our forecasting model.

Figure C.45
Global Pandemic: Forecasts of Demands for U.S. Ground Stability Operations Forces, 2017–2040

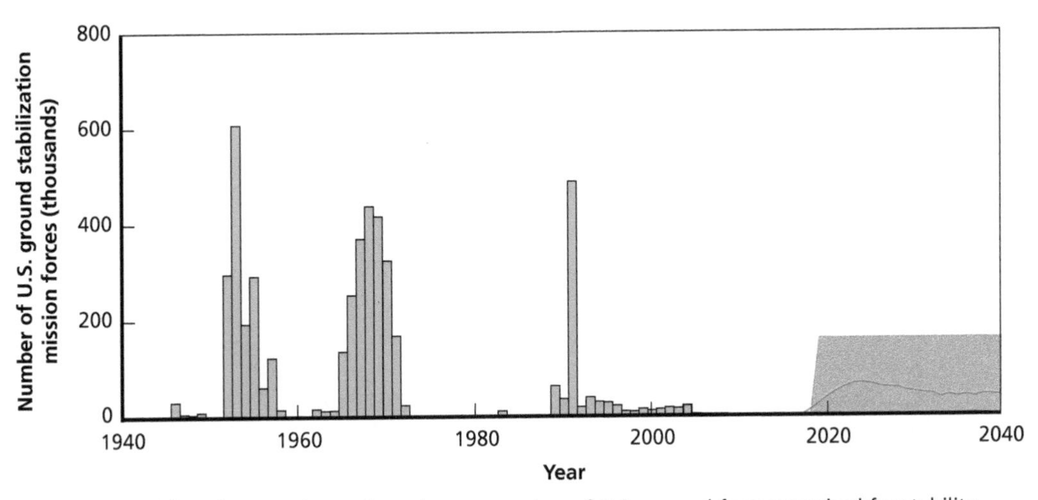

NOTES: The red line denotes the projected mean number of U.S. ground forces required for stability operation interventions each year, based on 500 iterations of our forecasting model. The gray shaded area represents the range of forecasts bounded by the 10th and 90th percentiles of U.S. ground forces required for stability operation interventions each year, based on 500 iterations of our forecasting model.

Figure C.46
Global Pandemic: Forecasts of U.S. Ground Deterrent Interventions, 2017–2040

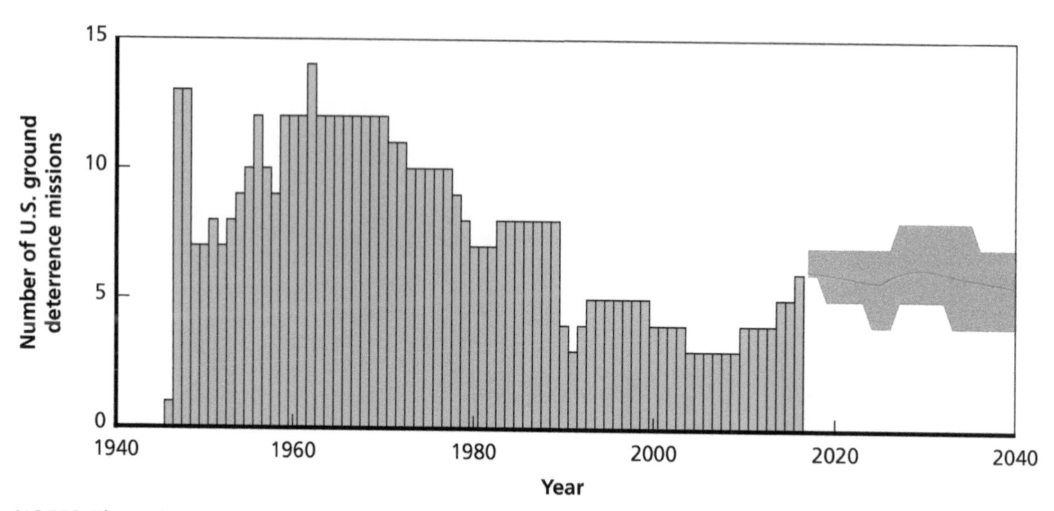

NOTES: The red line denotes the projected mean number of total U.S. ground deterrence missions each year, based on 500 iterations of our forecasting model. The gray shaded area represents the range of forecasts bounded by the 10th and 90th percentiles of U.S. ground deterrence missions each year, based on 500 iterations of our forecasting model.

Figure C.47
Global Pandemic: Forecasts of Demands for U.S. Ground Deterrent Intervention Forces, 2017–2040

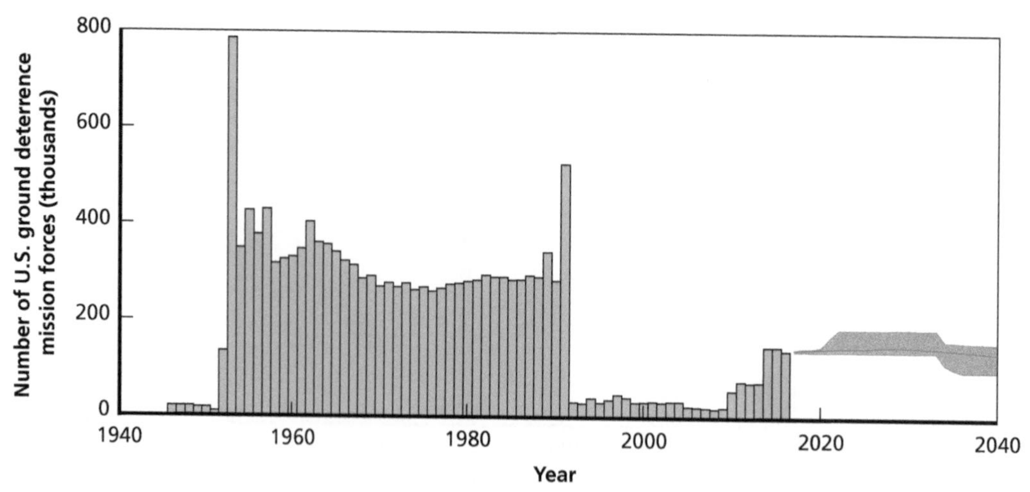

NOTES: The red line denotes the projected mean number of U.S. ground forces required for deterrence missions each year, based on 500 iterations of our forecasting model. The gray shaded area represents the range of forecasts bounded by the 10th and 90th percentiles of U.S. ground forces required for deterrence missions each year, based on 500 iterations of our forecasting model.

Figure C.48
Global Pandemic: Forecasts of Demands for Heavy U.S. Ground Intervention Forces, 2017–2040

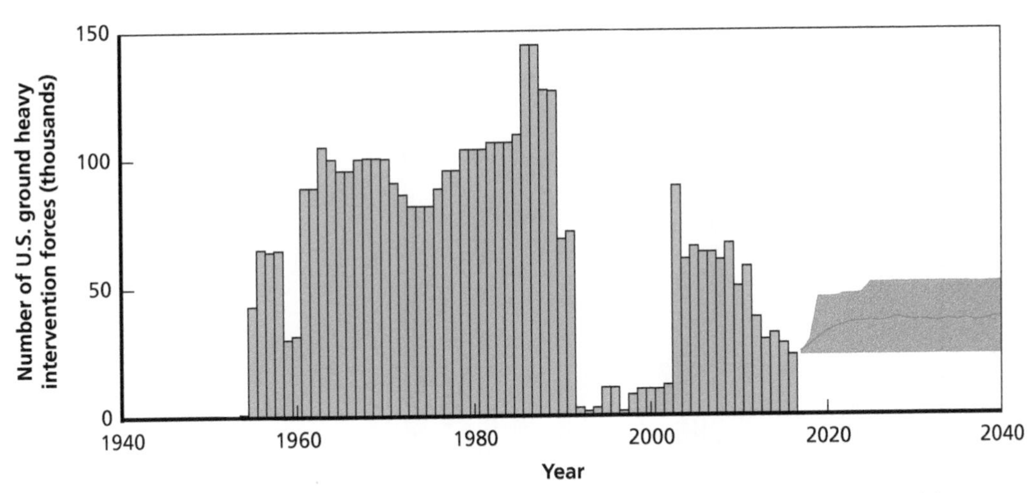

NOTES: The red line denotes the projected mean number of heavy U.S. ground forces required for interventions each year, based on 500 iterations of our forecasting model. The gray shaded area represents the range of forecasts bounded by the 10th and 90th percentiles of heavy U.S. ground forces required for interventions each year, based on 500 iterations of our forecasting model.

Forecasting Model Results for Alternative Scenario 4: U.S. Isolationism

Figure C.49
U.S. Isolationism: Forecasts of Interstate War Onsets, 2017–2040

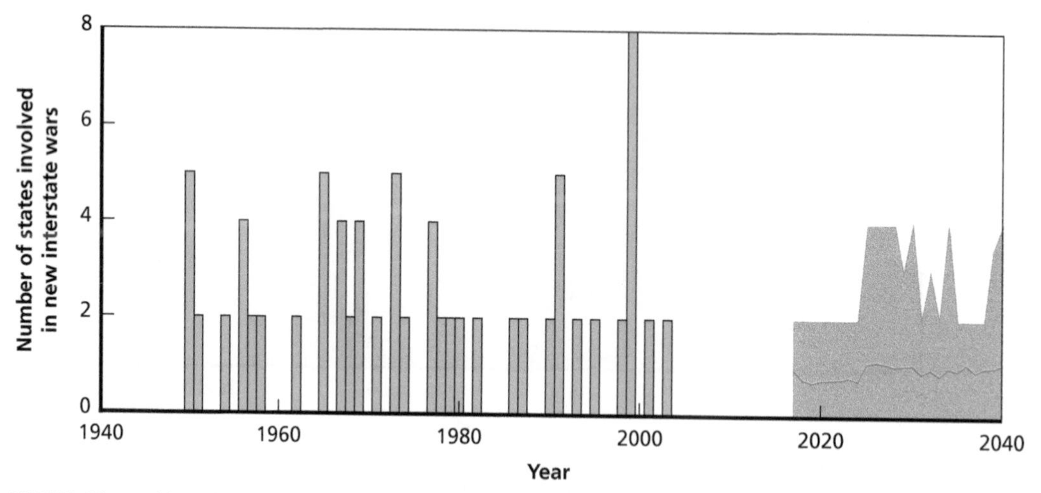

NOTES: The red line denotes the projected mean number of states involved in new interstate wars each year, based on 500 iterations of our forecasting model. The gray shaded area represents the range of forecasts bounded by the 10th and 90th percentiles of state involvement in interstate war onsets each year, based on 500 iterations of our forecasting model.

Figure C.50
U.S. Isolationism: Forecasts of Interstate War Occurrence, 2017–2040

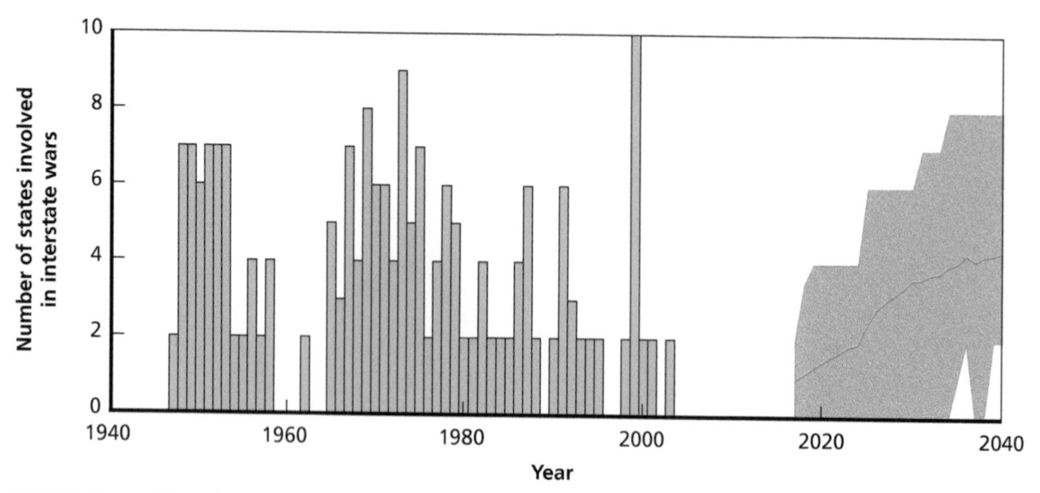

NOTES: The red line denotes the projected mean number of states involved in interstate war each year, based on 500 iterations of our forecasting model. The gray shaded area represents the range of forecasts bounded by the 10th and 90th percentiles of states involved in interstate war each year, based on 500 iterations of our forecasting model.

Figure C.51
U.S. Isolationism: Regional Forecasts of Interstate War Occurrence, 2017–2040

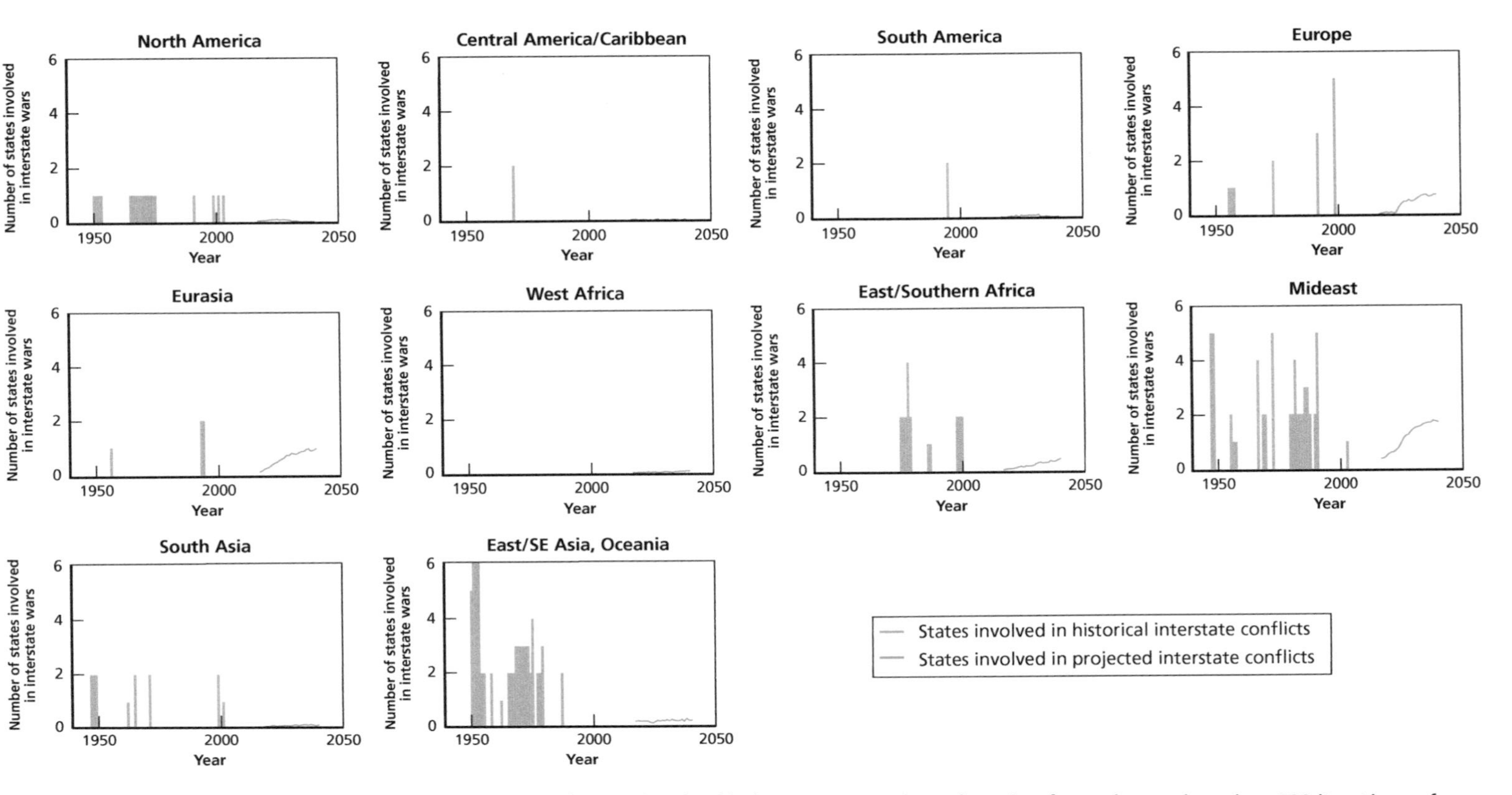

NOTE: The red lines denote the projected mean number of states involved in interstate wars in each region for each year, based on 500 iterations of our forecasting model.

Figure C.52
U.S. Isolationism: Forecasts of Intrastate Conflict Onsets, 2017–2040

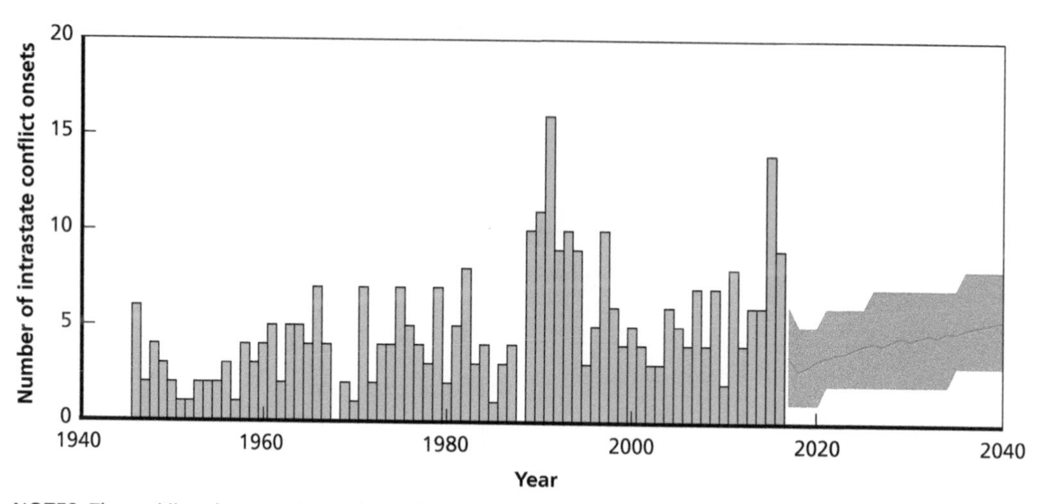

NOTES: The red line denotes the projected mean number of new intrastate conflict onsets each year, based on 500 iterations of our forecasting model. The gray shaded area represents the range of forecasts bounded by the 10th and 90th percentiles of intrastate conflict onsets each year, based on 500 iterations of our forecasting model.

Figure C.53
U.S. Isolationism: Forecasts of Intrastate Conflict Occurrence, 2017–2040

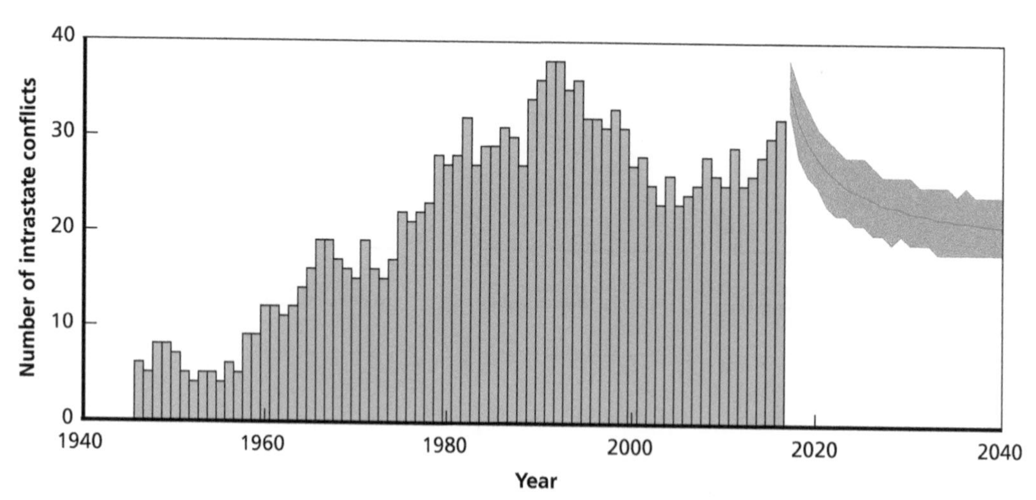

NOTES: The red line denotes the projected mean number of intrastate conflicts each year, based on 500 iterations of our forecasting model. The gray shaded area represents the range of forecasts bounded by the 10th and 90th percentiles of intrastate conflicts each year, based on 500 iterations of our forecasting model.

Figure C.54
U.S. Isolationism: Regional Forecasts of Intrastate Conflict Occurrence, 2017–2040

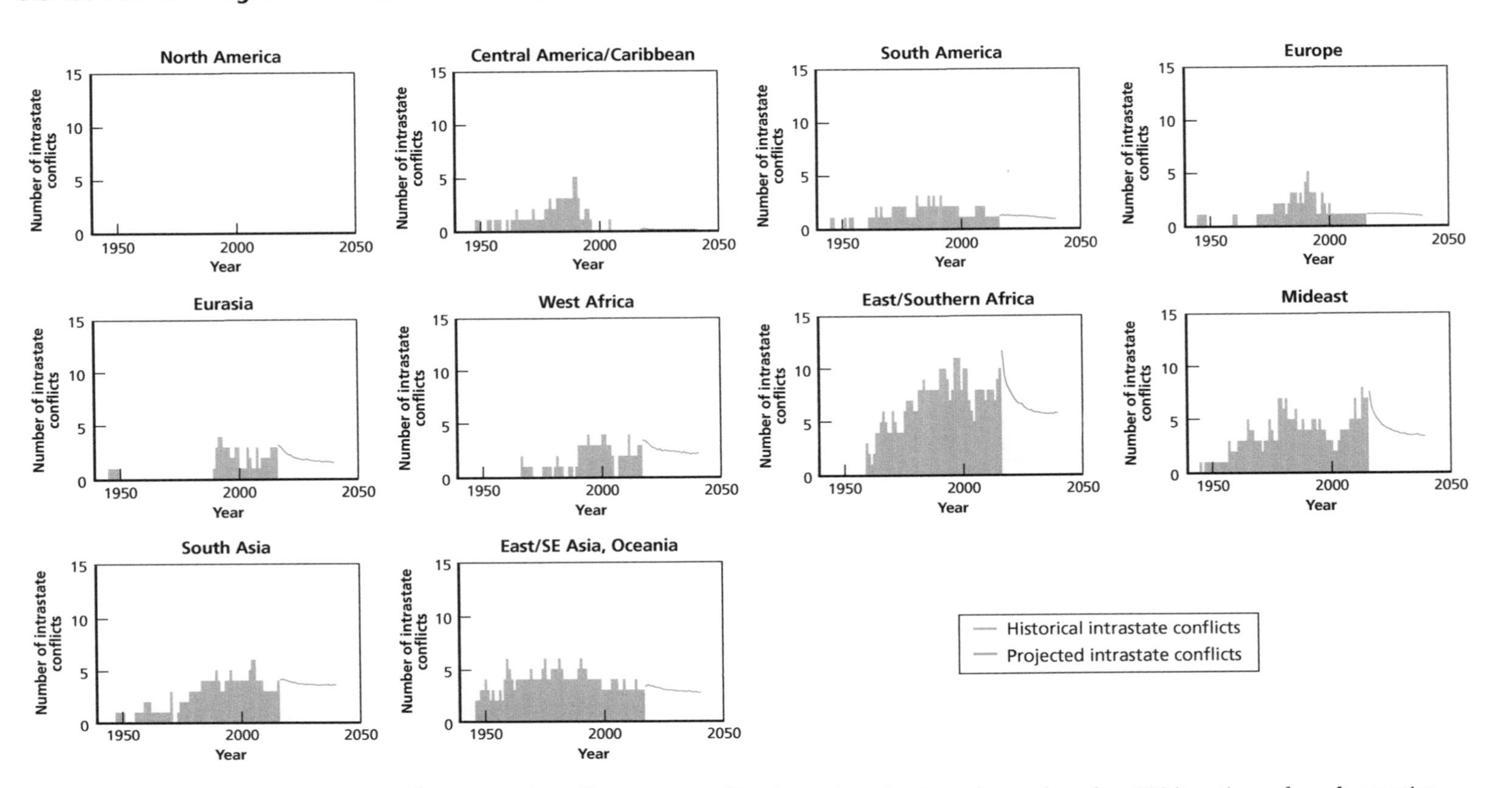

NOTE: The red lines denote the projected mean number of intrastate conflicts for each region in each year, based on 500 iterations of our forecasting model.

Figure C.55
U.S. Isolationism: Forecasts of Total U.S. Ground Interventions, 2017–2040

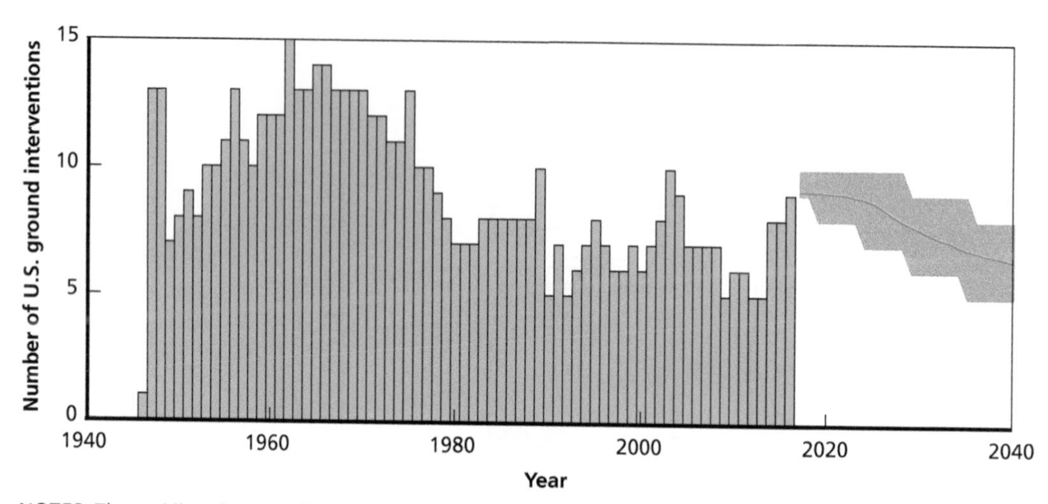

NOTES: The red line denotes the projected mean number of total U.S. ground interventions each year, based on 500 iterations of our forecasting model. The gray shaded area represents the range of forecasts bounded by the 10th and 90th percentiles of U.S. ground interventions each year, based on 500 iterations of our forecasting model.

Figure C.56
U.S. Isolationism: Forecasts of Demands for U.S. Ground Intervention Forces, 2017–2040

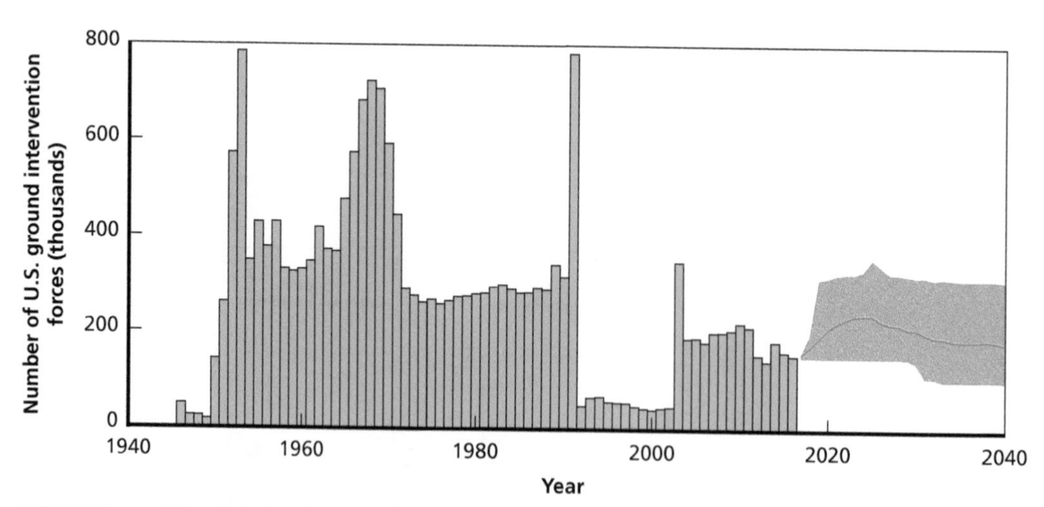

NOTES: The red line denotes the projected mean number of U.S. ground forces required for interventions each year, based on 500 iterations of our forecasting model. The gray shaded area represents the range of forecasts bounded by the 10th and 90th percentiles of U.S. ground forces required for interventions each year, based on 500 iterations of our forecasting model.

Figure C.57
U.S. Isolationism: Regional Forecasts of Demands for U.S. Ground Intervention Forces, 2017–2040

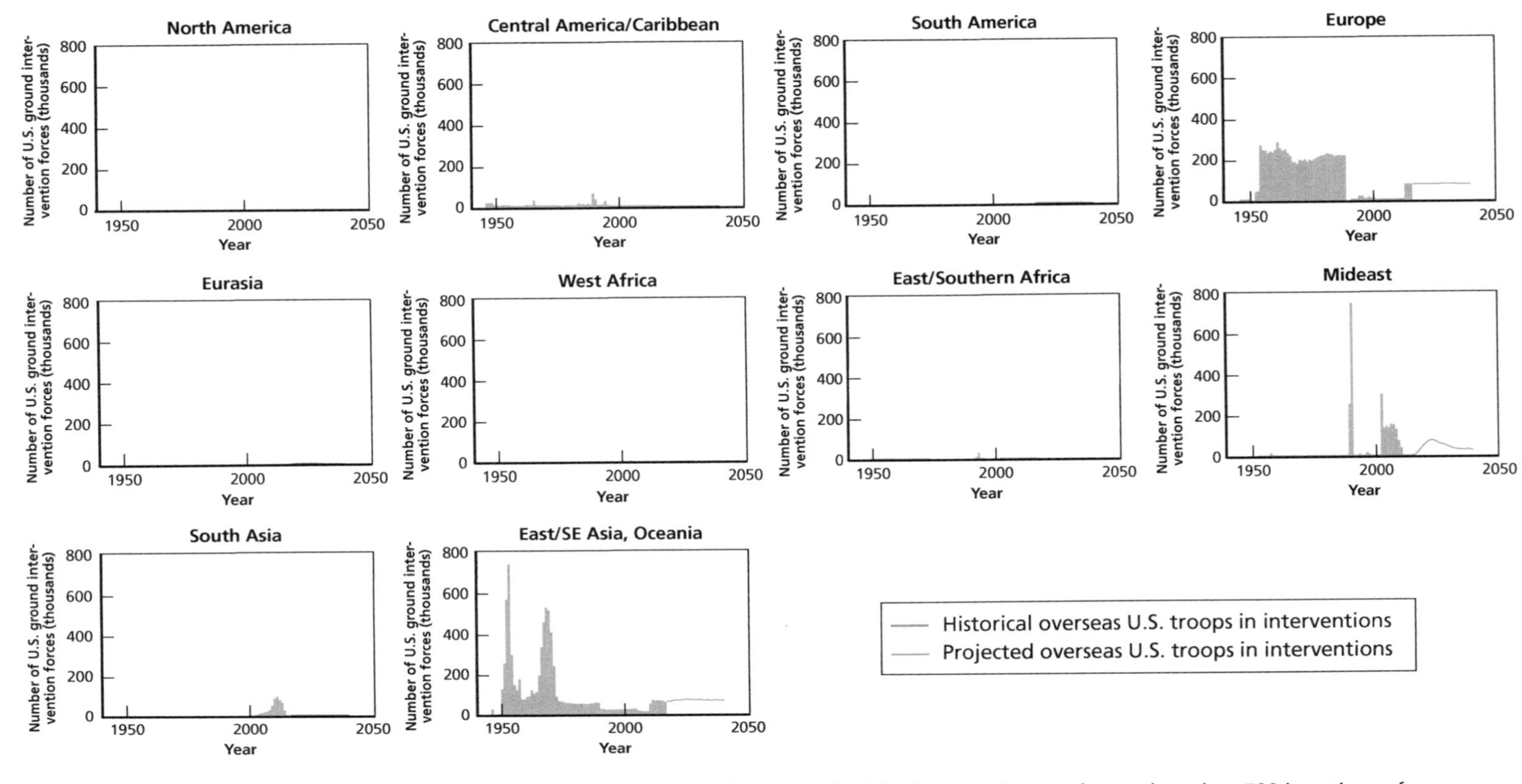

NOTE: The red line denotes the projected mean number of U.S. ground forces required for interventions each year, based on 500 iterations of our forecasting model.

Figure C.58
U.S. Isolationism: Forecasts of U.S. Ground Interventions into Armed Conflicts, 2017–2040

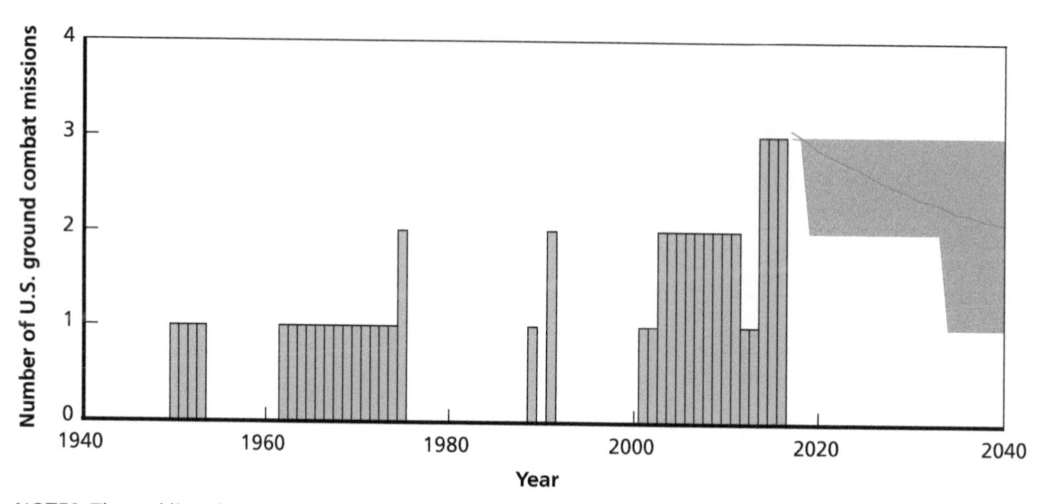

NOTES: The red line denotes the projected mean number of total U.S. ground armed conflict interventions each year, based on 500 iterations of our forecasting model. The gray shaded area represents the range of forecasts bounded by the 10th and 90th percentiles of U.S. ground armed conflict interventions each year, based on 500 iterations of our forecasting model.

Figure C.59
U.S. Isolationism: Forecasts of Demands for U.S. Ground Combat Mission Forces, 2017–2040

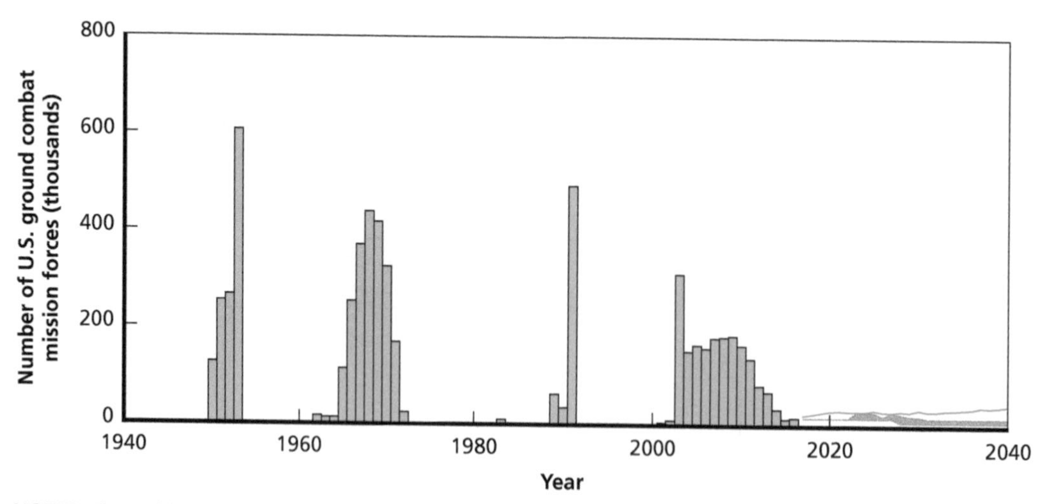

NOTES: The red line denotes the projected mean number of U.S. ground forces required for armed conflict interventions each year, based on 500 iterations of our forecasting model. The gray shaded area represents the range of forecasts bounded by the 10th and 90th percentiles of U.S. ground forces required for armed conflict interventions each year, based on 500 iterations of our forecasting model.

Figure C.60
U.S. Isolationism: Forecasts of U.S. Ground Stability Operations, 2017–2040

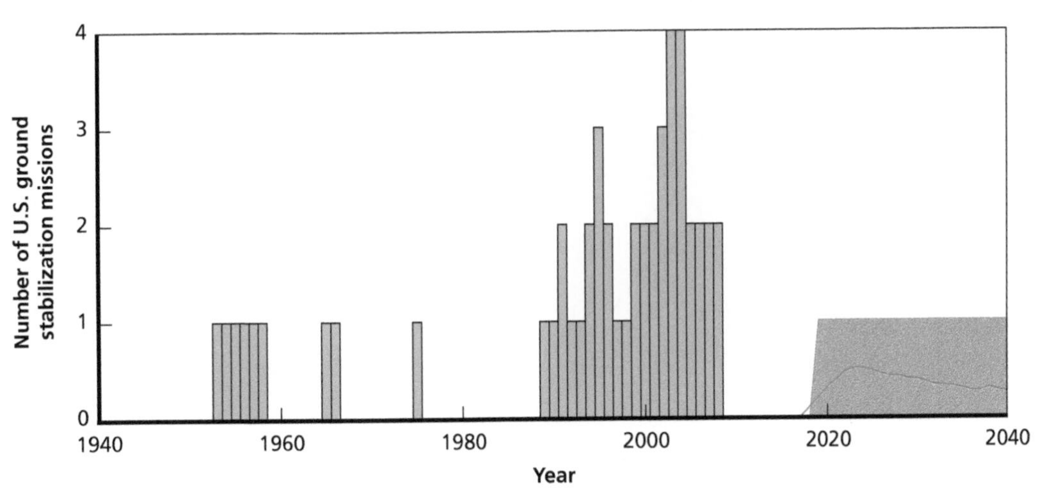

NOTES: The red line denotes the projected mean number of total U.S. ground stability operations each year, based on 500 iterations of our forecasting model. The gray shaded area represents the range of forecasts bounded by the 10th and 90th percentiles of U.S. ground stability operations each year, based on 500 iterations of our forecasting model.

Figure C.61
U.S. Isolationism: Forecasts of Demands for U.S. Ground Stability Operations Forces, 2017–2040

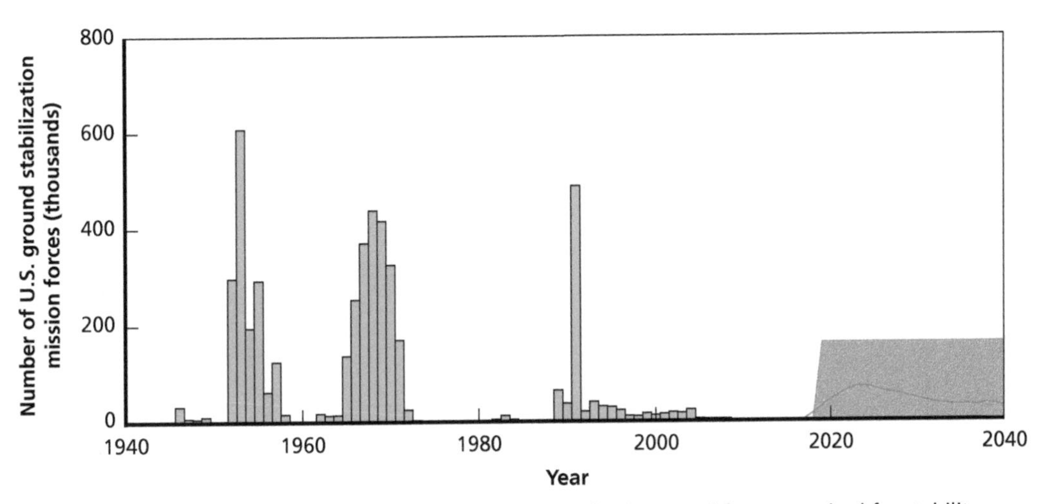

NOTES: The red line denotes the projected mean number of U.S. ground forces required for stability operations each year, based on 500 iterations of our forecasting model. The gray shaded area represents the range of forecasts bounded by the 10th and 90th percentiles of U.S. ground forces required for stability operations each year, based on 500 iterations of our forecasting model.

Figure C.62
U.S. Isolationism: Forecasts of U.S. Ground Deterrent Interventions, 2017–2040

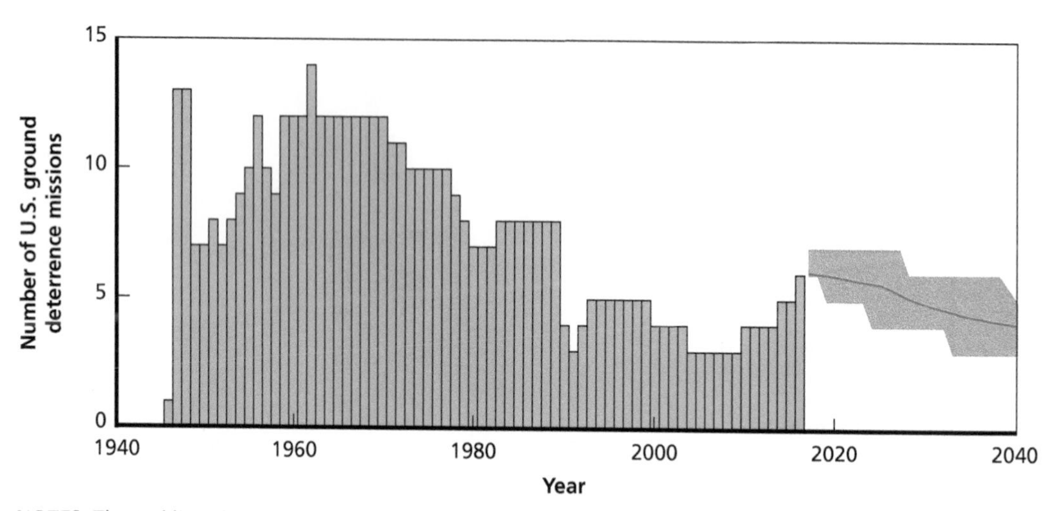

NOTES: The red line denotes the projected mean number of total U.S. ground deterrence missions each year, based on 500 iterations of our forecasting model. The gray shaded area represents the range of forecasts bounded by the 10th and 90th percentiles of U.S. ground deterrence missions each year, based on 500 iterations of our forecasting model.

Figure C.63
U.S. Isolationism: Forecasts of Demands for U.S. Ground Deterrent Intervention Forces, 2017–2040

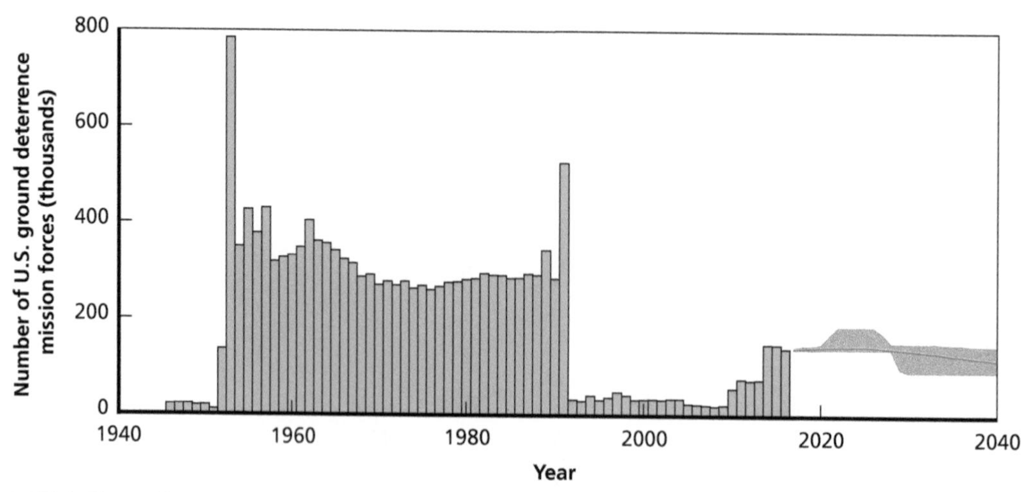

NOTES: The red line denotes the projected mean number of U.S. ground forces required for deterrence missions each year, based on 500 iterations of our forecasting model. The gray shaded area represents the range of forecasts bounded by the 10th and 90th percentiles of U.S. ground forces required for deterrence missions each year, based on 500 iterations of our forecasting model.

Figure C.64
U.S. Isolationism: Forecasts of Demands for Heavy U.S. Ground Intervention Forces, 2017–2040

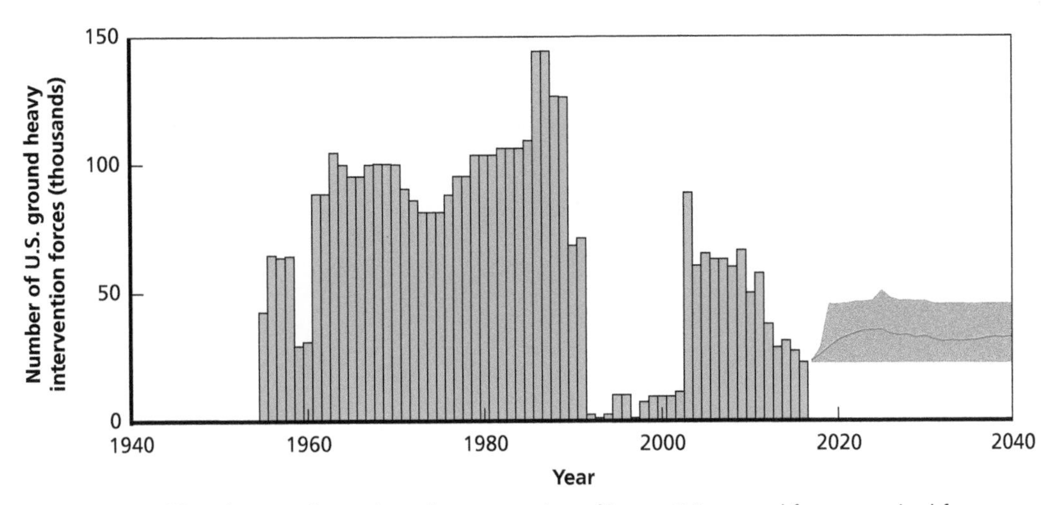

NOTES: The red line denotes the projected mean number of heavy U.S. ground forces required for interventions each year, based on 500 iterations of our forecasting model. The gray shaded area represents the range of forecasts bounded by the 10th and 90th percentiles of heavy U.S. ground forces required for interventions each year, based on 500 iterations of our forecasting model.

References

Allansson, Marie, Erik Melander, and Lotta Themnér, "Organized Violence, 1989–2016," *Journal of Peace Research*, Vol. 54, No. 4, 2017.

Ambrose, Stephen E., *Rise to Globalism: American Foreign Policy Since 1938*, sixth edition, New York: Penguin Books, 1991.

Army Futures Command Task Force, "Army Futures Command," March 28, 2018. As of May 1, 2019:
https://www.army.mil/standto/2018-03-28

Aydin, Aysegul, "Where Do States Go? Strategy in Civil War Intervention," *Conflict Management and Peace Science*, Vol. 27, No. 1, 2010.

Azar, Edward E., "The Conflict and Peace Databank (COPDAB) Project," *Journal of Conflict Resolution*, Vol. 24, No. 1, 1980.

Azar, Edward E., R. D. McLaurin, Thomas Havener, Craig Murphy, Thomas Sloan, and Charles H. Wagner, "A System for Forecasting Strategic Crises: Findings and Speculations About Conflict in the Middle East," *International Interactions*, Vol. 3, No. 3, 1977.

Bagozzi, Benjamin E., "Forecasting Civil Conflict with Zero-Inflated Count Models," *Civil Wars*, Vol. 17, No. 1, 2015.

Barbieri, Katherine, and Omar Keshk, *Correlates of War Project Trade Data Set Codebook*, Correlates of War Project, Version 3.0, 2012.

Beardsley, Kyle, "Peacekeeping and the Contagion of Armed Conflict," *Journal of Politics*, Vol. 73, No. 4, 2011.

Beck, Nathaniel, and Jonathan N. Katz, "Modeling Dynamics in Time-Series-Cross-Section Political Economy Data," *Annual Review of Political Science*, Vol. 14, 2011.

Beck, Nathaniel, Jonathan N. Katz, and Richard Tucker, "Taking Time Seriously: Time-Series-Cross-Section Analysis with a Binary Dependent Variable," *American Journal of Political Science*, Vol. 42, No. 4, 1998.

Bernanke, Ben S., "The Great Moderation," remarks delivered at the meetings of the Eastern Economic Association, Washington, D.C., February 20, 2004.

Blainey, Geoffrey, *The Causes of War*, London, UK: Macmillan, 1973.

Bolt, Jutta, and Jan Luiten van Zanden, "The Maddison Project: Collaborative Research on Historical National Accounts," *The Economic History Review*, Vol. 67, No. 3, 2014.

Bolt, Jutta, Robert Inklaar, Herman de Jong, and Jan Luiten van Zanden, "Rebasing 'Maddison': New Income Comparisons and the Shape of Long-Run Economic Development," University of Groningen, GGDC Research Memorandum 174, January 2018. As of May 1, 2019: https://www.rug.nl/ggdc/html_publications/memorandum/gd174.pdf

Brainerd, Elizabeth and Mark Siegler, *The Economic Effects of the 1918 Influenza Epidemic*, Paris, France: Centre for Economic and Policy Research, Discussion Paper No. 3791, 2003.

Braithwaite, Alex, "Resisting Infection: How State Capacity Conditions Conflict Contagion," *Journal of Peace Research*, Vol. 47, No. 3, 2010.

Brands, H. W., Jr., "Decisions on American Armed Intervention: Lebanon, Dominican Republic, and Grenada," *Political Science Quarterly*, Vol. 102, No. 4, Winter 1987–1988.

Brandt, Patrick T., John R. Freeman, and Philip A. Schrodt "Real Time, Time Series Forecasting of Inter- and Intra-State Political Conflict," *Conflict Management and Peace Science*, Vol. 28, No. 1, 2011.

Brooks, Stephen G., G. John Ikenberry, and William C. Wohlforth, "Don't Come Home, America: The Case Against Retrenchment." *International Security*, Vol. 37, No. 3, 2013, pp. 7–51.

Brown, Ryan, and Duncan Thomas, "On the Long-Term Effects of the 1918 U.S. Influenza Pandemic," unpublished manuscript, 2011.

Bueno de Mesquita, Bruce, *Predicting Politics*, Columbus, Ohio: Ohio State University Press, 2002.

Bueno de Mesquita, Bruce, James D. Morrow, Randolph M. Siverson, and Alastair Smith, "An Institutional Explanation of the Democratic Peace," *American Political Science Review*, Vol. 93, No. 4, 1999.

Buhaug, Halvard, Lars-Erik Cederman, and Jan Ketil Rød, "Disaggregating Ethno-Nationalist Civil Wars: A Dyadic Test of Exclusion Theory," *International Organization*, Vol. 62, No. 3, 2008.

Buhaug, Halvard, Lars-Erik Cederman, and Kristian Skrede Gleditsch, "Square Pegs in Round Holes: Inequalities, Grievances, and Civil War," *International Studies Quarterly*, Vol. 58, No. 2, 2014.

Buhaug, Halvard, and Kristian Skrede Gleditsch, "Contagion or Confusion? Why Conflicts Cluster in Space," *International Studies Quarterly*, Vol. 52, No. 2, 2006.

Burns, Andrew, Dominque van der Mensbrugghe, and Hans Timmer, *Evaluating the Economic Consequences of Avian Influenza*, World Bank, 2006.

Carpenter, Ted Galen, *A Search for Enemies: America's Alliances After the Cold War*, Washington, D.C.: Cato Institute, 1992.

Carter, David B., and Curtis S. Signorino, "Back to the Future: Modeling Time Dependence in Binary Data," *Political Analysis*, Vol. 18, No. 3, 2010.

Cederman, Lars-Erik, Nils B. Weidmann, and Kristian Skrede Gleditsch, "Horizontal Inequalities and Ethnonationalist Civil War: A Global Comparison," *American Political Science Review*, Vol. 105, No. 3, 2011.

Cederman, Lars-Erik, and Nils B. Weidmann, "Predicting Armed Conflict: Time to Adjust Our Expectations?" *Science*, Vol. 355, No. 6324, 2017.

Cederman, Lars-Erik, Andreas Wimmer, and Brian Min, "Why Do Ethnic Groups Rebel? New Data and Analysis," *World Politics*, Vol. 62, No. 1, 2010.

Centers for Disease Control and Prevention, "History of the 1918 Flu Pandemic," 2018. As of May 1, 2018:
https://www.cdc.gov/flu/pandemic-resources/1918-commemoration/1918-pandemic-history.htm

Chadefaux, Thomas, "Early Warning Signals for War in the News," *Journal of Peace Research*, Vol. 51, No. 1, 2014.

Cohen, Raphael S., *The History and Politics of Defense Reviews*, Santa Monica, Calif.: RAND Corporation, RR-2278-AF, 2018.

Colaresi, Michael, and Zuhaib Mahmood, "Do the Robot: Lessons from Machine Learning to Improve Conflict Forecasting," *Journal of Peace Research*, Vol. 54, No. 2, 2017.

Collier, Paul, and Anke Hoeffler, "Greed and Grievance in Civil War," *Oxford Economic Papers*, Vol. 56, No. 4, 2004.

Congressional Budget Office, *The U.S. Military's Force Structure: A Primer*, Washington, D.C.: July 2016.

Correlates of War Project, website, undated. As of May 3, 2019:
http://www.correlatesofwar.org/

———, "Direct Contiguity Data, 1816–2006," Version 3.1, 2016.

Davis, Paul K., *Capabilities for Joint Analysis in the Department of Defense: Rethinking Support for Strategic Analysis*, Santa Monica, Calif.: RAND Corporation, RR-1469-OSD, 2016. As of May 1, 2019:
https://www.rand.org/pubs/research_reports/RR1469.html

Defense Manpower Data Center, "DoD Personnel, Workforce Reports, & Publications," website, undated. As of May 6, 2019:
https://www.dmdc.osd.mil/appj/dwp/dwp_reports.jsp

———, Historical Report—Military Only (aggregated data 1950–current), Washington, D.C., 2016.

———, Worldwide Manpower Distribution by Geographical Area, Washington, D.C., multiple years.

Department of Defense Directive 8260.1, *Data Collection, Development, and Management in Support of Strategic Analysis*, Washington, D.C.: U.S. Department of Defense, December 6, 2002.

Department of Defense Directive 8260.05, *Support for Strategic Analysis (SSA)*, Washington, D.C.: U.S. Department of Defense, July 7, 2011.

DMDC—*See* Defense Manpower Data Center.

Dobbins, James, Seth G. Jones, Keith Crane, and Beth Cole DeGrasse, *The Beginner's Guide to Nation-Building*, Santa Monica, Calif.: RAND Corporation, MG-557-SRF, 2007. As of May 1, 2019:
https://www.rand.org/pubs/monographs/MG557.html

Doyle, Andy, Graham Katz, Kristen Summers, Chris Ackermann, Ilya Zavorin, Zunsik Lim, Sathappan Muthiah, Patrick Butler, Nathan Self, Liang Zhao, Chang-Tien Lu, Rupinder Paul Khandpur, Youssef Fayed, and Naren Ramakrishnan, "Forecasting Significant Societal Events Using the Embers Streaming Predictive Analytics System," *Big Data*, Vol. 2, No. 4, 2014.

Drezner, D. W., "Military Primacy Doesn't Pay (Nearly as Much as You Think)," *International Security*, Vol. 38, No. 1, 2013, pp. 52–79.

Enterline, Andrew J., and J. Michael Greig, "Beacons of Hope? The Impact of Imposed Democracy on Regional Peace, Democracy, and Prosperity," *Journal of Politics*, Vol. 67, No. 4, 2005.

Fazal, Tanisha M., "Dead Wrong? Battle Deaths, Military Medicine, and Exaggerated Reports of War's Demise," *International Security*, Vol. 39, No. 1, 2014.

Fearon, James D., and David D. Laitin, "Ethnicity, Insurgency, and Civil War," *American Political Science Review*, Vol. 97, No. 1, 2003.

Field Manual 3-0, *Operations*, Washington, D.C.: Department of the Army, October 2017.

Findley, Michael G., and Tze Kwang Teo, "Rethinking Third-Party Interventions into Civil Wars: An Actor-Centric Approach," *Journal of Politics*, Vol. 68, No. 4, 2006.

Finnemore, Martha, *The Purpose of Intervention: Changing Beliefs About the Use of Force*, Ithaca, N.Y.: Cornell University Press, 2003.

Fjelde, Hanne, "Generals, Dictators, and Kings: Authoritarian Regimes and Civil Conflict, 1973–2004," *Conflict Management and Peace Science*, Vol. 27, No. 3, 2010.

Frederick, Bryan A., *The Sources of Territorial Stability*, dissertation, Washington, D.C.: Johns Hopkins University School of Advanced International Studies, 2012.

Frederick, Bryan A., Paul R. Hensel, and Christopher Macaulay, "The Issue Correlates of War Territorial Claims Data, 1816–2001," *Journal of Peace Research*, Vol. 54, No. 1, 2017.

Frederick, Bryan, Stephen Watts, Matthew Lane, Abby Doll, Ashley L. Rhoades, and Meagan L. Smith, *Understanding the Deterrent Impact of U.S. Overseas Forces*, Santa Monica, Calif.: RAND Corporation, RR-2533-A, 2020. As of March 5, 2020:
https://www.rand.org/pubs/research_reports/RR2533.html

Frederick S. Pardee Center for International Futures, website, undated. As of May 3, 2019:
https://pardee.du.edu/

Friedman, Jeffrey A., "Manpower and Counterinsurgency: Empirical Foundations for Theory and Doctrine," *Security Studies*, Vol. 20, No. 4, 2011.

Garrett, Thomas, *Economic Effects of the 1918 Influenza Pandemic: Implications for a Modern-Day Pandemic*, St. Louis, Mo.: Federal Reserve Bank of St. Louis, November 2007.

Gartzke, Erik, "Kant We All Just Get Along? Opportunity, Willingness, and the Origins of the Democratic Peace," *American Journal of Political Science*, Vol. 42, No. 1, 1998.

Gartzke, Erik, and Matthew Kroenig, "A Strategic Approach to Nuclear Proliferation," *Journal of Conflict Resolution*, Vol. 53, No. 2, 2009.

George, Alexander L., and Andrew Bennett, *Case Studies and Theory Development in the Social Sciences*, Cambridge, Mass.: MIT Press, 2005.

Gibler, Douglas M., *International Military Alliances, 1648–2008*, Washington, D.C.: CQ Press, 2009.

———, *The Territorial Peace: Borders, State Development, and International Conflict*, Cambridge, UK: Cambridge University Press, 2012.

———, Correlates of War Formal Alliances Dataset (v4.1), 2014.

Gilpin, Robert, *War and Change in World Politics*, Cambridge, UK: Cambridge University Press, 1981.

Gleditsch, Nils Petter, Peter Wallensteen, Mikael Eriksson, Margareta Sollenberg, and Håvard Strand, "Armed Conflict, 1946–2001: A New Dataset," *Journal of Peace Research*, Vol. 39, No. 5, 2002.

Goldstone, Jack A., "Population and Security: How Demographic Change Can Lead to Violent Conflict," *Journal of International Affairs*, Vol. 56, No. 1, 2002.

Goldstone, Jack A., Robert H. Bates, David L. Epstein, Ted Robert Gurr, Michael B. Lustik, Monty G. Marshall, Jay Ulfelder, and Mark Woodward, "A Global Model for Forecasting Political Instability," *American Journal of Political Science*, Vol. 54, No. 1, 2010.

Greig, J. Michael, and Andrew J. Enterline, "Correlates of War Project: National Material Capabilities (NMC) Data Documentation, Version 5.0 (Period Covered: 1816–2012)," February 1, 2017. As of June 8, 2018:
https://correlatesofwar.org/data-sets/national-material-capabilities

Hegre, Håvard, Halvard Buhaug, Katherine V. Calvin, Jonas Nordkvelle, Stephanie T. Waldhoff, and Elisabeth Gilmore, "Forecasting Civil Conflict Along the Shared Socioeconomic Pathways," *Environmental Research Letters*, Vol. 11, No. 5, 2016.

Hegre, Håvard, Tanja Ellingsen, Scott Gates, and Nils Petter Gleditsch, "Toward a Democratic Civil Peace? Democracy, Political Change, and Civil War, 1816–1992," *American Political Science Review*, Vol. 95, No. 1, 2001.

Hegre, Håvard, Joakim Karlsen, Håvard Mokleiv Nygård, Håvard Strand, and Henrik Urdal, "Predicting Armed Conflict, 2010–2050," *International Studies Quarterly*, Vol. 57, No. 2, 2013.

Hegre, Håvard, Nils W. Metternich, Håvard Mokleiv Nygård, and Julian Wucherpfennig, "Introduction: Forecasting in Peace Research," *Journal of Peace Research*, Vol. 54, No. 2, 2017.

Hegre, Håvard, Marie Allansson, Matthias Basedau, Michael Colaresi, Mihai Croicu, Hanne Fjelde, Frederick Hoyles, Lisa Hultman, Stina Högbladh, Remco Jansen, Naima Mouhleb, Sayyed Auwn Muhammad, Desirée Nilsson, Håvard Mokliev Nygård, Gudlaug Olafsdottir, Kristina Petrova, David Randahl, Espen Geelmuyden Rød, Gerald Schneider, Nina von Uexkull, and Jonas Vestby, "ViEWS: A Political Violence Early-Warning System," *Journal of Peace Research*, Vol. 56, No. 2, 2019.

Hensel, Paul R., "Charting a Course to Conflict: Territorial Issues and Interstate Conflict, 1816–1992," *Conflict Management and Peace Science*, Vol. 15, No. 1, 1996.

———, Multilateral Treaties of Pacific Settlement (MTOPS) Data Set, Version 1.4, 2005.

Hollis, Duncan B., "Private Actors in Public International Law: Amicus Curiae and the Case for the Retention of State Sovereignty," *Boston College International and Comparative Law Review*, Vol. 25, 2002.

Horowitz, Michael C., *The Diffusion of Military Power: Causes and Consequences for International Politics*, Princeton, N.J.: Princeton University Press, 2010.

Huntington, Samuel P., "Why International Primacy Matters," *International Security*, Vol. 17, No. 4, 1993, pp. 68–83.

Huth, Paul K., "Extended Deterrence and the Outbreak of War," *American Political Science Review*, Vol. 82, No. 2, 1988.

———, "Major Power Intervention in International Crises, 1918–1988," *Journal of Conflict Resolution*, Vol. 42, No. 6, 1998.

———, *Standing Your Ground: Territorial Disputes and International Conflict*, Ann Arbor, Mich.: University of Michigan Press, 1998.

Huth, Paul K., D. Scott Bennett, and Christopher Gelpi, "System Uncertainty, Risk Propensity, and International Conflict Among the Great Powers," *Journal of Conflict Resolution*, Vol. 36, No. 3, 1992.

Huth, Paul K., Christopher Gelpi, and D. Scott Bennett, "The Escalation of Great Power Militarized Disputes: Testing Rational Deterrence Theory and Structural Realism," *American Political Science Review*, Vol. 87, No. 3, 1993.

International Futures Database, version 7.31, undated.

International Institute for Strategic Studies, *The Military Balance*, Washington, D.C., multiple years.

International Law and Policy Institute, "The Nuclear Umbrella States," ILPI Nuclear Weapons Project Nutshell Paper No. 5, 2012.

James, Patrick, and John O'Neal, "The Influence of Domestic and International Politics on the President's Use of Force," *Journal of Conflict Resolution*, Vol. 35, No. 2, 1991.

Judson, Jen, "Deterring Russia: U.S. Army Hones Skills to Mass Equipment, Troops in Europe," *Defense News*, March 17, 2017.

Kahneman, Daniel, *Thinking, Fast and Slow*, New York, N.Y.: Farrar, Straus and Giroux, 2011.

Kavanagh, Jennifer, Bryan Frederick, Matthew Povlock, Stacie L. Pettyjohn, Angela O'Mahony, Stephen Watts, Nathan Chandler, John Speed Meyers, and Eugeniu Han, *The Past, Present, and Future of U.S. Ground Interventions: Identifying Trends, Characteristics, and Signposts*, Santa Monica, Calif.: RAND Corporation, RR-1831-A, 2017. As of May 1, 2019: https://www.rand.org/pubs/research_reports/RR1831.html

Keohane, Robert O., *After Hegemony: Cooperation and Discord in the World Political Economy*, Princeton, N.J.: Princeton University Press, 1984.

Klare, Michael T., *Beyond the "Vietnam Syndrome": U.S. Interventionism in the 1980s*, Washington, D.C.: Institute for Policy Studies, 1981.

Kristian, Bonnie, "Trump is Right: It's time to rethink NATO," *Politico*, August 3, 2016.

Lagon, Mark P., "The International System and the Reagan Doctrine: Can Realism Explain Aid to 'Freedom Fighters'?" *British Journal of Political Science*, Vol. 22, No. 1, 1992.

Larson, Eric V., Derek Eaton, Michael E. Linick, John E. Peters, Agnes Gereben Schaefer, Keith Walters, Stephanie Young, H. G. Massey, and Michelle Darrah Ziegler, *Defense Planning in a Time of Conflict: A Comparative Analysis of the 2001–2014 Quadrennial Defense Reviews, and Implications for the Army*, Santa Monica, Calif.: RAND Corporation, RR-1309-A, 2018. As of May 1, 2019: https://www.rand.org/pubs/research_reports/RR1309.html

Layne, Christopher, "From Preponderance to Offshore Balancing: America's Future Grand Strategy," *International Security*, Vol. 22, No. 1, 1997.

———, *The Peace of Illusions: American Grand Strategy from 1940 to the Present*, Ithaca, N.Y.: Cornell University Press, 2006.

Lemke, Douglas, "The Tyranny of Distance: Redefining Relevant Dyads," *International Interactions*, Vol. 21, No. 1, 1995.

———, "The Continuation of History: Power Transition Theory and the End of the Cold War," *Journal of Peace Research*, Vol. 34, No. 1, 1997.

MacDonald, Paul K., and Joseph M. Parent, "Graceful Decline? The Surprising Success of Great Power Retrenchment," *International Security*, Vol. 35, No. 4, 2011, pp. 7–44.

Manaugh, Geoff, and Nicola Twilley, "It's Artificial Afghanistan: A Simulated Battlefield in the Mojave Desert," *The Atlantic*, May 18, 2013.

Maoz, Zeev, and Bruce M. Russett, "Normative and Structural Causes of Democratic Peace, 1946–1986," *American Political Science Review*, Vol. 87, No. 3, 1993.

Marshall, Monty G., Ted Robert Gurr, and Keith Jaggers, *POLITY IV Project: Political Regime Characteristics and Transitions, 1800–2016—Dataset Users' Manual*, Vienna, Va.: Center for Systemic Peace, 2017.

Mazzetti, Mark, *The Way of the Knife: The CIA, a Secret Army, and a War at the Ends of the Earth*, New York, N.Y.: Penguin Books, 2013.

McClelland, Charles A., and Gary D. Hoggard, *Conflict Patterns in the Interactions Among Nations*, Los Angeles, Calif.: University of Southern California Press, 1968.

Mearsheimer, John J., *Conventional Deterrence*, Ithaca, N.Y.: Cornell University Press, 1983.

Meernik, James, "United States Military Intervention and the Promotion of Democracy," *Journal of Peace Research*, Vol. 33, No. 4, 1996.

Metz, Steven, *Learning from Iraq: Counterinsurgency in American Strategy*, Carlisle, Pa.: U.S. Army War College, Strategic Studies Institute, January 2007.

Mitre, James R., *Force Planning and Scenario Development*, Washington, D.C.: Office of the Secretary of Defense (Policy), May 5, 2015, not available to the general public.

Muchlinski, David, David Siroky, Jingrui He, and Matthew Kocher, "Comparing Random Forest with Logistic Regression for Predicting Class-Imbalanced Civil War Onset Data," *Political Analysis*, Vol. 24, No. 1, 2016.

National Intelligence Council, *Global Trends 2030: Alternative Worlds*, Washington, D.C.: Office of the Director of National Intelligence, December 2012.

O'Brien, Sean P., "Crisis Early Warning and Decision Support: Contemporary Approaches and Thoughts on Future Research," *International Studies Review*, Vol. 12, No. 1, 2010.

O'Mahony, Angela, Miranda Priebe, Bryan Frederick, Jennifer Kavanagh, Matthew Lane, Trevor Johnston, Thomas S. Szayna, Jakub P. Hlavka, Stephen Watts, and Matthew Povlock, *U.S. Presence and the Incidence of Conflict*, Santa Monica, Calif.: RAND Corporation, RR-1906-A, 2018. As of May 1, 2019:
https://www.rand.org/pubs/research_reports/RR1906.html

Oneal, John R., and Bruce Russett, "The Classical Liberals Were Right: Democracy, Interdependence, and Conflict, 1950–1985," *International Studies Quarterly*, Vol. 41, 1997.

———, Triangulating Peace: Democracy, Interdependence, and International Organizations, New York, N.Y.: W.W. Norton and Company, 2001.

Organski, A. F. K., and Jacek Kugler, *The War Ledger*, Chicago, Ill.: University of Chicago Press, 1981.

Overholt, William H., *Asia, America, and the Transformation of Geopolitics*, New York, N.Y.: Cambridge University Press, 2008.

Pamlin, Dennis, and Stuart Armstrong, *Global Challenges: 12 Risks that Threaten Human Civilization*, Global Challenges Foundation, 2015.

Parent, Joseph, and Paul MacDonald, "The Wisdom of Retrenchment: America Must Cut Back to Move Forward," *Foreign Affairs*, Vol. 90, No. 6, 2011.

Pearson, Frederic S., and Robert A. Baumann, "Foreign Military Intervention and Changes in United States Business Activity," *Journal of Political and Military Sociology*, Vol. 5, No. 1, 1977.

Peck, Michael, "Don't Hold Your Breath for 'Sim Afghanistan,'" *Wired*, October 1, 2009.

Pinker, Steven, *The Better Angels of Our Nature: The Decline of Violence in History and It's Causes*, London, UK: Penguin Books, 2011.

Posen, B. R., "Pull Back: The Case for a Less Activist Foreign Policy," *Foreign Affairs*, Vol. 92, 2013.

Quinlivan, James T., "Force Requirements in Stability Operations," *Parameters: U.S. Army War College Quarterly*, Vol. 25, No. 4, Winter 1995/96.

Reardon, Mark J., "Chasing a Chameleon: The U.S. Army Counterinsurgency Experience in Korea, 1945–1952," in Richard Davis, ed., *The U.S. Army and Irregular Warfare, 1775–2007: Selected Papers from the 2007 Conference of Army Historians*, Washington, D.C.: Center of Military History, U.S. Army, 2008, pp. 213–228.

Rost, Nicolas, and J. Michael Greig, "Taking Matters into Their Own Hands: An Analysis of the Determinants of State-Conducted Peacekeeping in Civil Wars," *Journal of Peace Research*, Vol. 48, No. 2, 2011.

Rubin, Harvey, *Future Global Shocks: Pandemics*, Organisation for Economic Development/ International Futures Programme on Future Global Shocks, January 2011.

Sarkees, Meredith Reid, and Frank Whelon Wayman, *Resort to War: 1816–2007*, Washington, D.C.: CQ Press, 2010.

Schrodt, Philip A., Shannon G. Davis, and Judith L. Weddle, "Political Science: KEDS—A Program for the Machine Coding of Event Data," *Social Science Computer Review*, Vol. 12, No. 4, 1994.

Schrodt, Philip A., and Deborah J. Gerner, "Cluster-Based Early Warning Indicators for Political Change in the Contemporary Levant," *American Political Science Review*, Vol. 94, No. 4, 2000.

Schultz, Tammy, *Tool of Peace and War: Save the Peacekeeping and Stability Operations Institute*, Washington, D.C.: Council on Foreign Relations, July 31, 2018. As of May 1, 2019: https://www.cfr.org/blog/tool-peace-and-war-save-peacekeeping-and-stability-operations-institute

Sidorenko, Alexandra, and Warwick McKibbin, *What a Flu Pandemic Could Cost the World*, Washington, D.C.: Brookings Institution, April 28, 2009.

Singer, J. David, "Reconstructing the Correlates of War Dataset on Material Capabilities of States, 1816–1985," *International Interactions*, Vol. 14, No. 2, 1988.

Singer, J. David, Stuart Bremer, and John Stuckey, "Capability Distribution, Uncertainty, and Major Power War, 1820–1965," in Bruce Russett, ed., *Peace, War, and Numbers*, Beverly Hills: Sage, 1972.

Singer, J. David, and Michael D. Wallace, eds., *To Augur Well: Early Warning Indicators in World Politics*, Beverly Hills, Calif.: Sage, 1979.

Siroky, David S., "Navigating Random Forests and Related Advances in Algorithmic Modeling," *Statistics Surveys*, Vol. 3, 2009.

Smith, R. Jeffrey, "Military Admits Major Mistakes in Iraq and Afghanistan," *The Atlantic*, June 11, 2012.

Stinnett, Douglas M., Jaroslav Tir, Philip Schafer, Paul F. Diehl, and Charles Gochman, "The Correlates of War Project Direct Contiguity Data, Version 3," *Conflict Management and Peace Science*, Vol. 19, No. 2, 2002.

Stock, James H., and Mark W. Watson, "Has the Business Cycle Changed and Why?" *NBER Macroeconomics Annual 2002*, Vol. 17, Cambridge, Mass.: MIT Press, 2003.

Summers, Lawrence H., "U.S. Economic Prospects: Secular Stagnation, Hysteresis, and the Zero Lower Bound," *Business Economics*, Vol. 49, No. 2, 2014.

Swanson, Ana, "Once the WTO's Biggest Supporter, U.S. Is Its Biggest Skeptic," *New York Times*, December 10, 2017.

Szayna, Thomas S., Angela O'Mahony, Jennifer Kavanagh, Stephen Watts, Bryan Frederick, Tova C. Norlen, and Phoenix Voorhies, *Conflict Trends and Conflict Drivers: An Empirical Assessment of Historical Conflict Patterns and Future Conflict Projections*, Santa Monica, Calif.: RAND Corporation, RR-1063-A, 2017. As of May 1, 2019:
https://www.rand.org/pubs/research_reports/RR1063.html

Tetlock, Philip E., *Expert Political Judgment: How Good Is It? How Can We Know?* Princeton, N.J.: Princeton University Press, 2005.

Teulings, Coen, and Richard Baldwin, eds., *Secular Stagnation: Facts, Causes and Cures*, London, UK: Centre for Economic Policy Research, August 15, 2014.

Tilgham, Andrew, "More U.S. Troops Deploying to Europe in 2017," *Military Times*, February 2, 2016.

Uppsala University, ViEWS Project, website, undated. As of May 1, 2019:
http://www.pcr.uu.se/research/views/

Urdal, Henrik, "A Clash of Generations? Youth Bulges and Political Violence," *International Studies Quarterly*, Vol. 50, No. 3, 2006.

U.S. Army War College, *How the Army Runs: A Senior Leader Reference Handbook, 2015–2016*, Carlisle, Pa., 2015.

U.S. Department of Defense, *Quadrennial Defense Review Report*, Washington, D.C., February 6, 2006.

U.S. Department of Defense, *Summary of the 2018 National Defense Strategy of the United States of America*, Washington, D.C., 2018.

Vasquez, John A., and Marie T. Henehan, *Territory, War, and Peace*, New York, N.Y.: Routledge, 2010.

Vreeland, James Raymond, "The Effect of Political Regime on Civil War: Unpacking Anocracy," *Journal of Conflict Resolution*, Vol. 52, No. 3, 2008.

Walt, Stephen M., *Taming American Power: The Global Response to U.S. Primacy*, New York, N.Y.: Norton, 2006.

Walter, Barbara F., "Does Conflict Beget Conflict? Explaining Civil War Recurrence," *Journal of Peace Research*, Vol. 41, No. 3, 2004.

Ward, Michael D., and Andreas Beger, "Lessons from Near Real-Time Forecasting of Irregular Leadership Changes," *Journal of Peace Research*, Vol. 54, No. 2, 2017.

Ward, Michael D., Nils W. Metternich, Cassy L. Dorff, Max Gallop, Florian M. Hollenbach, Anna Schultz, and Simon Weschle, "Learning from the Past and Stepping into the Future: Toward a New Generation of Conflict Prediction," *International Studies Review*, Vol. 15, No. 4, 2013.

Warrick, Joby, *Black Flags: The Rise of Isis*, New York, N.Y.: Penguin Random House, 2015.

Wasserby, Daniel, "Pentagon budget 2019: US Army procurement shifts from aviation to artillery, combat vehicles," Jane's Defense Weekly, February 13, 2018.

Watts, Stephen, Bryan Frederick, Jennifer Kavanagh, Angela O'Mahony, Thomas S. Szayna, Matthew Lane, Alexander Stephenson, and Colin P. Clarke, *A More Peaceful World? Regional Conflict Trends and U.S. Defense Planning*, Santa Monica, Calif.: RAND Corporation, RR-1177-A, 2017a. As of May 1, 2019:
https://www.rand.org/pubs/research_reports/RR1177.html

Watts, Stephen, Patrick B. Johnston, Jennifer Kavanagh, Sean M. Zeigler, Bryan Frederick, Trevor Johnston, Karl P. Mueller, Astrid Stuth Cevallos, Nathan Chandler, Meagan L. Smith, Alexander Stephenson, and Julia A. Thompson, *Limited Intervention: Evaluating the Effectiveness of Limited Stabilization, Limited Strike, and Containment Operations*, Santa Monica, Calif.: RAND Corporation, RR-2037-A, 2017b. As of May 1, 2019:
https://www.rand.org/pubs/research_reports/RR2037.html

Watts, Stephen, Jennifer Kavanagh, Bryan Frederick, Tova C. Norlen, Angela O'Mahony, Phoenix Voorhies, and Thomas S. Szayna, *Understanding Conflict Trends: A Review of the Social Science Literature on the Causes of Conflict*, Santa Monica, Calif.: RAND Corporation, RR-1063/1-A, 2017c. As of May 1, 2019:
https://www.rand.org/pubs/research_reports/RR1063z1.html

Weidmann, Nils B., Doreen Kuse, and Kristian Skrede Gleditsch, "The Geography of the International System: The cShapes Dataset," *International Interactions*, Vol. 36, No. 1, 2010.

Weiss, Michael, and Hassan Hassan, *ISIS: Inside the Army of Terror*, New York, N.Y.: Reagan Arts, 2015.

"Why Soldiers Can Now Pretty Much Say Goodbye to Counter-Insurgency Training," We Are the Mighty, October 2017. As of May 1, 2019:
https://www.wearethemighty.com/news/why-soldiers-can-now-pretty-much-say-goodbye-to-counter-insurgency-trainin

Wilson, John B., *Maneuver and Firepower: The Evolution of Divisions and Separate Brigades*, Washington, D.C.: Center of Military History, U.S. Army, 1998.

Wimmer, Andreas, Lars-Erik Cederman, and Brian Min, "Ethnic Politics and Armed Conflict: A Configurational Analysis of a New Global Data Set," *American Sociological Review*, Vol. 74, No. 2, 2009.

Witmer, Frank D. W., Andrew M. Linke, John O'Loughlin, Andrew Gettelman, and Arlene Laing, "Subnational Violent Conflict Forecasts for Sub-Saharan Africa, 2015–65, Using Climate-Sensitive Models," *Journal of Peace Research*, Vol. 54, No. 2, 2017.

Wohlforth, William C., and Stephen G. Brooks, "American Primacy in Perspective," in David Skidmore, ed., *Paradoxes of Power*, New York: N.Y.: Routledge, 2015, pp. 29–38.

Wood, Reed M., Jacob D. Kathman, and Stephen E. Gent, "Armed Intervention and Civilian Victimization in Intrastate Conflicts," *Journal of Peace Research*, Vol. 49, No. 5, 2012.

World Bank, "World Development Indicators," 2012.

Yair, Omer, and Dan Miodownik, "Youth Bulge and Civil War: Why a Country's Share of Young Adults Explains Only Non-Ethnic Wars," *Conflict Management and Peace Science*, Vol. 33, No. 1, 2016.

Yeo, Mike, "Incoming US Pacific Command Chief Wants to Increase Presence Near China," *Defense News*, April 23, 2018.

Yoon, Mi Yung, "Explaining U.S. Intervention in Third World Internal Wars, 1945–1989," *Journal of Conflict Resolution*, Vol. 41, No. 4, 1997.

Zakaria, Fareed, *From Wealth to Power: The Unusual Origins of America's World Role*, Princeton, N.J.: Princeton University Press, 1999.

Lightning Source UK Ltd.
Milton Keynes UK
UKHW051045020922
408227UK00003B/14